Charge-Transfer Devices in Spectroscopy

Charge-Transfer Devices in Spectroscopy

Edited by
Jonathan V. Sweedler
Kenneth L. Ratzlaff
M. Bonner Denton

Editors:
Jonathan V. Sweedler
Department of Chemistry
University of Illinois
Urbana, IL 61801

Kenneth L. Ratzlaff
Instrumentation Design Lab
University of Kansas
Lawrence, KS 66045

M. Bonner Denton
Department of Chemistry
University of Arizona
Tucson, AZ 85721

This book is printed on acid-free paper. ♾

Library of Congress Cataloging-in-Publication Data

Charge-transfer devices in spectroscopy / edited by Jonathan V. Sweedler, Kenneth L. Ratzlaff, M. Bonner Denton.
p. cm.
Includes bibliographical references and index.
ISBN 1-56081-060-2
1. Spectrum analysis—Instruments. 2. Charge-transfer devices (Electronics) I. Sweedler, Jonathan V. II. Ratzlaff, Kenneth L. III. Denton, M. Bonner.
QC452.5.C45 1994
543′.0858—dc20 93-42138
CIP

Printed in the United States of America

ISBN 1-56081-060-2 VCH Publishers

Printing History
10 9 8 7 6 5 4 3 2 1

VCH Publishers, Inc.
220 East 23rd Street
New York, NY 10010-4606

VCH Verlagsgesellschaft mbH
P.O. Box 10 11 61
D-69451 Weinheim, Germany

VCH Publishers (UK) Ltd.
8 Wellington Court
Cambridge CB1 1 HZ
United Kingdom

Preface

Nothing begets good science as much as the development of a good instrument.
Sir Humphrey Davy (1778–1829)

This quotation, brought to our attention by Professor Gary Hieftje of Indiana University, emphasizes the pivotal importance of new ways of making measurements. The availability of new instrumental tools allows dramatic improvements in the acquisition of new scientific information. Data can be acquired at higher rates, at greater sensitivity, with higher resolution (spatial, temporal, etc.), and with lower noise (including interferences). Naturally, scientists would like to capture all of these improvements at once.

The photon is a powerful probe available to the chemist. Consider the following examples:

- Optical microscopy can allow a surface to be characterized with submicron resolution.
- Fluorescence spectroscopy can detect the presence of a single fluorophore under rigorously controlled situations.
- Raman spectroscopy can provide details concerning the identity and environment of a molecule.

The quality of information that can be extracted through these measurements depends on the quality of the optical detector.

Charge-transfer devices (CTDs) are a new class of multichannel detector array that not only improves the quality of existing optical techniques but also allows entirely new types of measurements. This book describes the characteristics and application of CTDs to a variety of analytical imaging and spectroscopic applications.

Two decades have now passed since the first commercial CTDs—the charge-coupled device (CCD) and the charge-injection device (CID)—became available. In this period nearly 1000 research articles have been published on characterizing, optimizing, and using these detectors. In addition, commercial analytical instrumentation has been introduced based on this new technology.

More significantly, over the last decade the electrooptical performance of CTDs has improved dramatically—prototype devices have been fabricated that approach the "perfect" photon detector to a remarkable degree, with quantum efficiencies of nearly 100%, dark currents of less than 1 electron/hour, and readout noises of less than 2 electrons! Instruments based on CTDs can now supply the resolution and sensitivity needed for dramatic improvements in good science, as per Sir Davy's dictum.

Although the diverse formats and characteristics of available CTDs reflect the wide variety of their intended uses, this diversity also indicates the need for a book that consolidates information from a variety of fields. This volume is a guide for the scientific user of CTDs; it both describes the devices and their operational parameters, and presents outstanding applications for them.

The first five chapters describe the operation, characteristics, and use of CTDs in a variety of spectroscopic and imaging applications. Two of these chapters describe nonsilicon array detectors and intensified CCDs. Each chapter in the second half of the book describes a specific application and reviews the use and impact of CTDs in the particular field. Not only do these chapters describe current technology and research, they emphasize the future trends in adapting CTDs as optical detectors to the particular requirements in each field.

The authors of these chapters are experts in their field. In many cases they are the researchers who first applied CTD technology to the particular application described. With 11 different contributors and 10 chapters, some overlap among chapters is inevitable. However, this overlap allows the chapters to be understood independently and reinforces key points.

Finally, we wish to express our appreciation to these authors for their cooperation and help in putting this book together.

Jonathan V. Sweedler
Urbana, IL

Kenneth L. Ratzlaff
Lawrence, KS

M. Bonner Denton
Tucson, AZ

November 1993

Contents

8 CTD Detection in Analytical Fluorescence Spectroscopy [illegible]

Robert S. Pomeroy

9 Two-Dimensional Array Detection for Plasma Diagnostics [illegible]

Cheryl L. S[illegible] and [illegible] Schefler

10 CTD Detection in Atomic Emission Spectroscopy [illegible]

Robert S. Pomeroy

Contributors

Mark Baker
Department of Chemistry
University of Arizona
Tucson, Arizona 85721

Robert B. Bilhorn
Analytical Technologies Division
Eastman Kodak Company
Rochester, New York 14652

Cheryl A. Bye
Department of Chemistry
University of Illinois
Urbana, Illinois 61801

Bruce Chase
Corporate Center for
Analytical Science
Dupont Experimental Station
Wilmington, Delaware 19880

M. Bonner Denton
Department of Chemistry
University of Arizona
Tucson, Arizona 85721

Richard L. McCreery
Department of Chemistry
Ohio State University
Columbus, Ohio 43210

Robert S. Pomeroy
Department of Chemistry
U.S. Naval Academy
Annapolis, Maryland 21402

Kenneth L. Ratzlaff
Instrumentation Design Laboratory
University of Kansas
Lawrence, Kansas 66045

Gary R. Sims
Spectral Instruments
303 E. Roger Rd.
Tucson, Arizona 85705

Alexander Scheeline
Department of Chemistry
University of Illinois
Urbana, Illinois 61801

Jonathan V. Sweedler
Department of Chemistry
University of Illinois
Urbana, Illinois 61801

Yair Talmi
Princeton Instruments, Inc
3660 Quakerbridge Rd.
Trenton, New Jersey 08619

List of Abbreviations

ADC	analog-to-digital converter
ADU	analog-to-digital unit
AES	atomic emission spectroscopy
AOTF	acoustooptic tunable filter
CCD	charge-coupled device
CCE	charge-collection efficiency
CDM	capacitive discharge mode
CE	capillary electrophoresis
CID	charge-injection device
CRAB	clocked recombination antiblooming
CTD	charge-transfer device
CTE	charge-transfer efficiency
CTF	contrast-transfer function
CW	continuous wave
DCP	direct current plasma
DMCP	double microchannel plate
DNA	deoxyribonucleic acid
EB	ethidium bromide
EBI	equivalent background illumination
EEM	excitation-emission matrix
EMA	extramural absorber
FET	field-effect transistor
FSF	fast spectral framing
FT	Fourier transform
FWHM	full width at half maximum
GC	gas chromatography
HPLC	high-performance liquid chromatography
HPTLC	high-performance thin-layer chromatography
ICCD	intensified charge-coupled device
ICP	inductively coupled plasma
IPDA	image-intensified photodiode array
IR	infrared
KTC	Boltzmann constant × temperature × capacitance (noise)
LED	light-emitting diode

LIF	laser-induced fluorescence
LDD	lightly doped diffusion
LN/CCD	liquid-nitrogen-cooled CCD
lp/mm	line pair/mm
LOD	limit of detection
LTE	local thermodynamic equilibrium
MAMA	multianode microchannel arrays
MC	multichannel
MCP	microchannel plate
MCT	mercury cadmium telluride
MDS	minimum detectable signal
MOS	metal oxide semiconductor
MPP	multipinned phase
MTF	modulation-transfer function
MX	multiplex
Nd:YAG	Neodymium:yttrium-aluminum garnet (laser)
NDRO	nondestructive readout
NEP	noise equivalent power
NIR	near infrared
ODA	oxydianiline
PAH	polycyclic aromatic hydrocarbon
PDA	photodiode array
PECVD	plasma-enhanced chemical vapor deposition
PMT	photomultiplier tube
QE	quantum efficiency
RA	resistive anode
RAI	random access integration
rf	radio frequency
RLD	reciprocal linear dispersion
rms	root mean square
RNA	ribonucleic acid
SC	single channel
SEM	scanning electron microscopy
SSC	simultaneous spectral coverage
SERS	surface-enhanced Raman spectroscopy
SIT	silicon intensified target
SNR	signal-to-noise ratio
SR	spectral resolution
SWIR	short wave infrared
TDI	time-delayed integration
TLC	thin-layer chromatography
UV	ultraviolet
VIS-NIR	visible-near-infrared
VIT	variable integration time
XIDS	Crossed interferometric dispersive spectrometer

1
A Brief History of Charge-Transfer Devices

Gary R. Sims

Spectral Instruments
Tucson, Arizona

1.1 Charge-Transfer Devices

Charge-transfer device (CTD) is a term given to two types of related solid-state optical detectors, the charge-injection device (CID) and the charge-coupled device (CCD), both of which were first described in the early 1970s. CTD technology moved rapidly from the research laboratory to the fabrication of practical, operating devices because of their potential as solid-state replacements for vidicon tubes for video image-pickup applications. Today the potential has become a reality and nearly every video camera made, including those for home video cassette recorders, uses a CTD as the image-pickup device.

At the same time that CTD technology was being perfected for video use, a small group of people in the scientific community began to install the devices into specially constructed, cooled, slow-scan cameras in attempts to replace photographic emulsions and intensified-vidicon tubes for high-precision scientific measurements. The attempts were, for the most part, extremely successful, and a focus on CTD technology specifically for scientific applications has evolved.

This chapter describes the inception of CTDs and gives an account of their early use as scientific imaging detectors. This account is by no means all-inclusive but tries to touch on the highlights in a variety of representative applications.

1.2 Invention of the Charge-Coupled Device

The CCD concept was invented in 1969 by Willard Boyle and George Smith of Bell Laboratories in Murray Hill, New Jersey. Boyle was executive director of the semiconductor division, and Smith was head of a department given the task of developing a silicon diode detector for a Picturephone. During a casual afternoon brainstorming session, the possibility of an electronic analog to the approach used to transport magnetic domains in magnetic bubble memory was discussed (bubble memory was in the forefront of electronic research activity at that time). It was known that charge could be stored in a semiconductor using the electric field generated by applying a voltage to a metal oxide semiconductor (MOS) capacitor. The idea of passing charge from one capacitor to another by changing an applied voltage in a coordinated fashion was devised, and the CCD was born!

The CCD concept was first described in an article authored by Boyle and Smith in the *Bell System Technical Journal* in 1970 [1]. In a separate article in the same volume, the experimental verification of the charge-coupled concept was announced by authors Smith, Gil Amelio, and Mike Tompsett [2]. The first device built and described in their publication was a simple linear eight-element surface-channel device.

The significance of the charge-coupled concept was quickly recognized throughout the semiconductor industry. In addition to being employed as an optical detector, it was used or considered for use as an analog signal delay line, digital logic, and digital memory. Research into the technology proceeded simultaneously in many industrial laboratories including RCA, Texas Instruments, and Fairchild Semiconductor. The first commercially available CCD imaging detector was introduced by Fairchild in 1973. A photograph of this 100 pixel × 100 pixel array is shown in Figure 1.1.

In 1976 the CCD was put to scientific use for the first time. A team including Jim Janesick from the Jet Propulsion Laboratory in Pasadena brought a system they were using to evaluate CCD technology for potential use in future interplanetary satellites to the 61-inch telescope on Mt. Bigalow outside of Tucson, Arizona. There, working in conjunction with Dr. Bradford Smith of the University of Arizona's Steward Observatory, they obtained the first astronomical images recorded with a CCD.

Through these types of experiments, it was recognized that CCDs had the potential to be superior to other imaging detectors as a replacement for film emulsions and vidicon-type detectors for scientific applications. The development of "scientific CCDs" was slow since the main incentive for industrial CCD development was to make superior video devices for commercial, consumer, and military applications; scientific CCDs are those intended for very-low-light-level, high-dynamic-range imaging by being operated at very low readout speeds and cooled to lower dark current. The one notable exception was a program funded by NASA beginning in 1973 and conducted by Texas Instruments to develop CCDs for use in the Galileo mission to Jupiter. Throughout the 1970s, the scientific use of CCDs was limited to the very few who were fortunate enough to acquire a device of usable quality.

Figure 1.1 Devices used in the pioneering work of evaluating CCDs. Left: Fairchild 100 pixel × 100 pixel array. Right: RCA SID501. Fairchild device courtesy of Jim Janesick, Jet Propulsion Laboratory.

In 1979, a 320 pixel × 512 pixel CCD made by RCA (RCA 501) became available to scientific users. This device, also shown in Figure 1.1, had many desirable properties for scientific applications including high quantum efficiency throughout the visible and near-ultraviolet spectrum. This was achieved by a new approach termed backside illumination. An equally important feature was that the device had buried-channel architecture, which resulted in good charge-transfer efficiency even at low light levels. Even though this device had a readout noise at least an order of magnitude above what is standard on new CCD designs today, many of the RCA CCDs are still in use.

Throughout the 1970s, the scientific use of CCDs was almost exclusively in the domain of astronomy. In 1979, Ken Ratzlaff, then at Northern Illinois University, described the use of a linear CCD for molecular absorption spectroscopy [3]. Otherwise, the reported nonastronomical applications of scientific CCDs began to appear in the early 1980s, with most appearing after 1983.

The decade of the 1980s was a tumultuous period for the supply of scientific grade CCDs. Most companies manufacturing CCDs either had severe production problems that prevented delivery of devices for long periods, or the companies withdrew from the business altogether. One of the biggest difficulties was that, for most CCD manufacturers, video quality devices were the principal product and scientific CCDs were a problematic sideline. Not until 1985 did one company, Tektronix, begin manufacturing CCDs exclusively for scientific use. Unfortunately, for the first several years production problems limited the supply of Tektronix devices.

A scientific CCD operates by moving a small packet of electrons across as much as several centimeters of silicon. Therefore, any minuscule imperfection in the device design or the fabrication process manifests itself as a defect or impairs performance. The decade of the 1980s was a time when many organizations worldwide, both public and private, focused on understanding the factors limiting CCD performance, the causes of defects, and how to successfully produce devices with near theoretically perfect performance.

In the 1990s, the scientific community is fortunate to have available a broad range of excellent scientific CCD devices from several manufacturers. While many advances are yet to be made in CCD technology, the devices available today are often the best imaging detectors for a wide range of applications. Conditions are now conducive for explosive expansion in the use of CCD detectors in scientific instrumentation.

1.3 Invention of the Charge-Injection Device

In 1972, Gerry Michon and Hugh Burke of the General Electric Company's corporate research center were working on the development of solid-state memory devices. They experienced difficulty during probe-testing the experimental wafers due to optical sensitivity. Consequently they decided to try to build a light detector that was intended to be light-sensitive and placed a single-element device consisting of two adjacent MOS capacitors as a test structure on the edge of a wafer of experimental memory devices.

This first device showed promise, so they next built the 32 pixel × 32 pixel array shown in Figure 1.2 and used it to acquire images. This first device had no on-chip multiplexer or scanners; it simply had 32 clocked lines and 32 lines to sense charge [4]. The commercial and military potential of this technology was recognized and led to the development of a much more sophisticated 100 pixel × 100 pixel device that incorporated the on-chip scanners. This device was used in a commercially available camera beginning in 1973, the same year that Fairchild introduced the first commercially available CCD imager.

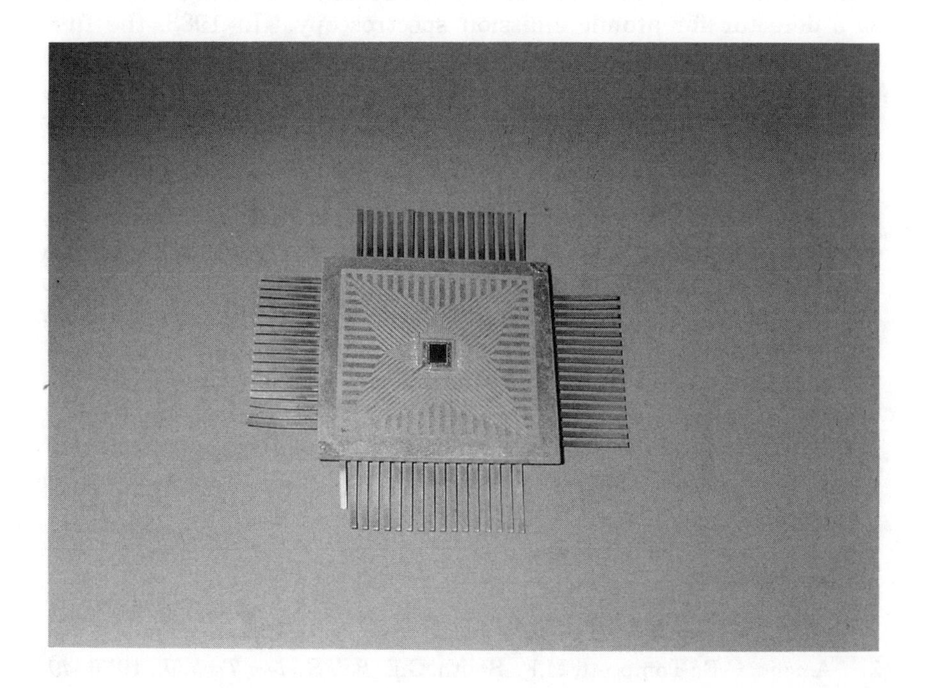

Figure 1.2 An early experimental 32 pixel × 32 pixel CID. Devices courtesy of Gerry Michon, General Electric Corporate Research Center.

In the early 1970s, CIDs were promising as scientific imaging detectors because they did not suffer from the charge-transfer inefficiencies exhibited by early CCDs. Also the CID was sensitive over a wider wavelength range and was less prone to having catastrophic defects. CIDs were reported to be used for astronomical measurements at the Kitt Peak National Observatory west of Tucson, Arizona as early as 1976 [5], the same year that the CCD was first used for astronomy. The CID was found to be better suited for higher-light-level applications such as solar astronomy where the high pixel-charge capacity was an advantage. The CIDs were not as well suited for very-low-light-level applications as CCDs because of higher readout noise.

In 1978, M. Bonner Denton of the University of Arizona learned of the properties of the CID while visiting the Steward Observatory laboratory for the newly built Multiple Mirror Telescope on Mt. Hopkins, south of Tucson. CID video cameras were being used there to optically align the six mirrors that made up the telescope. Soon his research group began to explore the possibilities of using this device as a detector for atomic emission spectroscopy. In 1983, the first spectroscopic results were reported, which demonstrated that a CID was a viable detector for multichannel emission spectroscopy [6]. This work has continued, resulting in an instrument that is now commercially available from Thermo Jarrell-Ash.

The scientific use of CIDs has been much more limited than that of CCDs. CIDs have many unique properties that make them the detector of choice in several applications, but they also suffer from a much higher readout noise than CCDs. Another factor that has affected the adoption of CIDs is that commercial devices are available from only one supplier (CIDTEC, a General Electric spinoff company) and in a limited selection of sizes and formats. Now that "CCD" has become a household word, the CID also suffers from an identity problem. Few scientists know what it is or of the properties that make it unique. Presently the CID is a mature and available technology looking for new applications.

References

1. Boyle, W.S.; Smith, G.E. *Bell System Tech. J.*, **1970**, *49*, 587.
2. Amelio, G.F.; Tompsett, M.F.; Smith, G.E. *Bell System Tech. J.*, **1970**, *49*, 593.
3. Ratzlaff, K.L.; Paul, S.L. *Appl. Spectrosc.*, **1979**, *33*, 240.
4. Michon, G.J.; Burke, H.K. *Digest IEEE Int. Solid-State Circuits Conf.*, **1973**, *16*, 138.
5. Aikens, R.S.; Nelson, R.E.; Lynds, C.R. *Proc. SPIE*, **1976**, *78*, 65.
6. Sims, G.R.; Denton, M.B. In *Multichannel Imaging Detectors*, Talmi, Y. Ed., ACS Symposium Series 236, Vol. 2. American Chemical Society, Washington, D.C., 1983.

Bibliography

Several excellent general books and review articles on CCD and CID technology have appeared in the literature. Some recommended treatments include the following:

Beynon, J.D.E., Lamb, D.R. *Charge-Coupled Devices and Their Applications*. McGraw-Hill, London, 1980.

Howes, M.J., Morgan, D.V. Eds. *Charge-Coupled Devices and Systems*. Wiley, New York, 1979.

Sequin, C.H., Tompsett, M.F. *Charge Transfer Devices*. Academic, New York, 1975.

Mclean, I.S. *Electronic and Computer-Aided Astronomy*. Ellis Horwood Limited, Chichester, 1989.

Amelio, G.F. "Charge-Coupled Devices," *Scientific American*, 296–304, February, 1974.

Janesick, J., Elliott, T. "History and Advancements of Large Area Array Scientific CCD Imagers." In *Astronomical Society of the Pacific Conference Series*, Tucson, Arizona, 1991.

2
Principles of Charge-Transfer Devices

Gary R. Sims

Spectral Instruments
Tucson, Arizona

2.1 Introduction

The purpose of this chapter is threefold: to introduce the terminology commonly used in describing CTDs, to outline the principles of CTD operation and the characteristics that pertain to their use in scientific applications, and to give a picture of the current state of the art in CTD technology. An attempt will also be made to describe the developments in CTD technology that are on the horizon.

Whereas CTDs were initially invented and designed for use as solid-state video pickup devices, their properties have proven so valuable as multichannel detectors for a variety of scientific measurement applications that an entirely unique class of devices, the scientific CCD and scientific CID, has emerged.

This chapter cannot possibly be a complete treatise on CTDs; even the subject of scientific CTDs would fill many volumes if treated adequately. Consequently, the reader is encouraged to consult the many references cited for a thorough understanding of the physics of CTDs, especially the semiconductor physics literature.

This chapter encompasses only silicon CTDs. Many CTDs are made from other materials, including Ge, PtSi, InGaAs (for use in the IR), but a discussion of this technology requires a separate treatment and is the subject of Chapter 4.

There are several excellent and quite detailed descriptions of metal oxide semiconductor (MOS) and CCD technology [1–3], and this chapter will not attempt to review all of the important issues. The discussion in this chapter is brief except when some particular aspect pertinent to scientific application has not been well documented elsewhere.

An excellent overview of scientific CCDs by James Janesick of California Institute of Technology and Morley Blouke of Tektronix appeared in *Sky and Telescope* [4]; to explain CCD operation, they used the analogy of measuring the spatial distribution of rainfall over a field using a system of conveyors. In this analogy, a grid of buckets is closely arrayed on a conveyor as in Figure 2.1. These buckets collect the rainfall, and when the rain has ceased, the conveyor is moved such that one line of buckets dumps into another single line of buckets on a conveyor at right angles to the first. The second conveyor moves the water from buckets serially to a graduated measuring vessel where the water is measured, then dumped in preparation for measuring the next bucket.

In CTD terms, the rainfall is equivalent to incident photons; the buckets, which are discrete sampling points, are called pixels, which is a condensation of the words *picture* and *elements* and is a term borrowed from television technology. When the photons strike the silicon, they are converted into charge that can be transported and measured by electronic structures built on a monolithic silicon chip that are analogous to the conveyors and measuring vessels.

Figure 2.1 The analogy of measuring rainfall on a field with buckets.

An analogy similar to the one used to describe CCD operation can be made for CIDs. In this case, buckets are again used to collect rain falling over a field. When the rain stops, the rain from one line of buckets is dumped into adjacent, graduated, measuring vessels as shown in Figure 2.1. The sum of all water from one orthogonal line of measuring vessels is read to determine the amount of rain that fell into a specific bucket. Once the measurement is completed, the water can be either dumped out of the line of measuring vessels or returned to the original collection bucket.

This analogy points out the principal difference between CCDs and CIDs. With CCDs, the charge from each pixel is brought to a single point to be measured. The measuring device is very precise, and so the most critical operation is to make sure the charge gets to the measurement point without "spilling." With a CID, the charge does not move very far because each pixel contains its own measurement device; however, the measuring devices are less precise. The CID has the major advantage that the charge in a pixel can be measured and then either discarded or poured back into the collection bucket to gather more charge.

2.2 Charge Generation in Silicon

2.2.1 Origin of Signal Charge

The valence electrons in a semiconductor such as silicon normally are immobile since they reside in a valence band of the lattice where they are associated with a specific atom. If an electron acquires sufficient energy to be promoted to the conduction band, it becomes mobile and thus can be transported and detected. When a free electron is generated, a vacancy for an electron, a *hole*, is simultaneously created. The hole is mobile like the electron, and the two will recombine after approximately 100 μs unless they are separated.

The indirect bandgap energy (for which a change in electron momentum with the assistance of a lattice phonon is necessary) for pure, undoped silicon is 1.12 eV at 300 K and 1.17 eV at 0 K [1]. If the lattice absorbs a photon of this or higher energy, a mobile photoelectron and a vacancy in the lattice (a hole) will be created. The process requires energy from a photon with a wavelength of approximately 1100 nm or less. At the temperatures that scientific CCDs are often operated, -70 to -100°C, the necessary wavelength decreases to approximately 1090 nm. At longer wavelengths, a photon cannot interact with the lattice, and therefore the silicon is transparent.

The direct bandgap energy for silicon (for which no lattice phonon assistance is necessary for charge generation) is 3.65 eV. When a photon has sufficient energy to promote an electron to the conduction band by the direct bandgap and sufficient energy remains to promote an electron by the indirect bandgap, multiple electron-hole pairs may be created by a single photon.

In the soft x-ray energy range, the generation of multiple electron-hole pairs can be used to determine the incident x-ray energy, since one electron is generated per 3.65 eV energy. This property has been exploited to make imaging energy-dispersive x-ray spectrometers sensitive in the 0.1 to 10 keV energy range [5].

2.2.2 Origin of Dark Current

Undesirable charge as "dark current," or more accurately "dark charge," is also generated in CTDs as a result of the promotion of electrons to the conduction band by thermal energy. For pure defect-free silicon the rate of thermal charge generation is quite low. It would be equal to approximately 30 e^-/pixel/s in a CCD at room temperature as opposed to the 1500 to 15,000 e^-/pixel/s actually experienced. The reason that dark current is so high is that there are many defects and discontinuities in the silicon lattice that give rise to energy states that lie between the conduction and valence bands. If the residence time for an electron in one of these states is sufficiently long, the electron can be promoted in a fashion analogous to climbing a staircase from the valence to conduction band. Only enough energy is required to climb one step at a given time, and this range of energy is readily available thermally.

The surface of silicon is normally oxidized, and the region where the transition from silicon to silicon dioxide occurs gives rise to most of the intermediate states in a modern CTD. These "surface states" arise for a variety of reasons, including the presence of suboxides at the interface; they not only are extremely important for dark-current generation but also affect charge transport. They will be discussed again in section 2.3.1.4.

Other defects in the silicon such as impurity atoms and crystal-lattice dislocations give rise to localized regions of extremely high dark current. These so-called hot defects can be sufficiently detrimental to render a device unusable.

Dark current due to all mechanisms can be reduced to insignificant levels by cooling the CCD. The reduction factor is approximately exponential, decreasing the dark current by a factor of two for every 6 to 7°C drop in temperature. Cooling in the temperature range of -10 to -50°C, where dark currents are typically a few to a few tens of e^-/pixel/s is usually achieved using multiple-stage thermoelectric (Peltier effect) refrigerators. Lower temperatures are achieved using cryostats filled with liquid nitrogen. Cryogenically cooled CCDs are normally operated in a temperature range of -100 to -140°C where dark current can be less than 1 e^-/pixel/hr.

2.2.3 Anomalous Charge Spikes

When long optical integration times are used in CCD detection, it is common to observe that some pixels have anomalous charge spikes, commonly called cosmic ray spikes, upon readout. These spikes are randomly distributed in time and space and can consist of many hundreds or thousands of electrons either in single pixels or tightly clustered about a few neighboring pixels. These spikes are caused by high-energy particles or radiation from either terrestrial or extra-terrestrial sources. The extraterrestrial sources are primarily due to by-products (muons) of cosmic rays. Terrestrial sources are numerous but are mainly natural radioactivity from materials near the CCD.

It is not practical to shield the camera from cosmic-ray-induced spikes. They must be identified and corrected in the data analysis algorithm. On the other hand, careful selection of materials used in the CCD detector system can have a large impact on the number of spikes caused by natural radioactivity. Some materials that have been identified as having high radioactivity are ceramic materials (even those used in some CCD packages), optical glasses, and antireflection coatings. Some glasses use materials that have a high proportion of radioactive isotopes, such as uranium and thorium, to adjust the index of refraction. Likewise, similar materials are sometimes used as antireflection coatings. Even common optical glasses that do not intentionally use radioactive materials, such as BK-7, have been found to have radioactive impurities that lead to significantly higher anomalous spikes than high-purity fused silica [6].

2.3 Structure of Charge-Coupled Devices

2.3.1 Charge Storage and Transport in MOS Structures

The ability to store charge indefinitely in silicon using MOS structures is the enabling capability of many modern, solid-state electronic devices. The added capability to transfer charge quantitatively between adjacent MOS structures is the essential element of CCDs.

The charge-transport capability of modern CCDs is extraordinary. It is now routine practice to transport only a few electrons up to several centimeters from the point they were created to where they are detected without the discernible gain or loss of any electrons! This capability is the key that makes modern CCDs highly useful optical detectors for critical scientific applications.

A simple MOS structure is shown in Figure 2.2. The key elements are the *substrate*, *epitaxy*, *insulators*, and *gate*. The substrate is simply a mechanical support for the other elements of the structure and also often serves as a common electrical contact. It consists of highly *p*-doped (with boron) silicon and is approximately 0.5 mm thick. The epitaxy, or *epi*, is the photoactive portion of the CCD. It is typically 10 to 20 μm thick and consists of lightly *p*-doped silicon. CCDs use *p*-doped silicon for substrate and epi and collect photogenerated electrons. Electrons are more mobile than holes in silicon; collecting and moving electrons rather than holes is preferable in CCDs because faster charge-transfer rates and higher readout rates are possible.

The insulators typically consist of two layers: silicon dioxide and silicon nitride. The silicon dioxide is grown by oxidizing the surface of the silicon to a depth of approximately 200 angstroms. The conditions used for this oxidation step are critical because these conditions affect the density of surface states and thus the dark current. The silicon nitride is deposited by chemical vapor deposition to a thickness of approximately 200 Å. It is used to cover any small holes in the silicon dioxide and to ensure that there is an insulator between the epi and the gate.

The gate electrode is not normally metal in CCDs, as the name *metal oxide semiconductor* implies, but is instead highly doped polycrystalline silicon. This material, commonly called polysilicon, is partially optically transparent at wavelengths longer than 400 nm. It must be highly boron doped to increase its conductivity. A metal contact is made to the gate so that it can be connected to an external source of electrical potential.

When an electrical potential is applied to the gate, an electrical field is created in the silicon immediately below the gate. The field is strongest at the interface between the silicon and silicon dioxide. If the applied gate potential is positive with respect to the epi, a region where electrons can be stored is created under the gate. Electron-hole

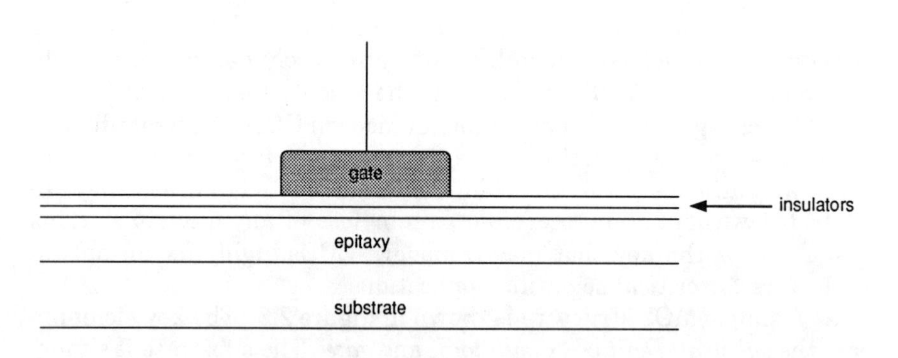

Figure 2.2 A simple MOS structure showing the substrate, epitaxy, insulators, and gate.

recombination will not occur in this area because the holes are forced out of the vicinity (they are normally conducted out of the device through the substrate). Commonly, the charge-storage capacity underneath gates is diagrammed as potential wells, as is shown in Figure 2.3. The depth of the well is a relative indication of the applied gate potential and the corresponding charge-storage capacity under the gate. The amount of charge in the potential well is indicated by the shaded area in the potential well. It must be remembered that while the potential well diagram shows the charge at the bottom of the potential wells, in reality the charge resides as near to the gate as possible, which places it at the silicon/silicon dioxide interface.

2.3.1.1 Charge Coupling

In Figure 2.4, a second gate has been added to the structure that partially overlaps the first gate. If the same positive potential is applied to both gates, charge can be stored under both. If one gate is biased negatively, the potential well under it collapses and the charge is stored under only one gate. Because the gates overlap, their potential wells overlap and charge can be moved from under one gate to under the other. This ability to move, or couple, the charge from one gate to another is the central concept behind the CCD.

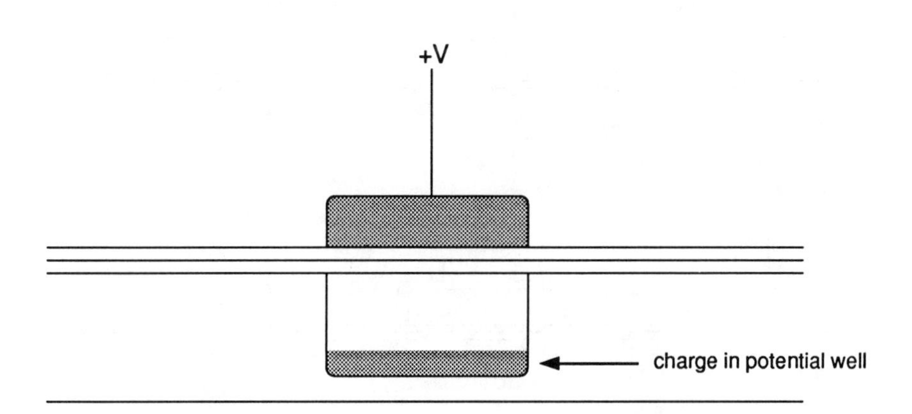

Figure 2.3 The potential well structure of a CTD, which provides the charge-storage capacity.

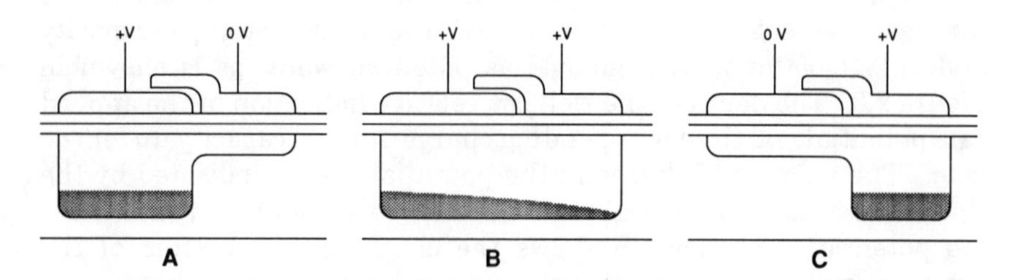

Figure 2.4 Charge coupling by the addition of a second gate. The charge can be moved by changing potentials on the gates.

If more electrodes are added to the first two, the potential wells can be changed over time in a coordinated manner such that the charge is made to move from one electrode to another in a constant direction as is shown in Figure 2.5. This is analogous to a digital shift register, and so the term *CCD register* is commonly used for this structure.

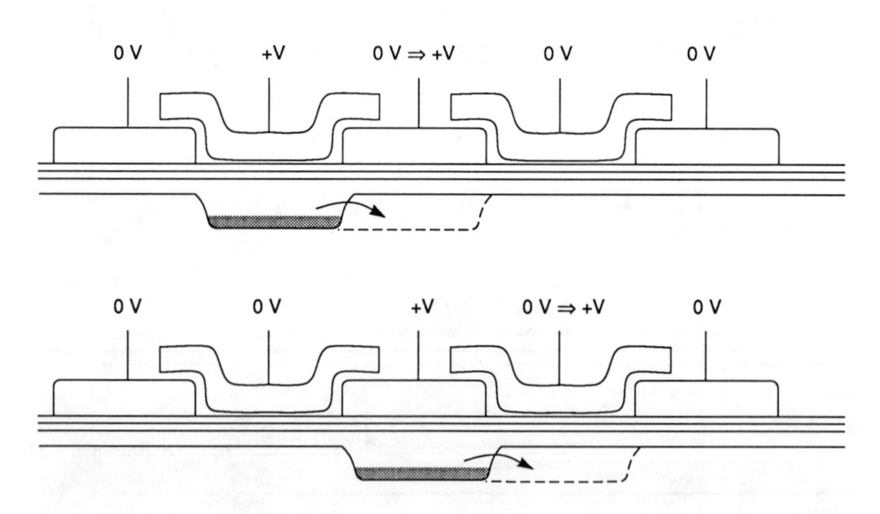

Figure 2.5 Multiple electrodes to create the CCD's analog shift register.

Electrodes can be grouped into sets of $n = 2, 3$, or 4 to create distinct charge packets that are moved in concert with one another. This division is how the register is divided into pixels and is shown in Figure 2.6. Every nth electrode is connected to the same source of potential, so that there are only n connections, or *phases*, to the register. When charge is being collected, the phases held positive are called the collecting phase and those held negative are called the barrier phases.

2.3.1.2 Charge-Collection Efficiency

It is important that the charge generated in a pixel be collected only in that same pixel. If charge is collected by an adjacent pixel, there is a loss in fidelity equivalent to a loss of spatial resolution. The measure of how effective pixels are at collecting their own charge is called charge-collection efficiency, abbreviated CCE.

CCE is strongly dependent on the epi thickness, the electric field penetration into the epi from the collection-phase gates, and the wavelength of incident light. In standard commercially available CCDs, the epi is 10 to 20 μm thick, but the field from the collection gates penetrates only a few microns, as is shown in Figure 2.7. The penetration depth is normally called the depletion depth, because the electric field reduces the population of holes where the field is present, thus depleting their population in the p-type silicon.

The depletion depth is dependent on the gate potential and the electrical permitivity of the silicon. In standard commercial CCDs, the depletion depth is only a few microns, yet the epi thickness is 10 to 20 μm. Photons absorbed in the field-free region below the absorption depth can either migrate until they fall into the depletion region and are collected, or drift laterally until they are collected in another pixel or recombine with a hole. For this reason, the spatial resolution of CCDs falls at near-infrared wavelengths.

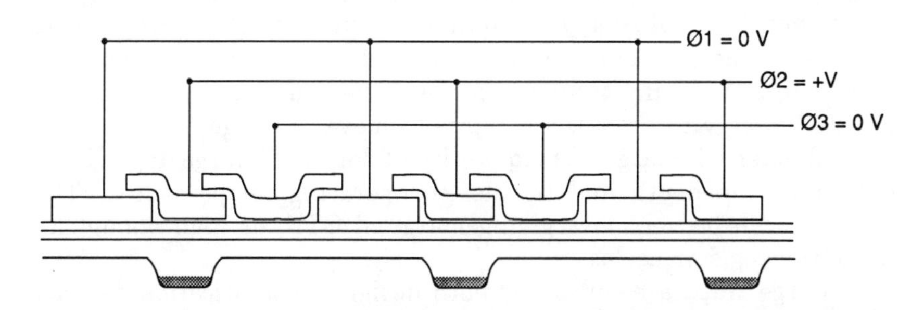

Figure 2.6 Creation of pixels via the electrode structure.

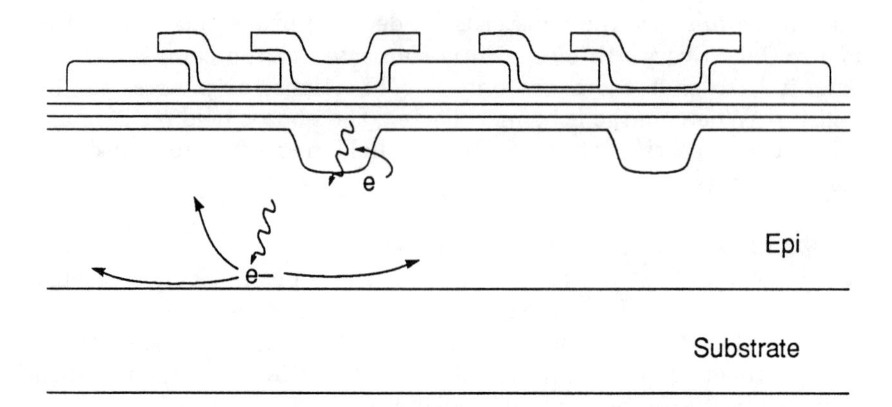

Figure 2.7 Effect of the absorption depth on the charge-collection efficiency.

2.3.1.3 Charge-Transfer Efficiency

In CCD registers, charge packets undergo thousands of transfers and travel across up to several centimeters of silicon. The process of transferring charge from one electrode to another must be extremely efficient, or the charge packet will become spread out. The term *charge-transfer efficiency*, or CTE, is used to describe the degree of charge transfer. A CTE of 1.0 denotes perfect transfer.

Charge-transfer efficiency is one of the most important performance measures of CCD registers. It is essential that when a charge packet is transferred, none is left behind. There are several possible causes for decreased CTE, but all can be reduced to wrinkles, or traps, in the potential wells, either under gates or between gates.

While the previous drawings of potential wells showed them to be perfectly smooth, in reality there can be imperfections in the bottoms of the wells and between wells. When charge is moved from one gate to another, thermal energy is required to cause the electrons to move laterally so they can fall into the deeper potential well as is illustrated in Figure 2.8. If any traps are present there may not be sufficient thermal energy to prompt all charge to move. Charge left behind is termed deferred charge. The quantity of deferred charge depends on both the trap depth and the device operating temperature. The quantity can be very low, giving marginal CTE, or high enough to make the register useless.

Charge traps are caused by both design and fabrication defects. Design defects are no longer common on commercially available CCDs, but until recent years this was not the case [7]. More often, traps are

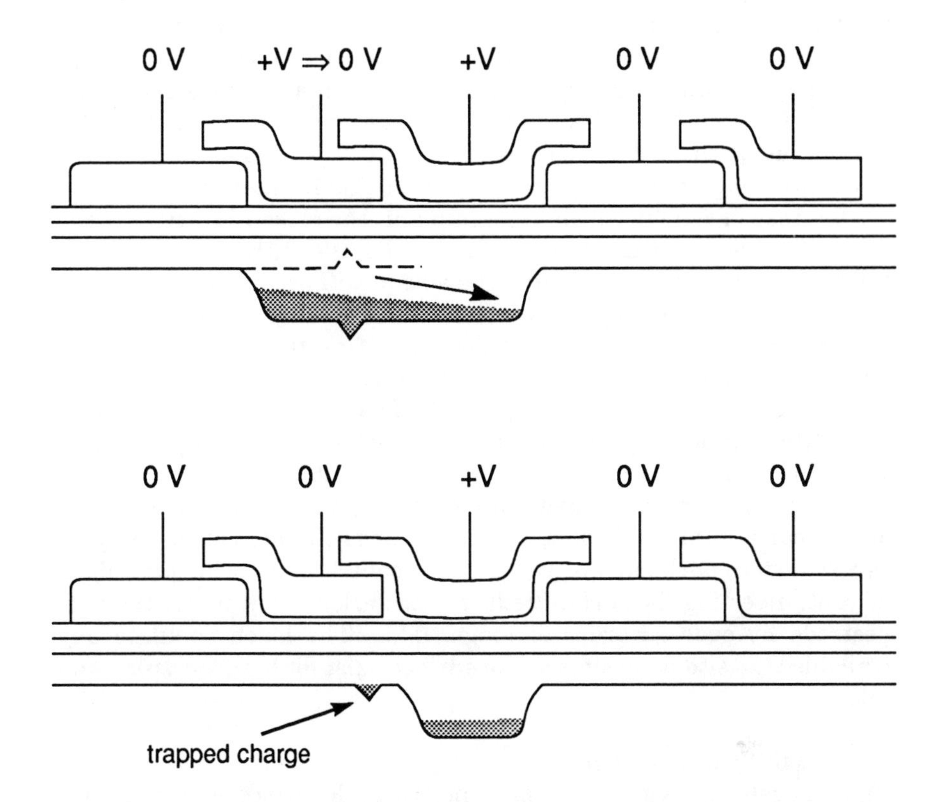

Figure 2.8 Traps in the potential wells caused by defects affect charge-transfer efficiency.

caused by defects in the silicon used to fabricate the CCD or by a defect introduced in the manufacturing process.

2.3.1.4 Surface States and Buried-Channel CCDs

When electrons are allowed to contact the silicon/silicon dioxide interface they interact with the physical defects that give rise to surface states. When an electron falls into a surface state, it becomes physically attached to a particular site, called an interface trap, analogous to the traps described in the previous section. This effect caused poor CTE in early CCDs, and thus the buried-channel CCD was introduced. The buried-channel CCD has an n-type region just under the surface of the silicon that is created by implanting phosphorus ions into the silicon. This buried-channel implant creates an electric field inside the silicon that causes the potential energy minimum for electrons to exist below the silicon surface. This means that for most normal applied gate potentials, the stored charge will not contact the silicon/silicon dioxide interface where it can interact with interface traps. If the gate potential becomes too high, the charge can be pulled to the surface, and this is intentionally done in a few rare cases that will be discussed later. The buried-channel approach yields CTE far above that of the older surface-channel devices, and thus all CCDs made after approximately 1980 have used this technology.

Because of the buried-channel implant, the silicon is n-type at the silicon dioxide interface. If a gate is biased sufficiently negative, holes will be drawn to the surface, and the silicon will become inverted to p-type. Once holes begin populating the surface, making the gate more negative will only attract more holes, offsetting the higher potential. The electric field in the CCD will not change. When this condition exists, the surface is said to be pinned. The gate potential necessary to pin the surface depends on the buried-channel implant dose and insulator thickness, but is typically -6 to -7 volts.

When the surface is inverted or pinned, the holes drawn to the surface can annihilate dark-current electrons generated at the surface. As the surface dark current is the major contributor to the total dark current, inverting the surface reduces the dark current under the gate electrode by over an order of magnitude. The holes populate any available trap site (trap sites at the surface exist for both electrons and holes).

2.3.1.5 Spurious Charge

When a gate is switched from a negative, inverted potential to a positive potential, the holes are forced away from the surface. Some remain stuck in interface trap sites for up to several milliseconds, where they can continue to annihilate dark electrons until exhausted. When holes are forced from the surface, they can reach high velocities that can cause impact ionization. Charge generated by clocking from a negative, inverted potential to a positive potential is termed spurious charge. It is detrimental in CCDs because the charge is indistinguishable from dark current. Spurious charge can be controlled either by not using voltages so negative as to cause inversion or by making the transition from negative to positive slowly so as not to accelerate the holes to too high a velocity.

2.3.1.6 Charge Capacity

The charge capacity (full-well capacity) of MOS storage structures is dictated by the volume of depleted silicon. This in turn is dictated by the surface area of the gate electrodes (which are positively biased), the positive potential applied, and the permitivity of the silicon, as was discussed in the section on CCE.

To increase the storage capacity of a gate, the potential is increased as much as possible. If the potential becomes too high, stored charge will leave the buried channel and reside at the silicon/silicon dioxide interface as the pixel fills with charge. This is equivalent to operating as a surface-channel CCD. As discussed before, this condition leads to poor charge-transfer efficiency and residual oxide charge, even though the charge capacity is higher. When precise measurements are needed from a CCD, surface-channel operation must be strictly avoided.

When the well capacity of a buried-channel CCD is exceeded, the charge spills into adjacent pixels, a phenomenon termed blooming. The goal in establishing the positive potentials of the register gate is to maximize the pixel charge capacity but at the same time to ensure that excess charge will stay in the buried channel while spilling to the adjacent pixels instead of interacting with the surface.

2.3.2 Register Architectures

A number of different approaches to CCD fabrication have evolved. They are functionally equivalent in that they move charge, but they differ in other characteristics.

The four-phase CCD architecture is one of the most simple both conceptually and in terms of necessary steps to fabricate. Every fourth gate is connected, as shown in Figure 2.9. Only four external connections are then necessary, and these are termed clock phases 1 through 4. The structure repeats itself every fourth gate, and there are four clock phases used; thus it is known as a four-phase CCD. Every repetition is a unique structure and constitutes a pixel.

When charge is collected, only one clock phase is held positive. By changing the clock phases through eight unique states, the charge is

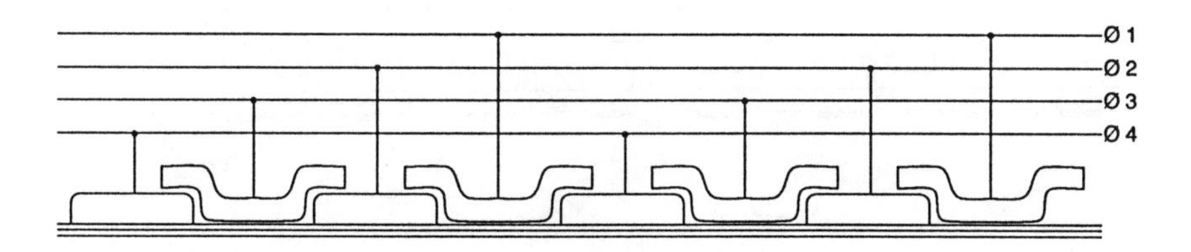

Figure 2.9 Four-phase architecture.

moved one pixel, as is shown in Figure 2.9. The direction of travel can be either left or right depending on the sequence of clocks.

The four-phase CCD architecture is simple because only two levels of polysilicon are required to create the necessary overlapping electrodes. A somewhat more complicated structure is the three-phase CCD, shown in Figure 2.6. This structure requires three levels of polysilicon. The reason the three-phase architecture is widely used in scientific CCDs is that the fabrication yields are generally higher than those of the four-phase architecture despite the additional fabrication steps necessary.

In operation the three-phase CCD is analogous to the four-phase approach. Only six unique clock states are needed to move the charge, and it can be moved either direction.

The two-phase CCD architecture is similar to the four-phase in that two polysilicon levels are used, but an added structure is needed to cause the charge to move only in one direction. The added structure is a shallow *p* implant placed between the *n* buried channel and the gate. The implant covers only the area of one-half of each gate, where it raises the potential well under each gate as shown in Figure 2.10. The stepped potential well causes the charge to move in one direction only when the two clock phases are simultaneously changed.

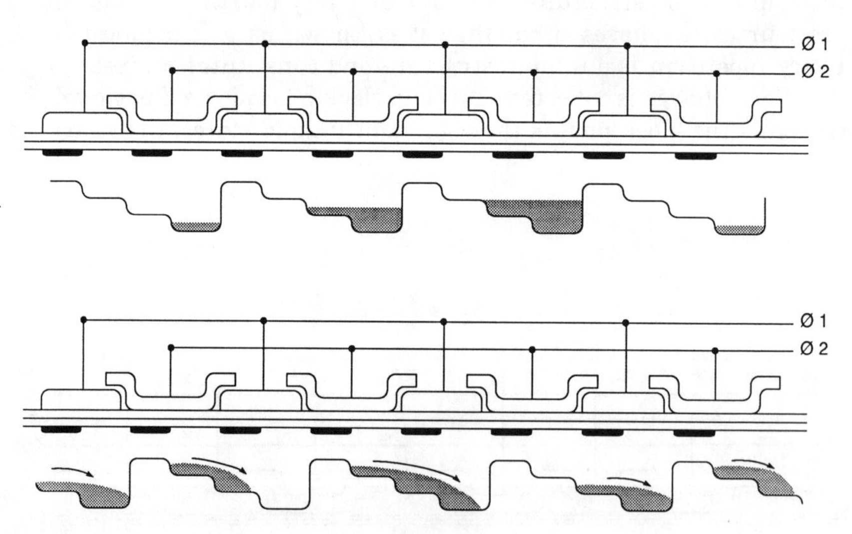

Figure 2.10 Two-phase architecture.

The two-phase architecture is very simple to operate because it requires only two complementary clock phases, but both clock phases must be clocked at precisely the same time. This architecture is commonly used in high-speed CCDs due to the clocking simplicity.

A fourth type of CCD architecture is uniphase or virtual-phase. This approach takes the two-phase design a step further. Several implants are used in a way that allows one of the two gates to be removed completely. A "virtual" electrode is constructed that consists only of implants; these shallow boron implants are used over the open area to pin the surface and eliminate surface dark current. The operation of the virtual-phase CCD is diagrammed in Figure 2.11. Like the two-phase CCD, the virtual-phase CCD can move charge in only one direction.

One of the main distinguishing features of the virtual-phase approach is that with one polysilicon electrode missing, photons can reach the epi without having to pass through a semitransparent gate. This causes an increased optical sensitivity, especially in the ultraviolet.

The disadvantages of the virtual-phase architecture are that the charge capacity per pixel area is small, charge-transfer efficiency can

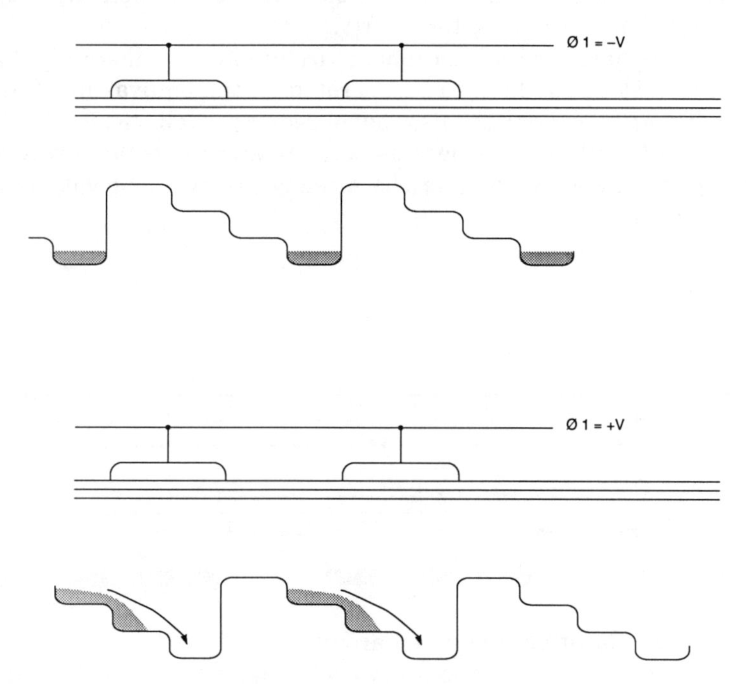

Figure 2.11 Uniphase or virtual-phase architecture.

be low when the CCD is operated at low temperatures, and dark-current spikes are commonplace. Well capacity is reduced by a factor of three to four compared with a multiphase CCD of the same pixel area. Another problem is that high spurious charge sets a high noise threshold. Spurious charge is a problem in virtual-phase CCDs because the lower clock level must be operated below -10 volts to achieve good CTE. This low potential fully populates the surface with holes, which are accelerated when the potential is changed to the normal positive potential of +1 to +2 volts.

2.3.2.1 MPP Three- and Four-Phase Architectures

In recent years, the three- and four-phase CCD architecture has been modified to permit operation in what has been termed multipinned phase or MPP mode. A shallow *p* implant is placed under one gate, which raises the potential energy only under that gate as is shown in Figure 2.12. A barrier phase is created, even when all gates are biased to the same potential. During charge integration, all gates can be biased negatively to cause surface inversion and to shut down dark current from all surface states. The dark current is typically lowered by a factor of 10 to 30 over that observed when only two of three (or three of four) gates are inverted. The barrier implant is not normally quite as efficient as biasing a phase positive, so full well capacity is reduced by as much as a factor of two.

The two-phase and virtual-phase architectures inherently have structures that allow them to be operated in the equivalent of MPP mode. Two-phase devices also have reduced well capacity when operated MPP, but virtual-phase devices are always operated with the clocked phase inverted, and so the already low well capacity is not further reduced.

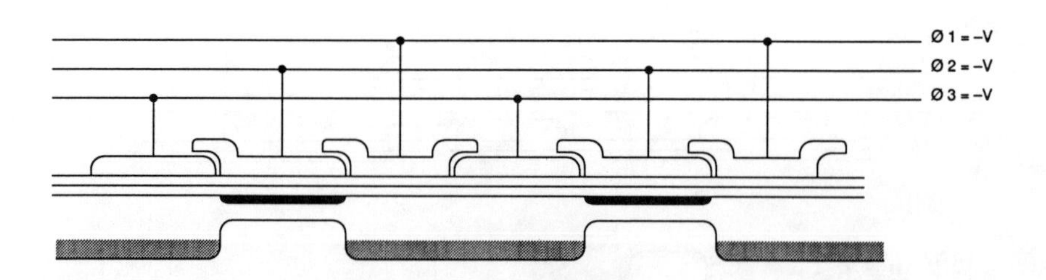

Figure 2.12 Multiphase pinned architecture.

2.3.2.2 Linear and Area Arrays

The register architectures discussed thus far have all been *linear arrays* in which there is a single line of pixels. These arrays are common and important for many commercial applications, such as for use in photocopiers and FAX machines. But for scientific use, most CCDs are *area arrays*. This is mainly because linear arrays with only a single line of pixels have extremely small area for photon collection and have spatial resolution in only one direction. The linear arrays that are commercially available today are mostly designed for high-speed readout, where low readout noise and dark current are not as important, and are not normally considered scientific devices.

Area arrays are much more widely used for scientific applications because of the larger photon collection area and the two-dimensional spatial resolution. Area arrays can be thought of as being constructed of a large number of linear registers. There are two CCD registers on an area array, the *serial register* and the *parallel register*. The parallel register is the photoactive register and consists of hundreds or thousands of linear registers aligned parallel with each other. The linear registers are separated from each other by *channel stops*, which are a region of the epi that has been highly *p*-doped and oxidized. This creates a strong, high-potential-energy region that prevents charge from one linear array from migrating to adjacent arrays, as is shown in Figure 2.13. The gates of the parallel register are common to all of the linear registers; that is, there is one polysilicon structure that extends through all of the linear registers and is connected to electrical contacts only at the top and bottom of the structure.

The parallel register normally has no provisions for charge readout but instead loads the serial register only with charge from the side. One pixel shift of the parallel register loads the serial register with a line of charge packets. The serial register is then operated so that the charge packets are shifted out one at a time to the output amplifier where the charge is measured. The serial register is almost always several pixels longer than the parallel array. This is necessary to have room for the output amplifier at the end of the register. The pixels between the parallel register and the output amplifier are called the serial register extension. This extension normally consists of between 6 and 50 pixels.

2.3.2.3 Parallel Register Shift Times

The parallel register gates have high capacitance due to their length of a centimeter or more. A typical parallel gate capacitance value for a 512 pixel $\times$ 512 pixel array is approximately 5000 pf. These gates also have appreciable resistance, on the order of several kiloohms. The combination of these two factors mean that when a gate is clocked, the clock signal lags by up to several microseconds at the center of the

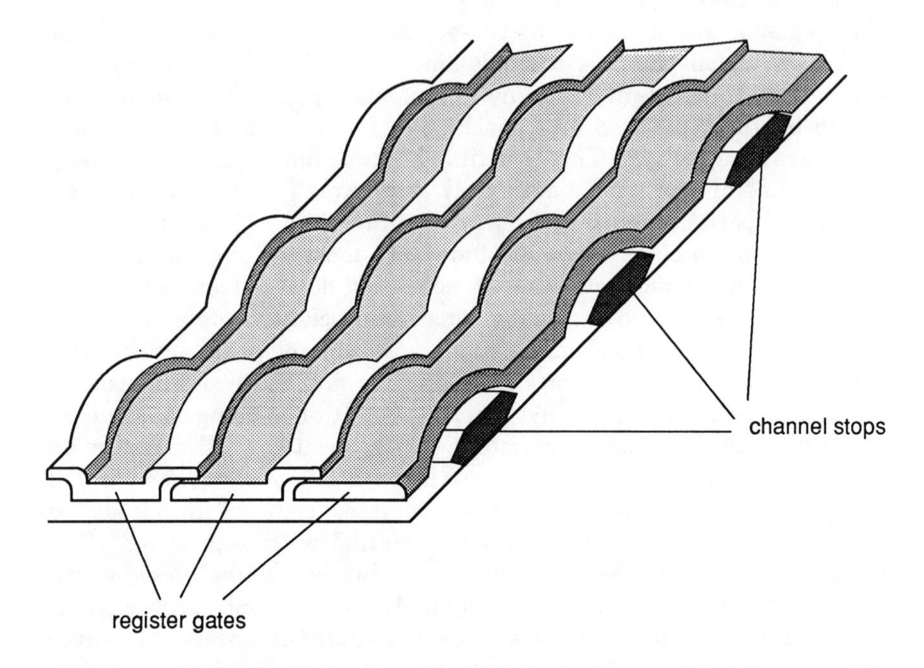

Figure 2.13 Diagram of linear registers showing the channel stops.

parallel array. If the array is shifted too fast, the CTE in the central region will be poor at high light levels because the gates in this area never experience the full potential swing applied to the ends of the gates.

The maximum shift speed varies widely in devices from different manufacturers because it depends highly on the gate insulator thickness (which affects gate capacitance) and the gate polysilicon doping level (which affects gate resistance). In general, even small arrays take 5 to 10 μs to shift a line, and larger arrays can take up to 100 μs. Experimental devices have been built in which the polysilicon gate resistance is decreased by strapping them with a narrow conductor of aluminum, but this reduces the optical sensitivity due to the opacity of the metal. No commercial devices of this type are available at this time.

2.3.2.4 Serial Register Shift Times

Serial register gates have much lower capacitance and resistance than parallel gates due to their small size. It is possible to shift serial registers at speeds as high at 0.1 μs/pixel and retain acceptable CTE. The peripheral electronics used to drive the gates begins to be an issue at these speeds since high-current, high-speed gate drivers are difficult to design. It is simpler from a system point of view to use virtual-phase or two-phase design for the serial registers on devices intended to be used at higher speeds, as fewer clock transitions are needed to move a pixel than with three- and four-phase register designs.

2.3.2.5 CCD Pixel Sizes

CCD pixels are normally between 10 and 30 μm on a side (though one device, the Kodak KAF1400, has 6.8-μm pixels). The pixel size is limited on the large end by the maximum size of individual gates and on the small end by photolithography constraints. The length of CCD gates is limited because charge transfer from under one gate to under another is assisted by the fringing electric fields between the two gates. If gate length becomes greater than approximately 12 μm, there is a region of flat potential profile at the center of the gate. The only mechanism to move the charge in this region is diffusion, and this process is too slow in most cases.

Pixels on scientific CCDs are made square so that the spatial sampling frequency in horizontal and vertical dimensions is equal. CCDs designed for television use often use asymmetric pixels with aspect ratio of approximately 1.7:1.

2.3.3 Quantum Efficiency

The quantum efficiency (QE) of the device depends on the number of photons that are absorbed in the silicon with sufficient energy to produce a photoelectron. In this section, several QE factors are described together with various solutions to limitations.

2.3.3.1 Optical Properties of Silicon

Silicon exhibits a high index of refraction throughout the near-IR, the visible, and the UV region of the spectrum. The high index of refraction leads to high reflectivity, approximately 36% reflection at 500 nm and 56% reflection at 375 nm, unless antireflection coatings are applied, as is shown in Figure 2.14.

The optical absorption length for silicon, arbitrarily defined as the thickness necessary to absorb 90% of incident radiation, varies widely with wavelength as shown in Figure 2.15. The length varies from several tens of microns in the near-IR to approximately 20 Å at 275 nm, which is the wavelength of minimum absorption length.

Both reflectivity and absorption length affect the QE of CTDs. The effects of reflectivity are apparent, but the obvious solution of using antireflection coatings is not always effective. Because a CCD is a solid-state integrated circuit, it is necessary to have specific materials, such as silicon dioxide, silicon nitride, and polysilicon, overlaying the photosensitive epitaxy. Because each of these necessary

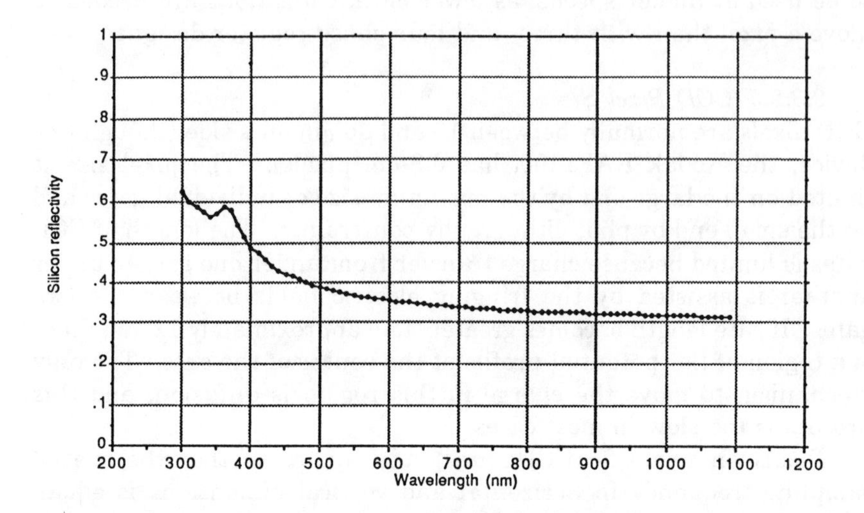

Figure 2.14 Reflection from silicon as a function of wavelength, computed from the index of refraction.

materials has a different index of refraction, their use highly restricts the benefit that can be gained by using antireflection coatings. The index of refraction of an antireflection coating, n, should be $n = n_{Si}^{1/2} \approx 3.5^{1/2}$. Silicon dioxide and silicon nitride have much lower indices of refraction, and while they reduce the reflection from silicon to a degree, they are far from optimum. An antireflection coating could be applied to the top of the passivation glass, but this would decrease the reflection by only about 4%. In fact, antireflection coatings are not typically used on CCDs except on thinned, backside-illuminated CCDs where it is not necessary to have other materials covering the epitaxy.

The silicon absorption length affects CCD QE in two ways. First, the photosensitive silicon layer is typically only 10 to 20 μm thick. At wavelengths where the absorption length is greater than this, the QE is reduced because some light passes through the photosensitive layer. At wavelengths where the absorption depth is very short, the QE is also reduced. This is primarily due to photon absorption in the polysilicon gate electrodes. This effect is especially detrimental in CCDs where the epitaxy is completely covered with polysilicon. Consequently, the QE of *frontside-illuminated* CCDs drops to essentially zero at wavelengths shorter than 400 nm. The effect is not as severe in virtual-phase CCDs where a significant amount of photosensitive silicon is not covered with polysilicon.

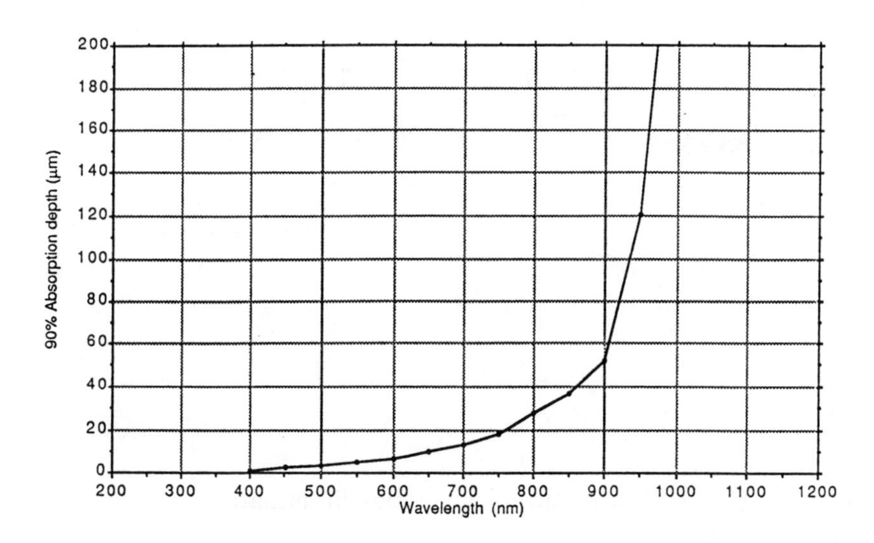

Figure 2.15 Silicon optical absorption depth.

2.3.3.2 Backside-Illuminated CCDs

An alternative to the frontside-illuminated CCD structure discussed thus far is the thinned, backside-illuminated CCD where, instead of causing incident photons to travel through the gate structure on the front of the device, the substrate is removed and photons are incident directly on the epitaxy from the backside. The substrate is normally removed after the CCD wafer has been completely processed and initial functional tests have been passed. The wafer is attached to a rigid substrate so that the back surface is exposed and the substrate is removed. This may involve mechanical lapping to thin the wafer to approximately 200 μm followed by chemical etching to the desired depth. A solution of acid mixtures is used to selectively etch the substrate. The etching reaction rate is orders of magnitude lower for the lightly doped epitaxy than for the highly doped substrate, so the device can be uniformly thinned to the epitaxy-substrate junction.

To make a common electrical contact, either some of the substrate is left on the back side at the periphery of the CCD die or a deep *p* diffusion is made on the front surface, as shown in Figure 2.16.

The thinned region of a CCD is a silicon membrane approximately 15 μm thick; it is very fragile. Two commercial approaches to supporting the membrane are to leave a border of substrate surrounding the photoactive region, the so-called window-frame thinned device, or to bond the membrane to a solid support from the front surface.

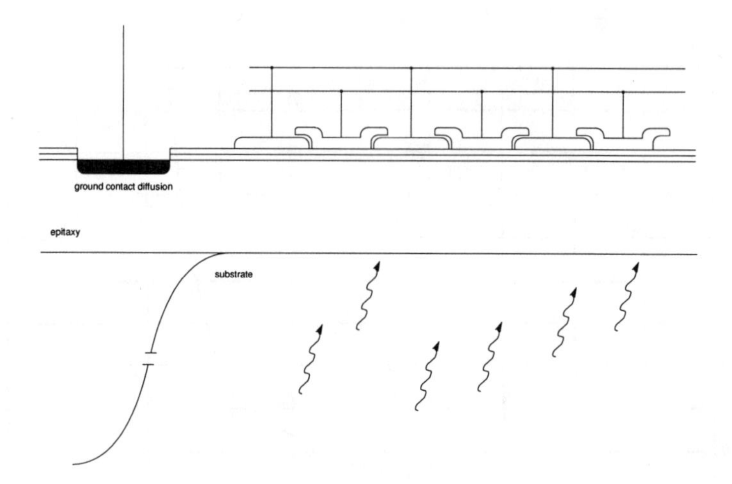

Figure 2.16 CCD structure for backside illumination.

The latter approach is more complicated in that to make electrical contact to the bonding pads on the front surface, all silicon over the bond pads must be etched away.

A newly etched silicon surface has an extremely high surface-state density due to the abrupt ending of the silicon lattice. The dangling bonds at the surface cause a high positive surface charge that traps electrons until they recombine. The net result is that newly thinned CCDs have QE no greater than frontside devices. The positive surface potential must be eliminated, and preferably negatively charged, to drive charge toward the buried channel on the frontside (see Figure 2.17).

Several approaches to backside charging are currently being explored, including low-temperature oxidation followed by UV illumination, exposure to chlorine or nitrous oxide gas, deposition of a thin platinum film, and shallow *p* implants. This is currently a very active research area, and new approaches are emerging; however, the commercially available backside devices use a shallow *p* (boron) implant on the back surface to create a zone of high potential energy for electrons just under the silicon surface. The potential-energy profile for a device of this type is diagrammed in Figure 2.17. Photons must penetrate far enough to be absorbed past the "hump" in potential energy, or they will be driven to the surface and lost. By keeping the implant as shallow as possible, the sensitivity can be extended into the UV. The photon absorption depth is so shallow at wavelengths less than 350 nm that the QE rapidly drops.

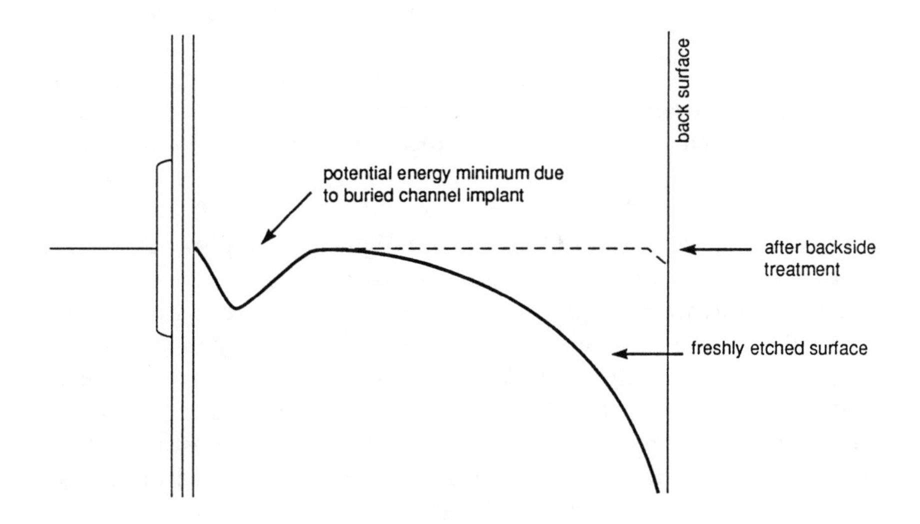

Figure 2.17 Graphical representation of electron potential energy profile as a function of depth in a backside-illuminated CCD.

Antireflection coatings are especially useful in conjunction with backside-illuminated devices. The high reflection loss of silicon can be reduced significantly over a wide range and to nearly zero at the design wavelength with use of appropriate materials and thicknesses. Both single- and two-layer coatings are presently used on commercial devices.

2.3.3.3 Organic Phosphor Coatings
The current method to achieve sensitivity throughout the majority of the UV and vacuum-UV for both frontside- and backside-illuminated CCDs is to use a thin coating of organic phosphor to convert the short wavelength photons to longer wavelengths. The phosphor absorbs the UV photons and re-emits by fluorescence at a longer wavelength. The important characteristics for the phosphor are high absorptivity in the UV, high fluorescence efficiency, and emission near the wavelength of peak sensitivity for the CCD.

The current method of choice is to deposit a thin (300–500 nm) layer of phosphor by vacuum sublimation. The materials used have nearly 100% efficiency, but since the emission is isotropic, half of the light is lost. Consequently the best UV sensitivity possible is equal to one-half the sensitivity of the CCD at the wavelength of phosphor emission, which is normally 500 to 550 nm. Because the phosphor coating is extremely thin, there is no appreciable loss of spatial resolution [8].

Representative QE curves for different CCD types, including frontside illuminated, backside illuminated, virtual phase, and phosphor coated, are given in Figure 2.18. While the backside-illuminated devices have the highest sensitivity, even frontside devices have much higher QE than photoemissive devices. This feature is one of the main attractions of CCDs for low-light-level applications.

CCDs also have sensitivity (QE greater than approximately 10%) in the soft x-ray wavelength range where the photon absorption depths are similar to those in the visible spectrum. This occurs in the energy range of approximately 0.2 to 10 keV. While devices are sometimes used for direct x-ray detection in this wavelength region, they are slowly damaged by the radiation. The damage manifests itself as higher dark current and reduced CTE caused by increased interband energy state traps in the bulk silicon.

2.3.4 Charge Quantitation

The quantity of charge in CCD pixels is measured by determining the voltage change on a MOS gate when the charge is inserted.

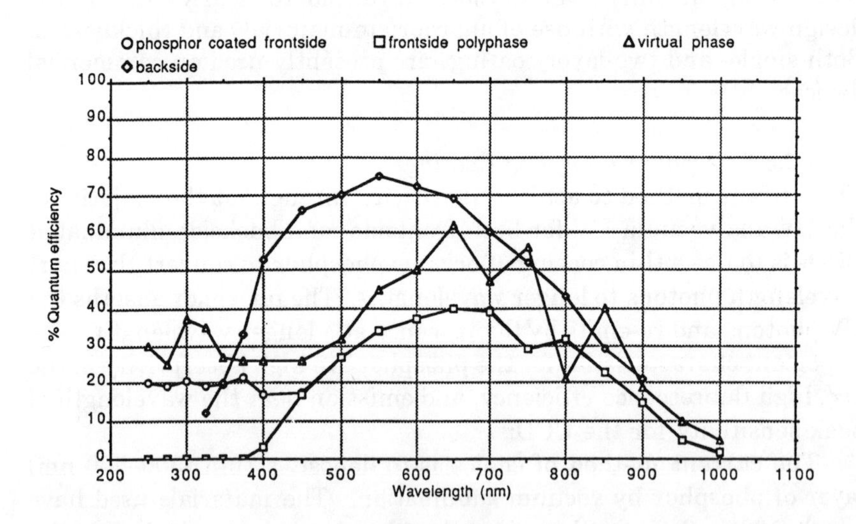

Figure 2.18 Representative QE curves for frontside-illuminated, backside-illuminated, virtual-phase, and phosphor-coated CCDs.

The voltage across a capacitor, V, is a function of the quantity of charge, q, in the capacitor and the capacitance, C:

$$q = CV \tag{2.1}$$

To measure the charge in CCD pixels, the charge packets are shifted serially into a capacitor where the voltage change is indirectly determined. The charge packet has then served its purpose, so it is destroyed in preparation for measuring the next packet. The charge is measured and destroyed with MOS field-effect transistors (FETs). This approach is used mainly because it is compatible with the MOS semiconductor processing used to fabricate the CCD register.

A cross-section diagram of a MOS-FET is shown in Figure 2.19. The device consists of many of the same structures used in a CCD. A gate insulated from the epi sits over a buried channel. Two *n* diffusions connect the *n* buried channel to the surface of the epi. The diffusions, termed the source and drain, are in turn contacted to metallizations that connect to pins on the CCD package or to other structures on the CCD. The source and drains are structurally and electrically equivalent. Their purpose is to carry current to and away from the MOS-FET buried channel. The buried channel acts as a resistor with resistance that varies from nearly infinity to approximately 1000 Ω depending on the potential applied to the gate.

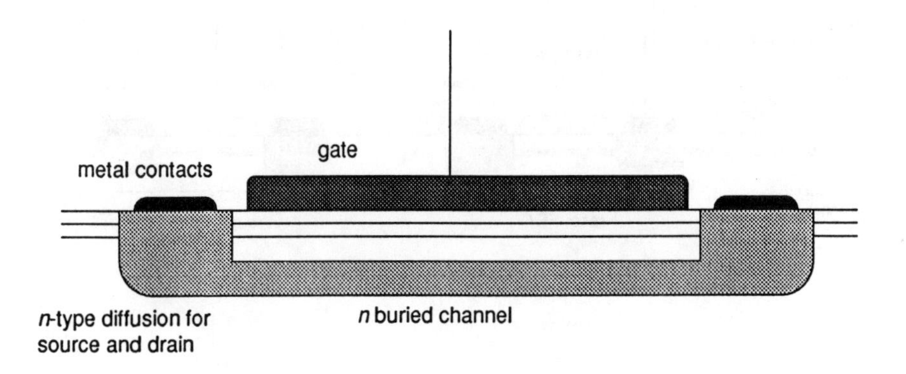

Figure 2.19 Cross section of a MOS-FET.

The most common output amplifier architecture for scientific CCDs is the floating diffusion output, diagrammed in Figure 2.20. The output diffusion is an *n* diffusion placed at the end of the CCD register. One FET, termed the reset FET, is used as an electrical switch to briefly connect the output diffusion to a potential source of +12 to +14 volts. The switch is on when the gate is biased to approximately +10 volts and off when at 0 volts. Setting the output diffusion to this high potential means that any charge that enters it from the register cannot be transferred back to the register. It can be conducted away only when the output diffusion is reset.

The output diffusion is also connected to the gate of the output FET. This FET is used in a source-follower-amplifier configuration [9] where the output of the amplifier is linearly proportional to the potential on the gate. The gain of the source follower is approximately 0.8 V/V. The linearity of the CCD, in terms of output signal related to photons in, is affected exclusively by the linearity of the output FET and supporting electronics. CCDs using source-follower outputs typically experience deviation from linearity of << 1% over the full dynamic range.

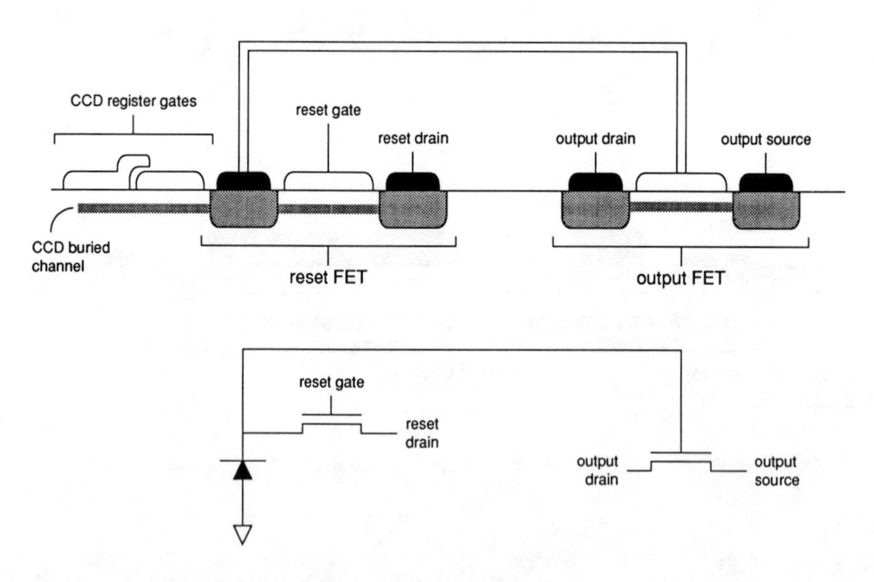

Figure 2.20 Floating diffusion CCD output.

The voltage that appears on the output FET gate is proportional to the product of the charge placed on the output diffusion with the sum of the capacitance of the output diffusion, the output FET gate, and the reset FET source, according to the relationship in equation 2.1. The capacitance normally experienced on the output node is 0.1 to 0.01 pF, which yields a signal of 1 to 10 μV per electron of charge.

The external circuitry used in conjunction with the output FET in the voltage-follower configuration is shown in Figure 2.21. The output drain is connected to a high positive potential of typically 19 to 25 volts. The current through the FET is limited by a load resistor to approximately 1 mA. The FET acts as a variable resistor with the resistance dictated by the voltage on the gate. As the source-to-drain resistance changes, the voltage at the FET source changes. The signal is capacitively coupled into stages of amplification and signal processing.

The limiting noise source in a scientific, cooled CCD detector system is, or should be, the noise from the output FET. Therefore,

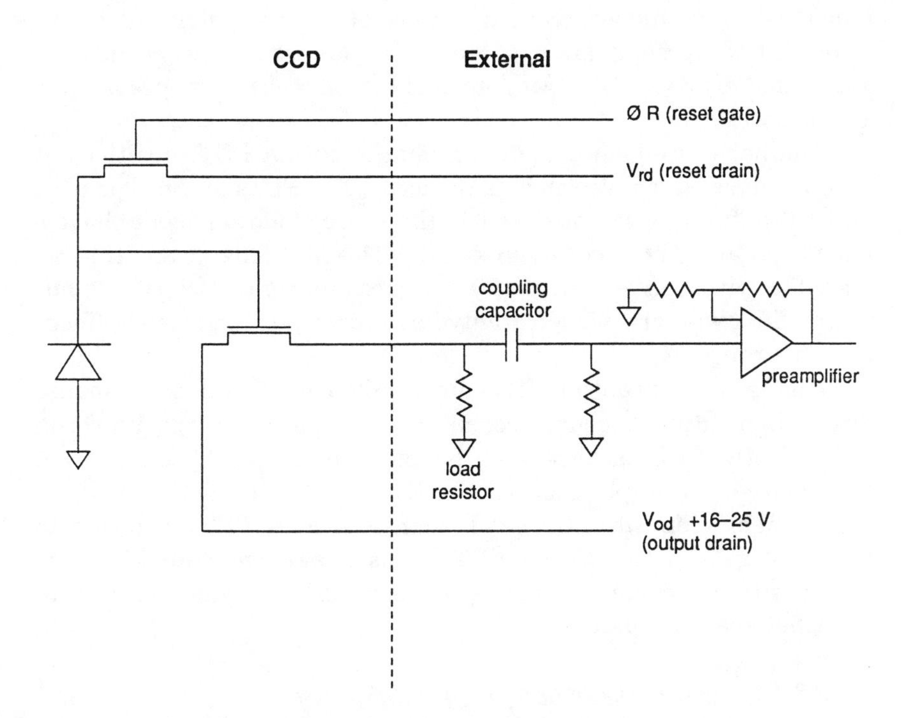

Figure 2.21 External signal circuitry for a CCD.

much work has gone into maximizing the signal-to-noise ratio (SNR) obtainable from this stage. One obvious approach to improving SNR is to increase the output sensitivity by minimizing the capacitance of the output node. The largest contributor to the output node capacitance is the output FET gate capacitance. In particular, the capacitance between the gate and the source and drain diffusions is parasitic in that it does not contribute to the performance of the output FET but decreases the signal level. Within the last 2 years, a few commercial devices have become available that use "lightly doped diffusion" or LDD technology. This greatly reduces the gate to source and drain capacitance, and yields approximately a factor of two improvement in SNR. This advancement has now made possible routine measurements with less than 5 e^- root mean square (RMS) readout noise using pixel readout times of 50 μs.

Another approach to increasing SNR is to reduce the area of the output FET gates to decrease their capacitance. This approach is used in some devices to increase the output sensitivity to approximately 10 $\mu V/e^-$. The SNR improvement is not straightforward, however, because making the gate smaller changes the noise-spectrum characteristics of the FET. The net result is that the achievable SNR is much better at higher readout speeds of approximately 10 to 0.2 μs/pixel, but worse at lower speeds. The lowest achievable noise is approximately 8 to 10 e^- RMS, but this can be realized at speeds up to 1 μs/pixel [10].

Another consequence of increasing the output FET sensitivity is that the linear dynamic range is reduced. The FETs of the type used on CCD outputs, when operated in the source-follower mode, have a linear operating range of approximately 1 volt. This means that an amplifier with 1 $\mu V/e^-$ sensitivity can measure up to 1×10^6 e^-, but an amplifier with greater sensitivity has a correspondingly lower linear dynamic range.

High-sensitivity output FETs have limited high frequency response due to high output impedance combined with parasitic capacitance on the CCD, the CCD package, and the external circuit. To circumvent this, a large, low-output-impedance FET is placed on the CCD to drive the signal off-chip, while the small, high-sensitivity FET needs only to drive the gate of the second FET. This dual-stage source-follower configuration is common on many modern CCDs designed to operate at higher readout speeds.

2.3.4.1 Floating Gate and Skipper Amplifiers

An alternate output-amplifier architecture is the floating gate amplifier. In this approach, diagrammed in Figure 2.22, the charge is sensed on a floating gate rather than a floating diffusion. The charge is transferred from the CCD register to a floating gate that was previously preset to a high positive potential through a gate-reset FET switch, similar to the reset FET on the floating diffusion CCD. When the charge is under the gate, its potential changes and is sensed by the output FET in the same manner as ocurs with the floating diffusion amplifier. After the charge is measured, it is shifted via an additional isolation gate to a diffusion, where it is conducted away.

The key feature of the floating gate amplifier is that measuring the charge does not destroy it, as is the case with the floating diffusion amplifier. This nondestructive readout mode has been exploited in the "skipper" amplifier architecture, which adds the capability to measure a charge packet several times before it is destroyed in order to lower effective readout noise by signal averaging [11]. With a skipper amplifier, a charge packet can be shifted under the floating gate and measured, and if the signal is extremely low, the packet can be reverse shifted back into the CCD register, then shifted back under the floating gate, and then remeasured. This process can be repeated any desired number of times.

The improvement in SNR using the skipper amplifier is a factor approximately equal to the square root of the number of reads averaged. Readout noise of less than 1 e^- RMS has been achieved using a skipper amplifier. The only penalties paid for this architecture are the added complexity of some additional gates and a separate reset FET for the floating gate, and the additional readout time of a pixel when multiple re-reads are made.

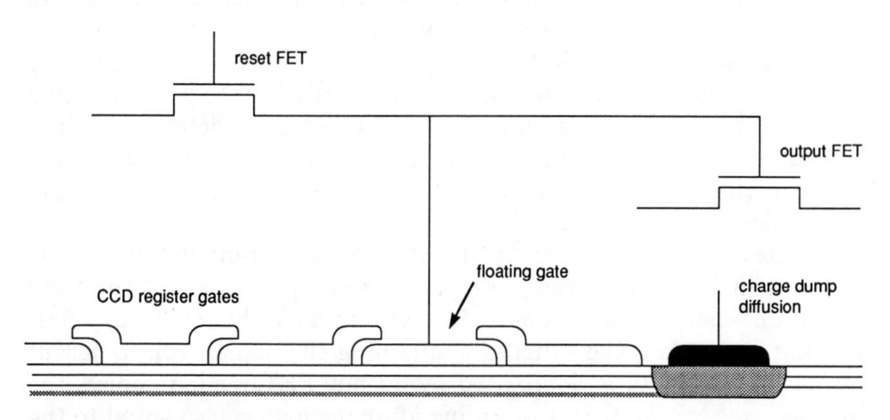

Figure 2.22 Floating gate amplifier.

No CCDs with skipper amplifiers are commercially available. Only a few experimental devices have been designed and tested.

2.3.4.2 KTC Noise and Correlated Double-Sample Readout

The floating diffusion and floating gate amplifiers depend on the reset switch briefly closing to set the potential on the gate or diffusion. Before the charge is measured, the switch must be opened. When this occurs, the potential left on the gate or diffusion is not exactly known, since the distribution of charge in the conductor at the moment that the switch opens has an inherent uncertainty. This uncertainty has been called thermodynamic noise, switch noise, or KTC noise. The term *KTC noise* is commonly used because the noise magnitude is proportional to $(KTC)^{1/2}$, where K = Boltzmann's constant, T = absolute temperature, and C = the capacitance of the node being reset. KTC noise will dominate the readout noise of the CCD to the level of approximately 100 to 500 e^- if not eliminated. Fortunately, KTC noise can be eliminated by measuring the exact potential at the output both before and after a charge packet has been shifted to the output. By subtracting the difference of the two measurements, the signal due only to the charge packet is obtained. This measurement technique, called correlated double sampling, can be achieved using a number of different analog signal processing approaches. The best approach is usually dictated by the readout speed required [12]. Correlated double sampling on readout is one of the key features that differentiates slow readout CCD systems for scientific use from CCD video cameras. Video rate readout is generally too fast to implement correlated double sampling, thus the readout noise is limited by KTC noise.

2.3.4.3 Binning

A common technique used with scientific CCD detector systems is to combine two or more charge packets before readout. This results in an increase in SNR and a loss of spatial resolution. In low-light-level applications where spatial resolution is not so important, combining charge by binning can result in increase of SNR up to a factor equal to the number of packets binned (see Chapter 3 for details). This is the case only when readout noise is dominant and when the time between the two output samples from the correlated-double-sample circuit does not increase.

If the noise is dominated by photon or dark-current shot noise, the increase in SNR by binning is equal to the square root of the total amount of charge. This increase in SNR could also be realized by first reading each pixel, then digitally summing the signal from adjacent pixels. If the noise is dominated by output FET readout noise, the increase in SNR by digital summing after readout is also equal to the square root of the number of pixels summed. If, however, the summation is performed in the CCD before readout by binning, the increase in SNR is equal to the number of pixels summed.

With many CCDs, binning must be performed in the following sequence: (1) reset the output node, (2) measure the output signal, (3) shift the register until the desired number of charge packets are delivered into the output node, and (4) make the second sample of the output. When this approach is used, the first and second samples of the output become separated in time to an appreciable extent. Since the serial register shift time is typically 1 to 2 μs/pixel, the samples can become separated by many tens of microseconds. When this occurs, the two samples are no longer correlated in time, and the system becomes sensitive to low-frequency electronic noise. This can significantly reduce the expected SNR improvement.

To circumvent this problem, many modern scientific CCDs now have a single gate, called a summing well, between the CCD register and the output. The function of this gate is to collect charge from several pixels when binning and hold it ready for the output amplifier. With this structure, readout starts by operating the CCD register until charge from the desired number of pixels are binned into the summing well. As soon as the first sample of the output is complete, the summing well dumps all of the binned charge into the output node, and the second sample is taken. With this approach, the two samples remain correlated in time, and the readout noise does not increase with increased binning.

2.3.5 Device Architectures

CCD area arrays are commercially available in many different pixel formats, sizes, and architectures. The simplest form, and the one most commonly used in scientific applications, is the *full frame* array. This is the same architecture that was described as an area array in an earlier section. There is only one parallel register, and the entire parallel register is photoactive.

Other variations of the full frame register are sometimes implemented to allow readout through more than one output amplifier. The parallel register can be split into two adjacent sections and two serial registers used, one adjacent to each split of the parallel array. The serial registers can likewise be split and output amplifiers placed on the four corners of the device, as is shown in Figure 2.23. It is common for this type of device to be built such that charge can be shifted across the parallel array and serial register boundaries as if there were no split. The direction of charge transfer can be independently determined for each section of the device if independent connections to the clock phases are available on the package. With this architecture, the device can be read out through one amplifier, two amplifiers in two different configurations, or four amplifiers.

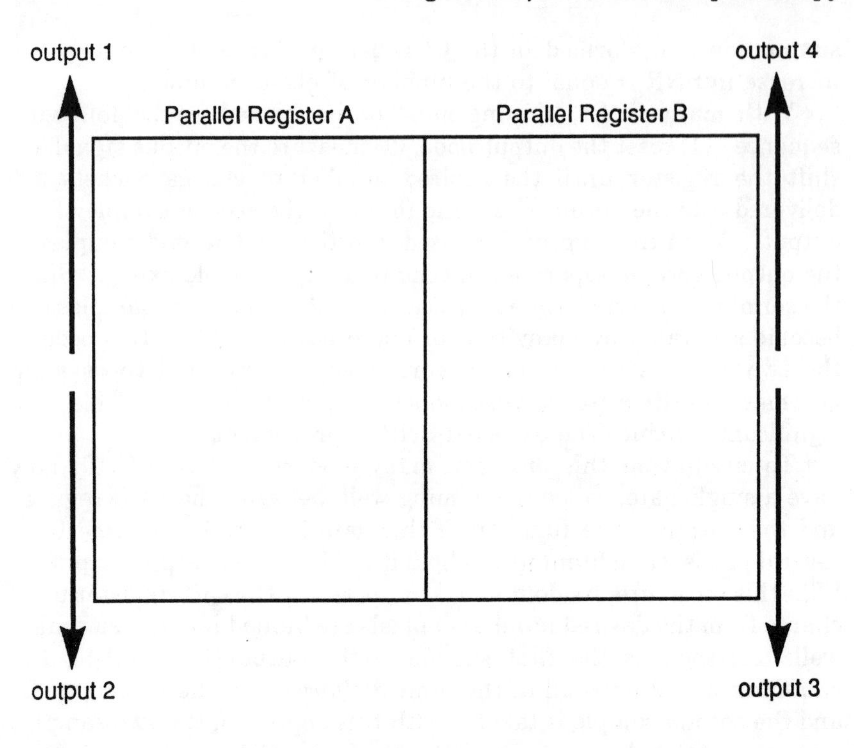

Figure 2.23 Architecture of a CCD with four output amplifiers.

Many other such variations with even more output amplifiers have been made for very large CCD devices and devices that need to read at very high speeds while maintaining low readout noise.

Full frame arrays are used with optical shutters so that no light will fall on the array during the readout, causing the image to be smeared. This is acceptable for applications where the optical integration time is very long compared with the readout time or where only one image is to be recorded. For applications where a sequence of images is to be taken and the integration time is short, the readout time is "dead time," during which no light can be integrated. In this case it is more appropriate to use a *frame-transfer* array. This architecture has two independent parallel arrays of equal size, side by side. The array closest to the serial register is optically masked by depositing a thin aluminum film. This register is not photoactive but simply serves as a *storage array* for the *imaging array*. During operation, the image array collects light for the desired time, then rapidly shifts the charge pattern to the storage array, as shown in Figure 2.24. While the next image is integrating on the image array, the storage array is being read out. There is a small amount of image smear due to light falling on the array during shift from image to storage array, but this is usually insignificant. If the smear is a significant problem, it can be avoided by closing an electromechanical shutter during the shift time.

Another device architecture commonly used for television applications is the interline-transfer array, shown in Figure 2.25. In this approach, the parallel array is divided into interdigitized lines of photoactive regions and masked registers. The photoactive regions integrate light and then pass the charge to the adjacent masked registers. The registers are read during the next integration period in a fashion analogous to the action of the frame-transfer architecture. The advantage of the interline transfer approach is that there is virtually no optical smearing during transfer. The disadvantage is that the optical sensitivity is low because only one-half of the array is photoactive, and the spatial resolution and contrast-transfer function

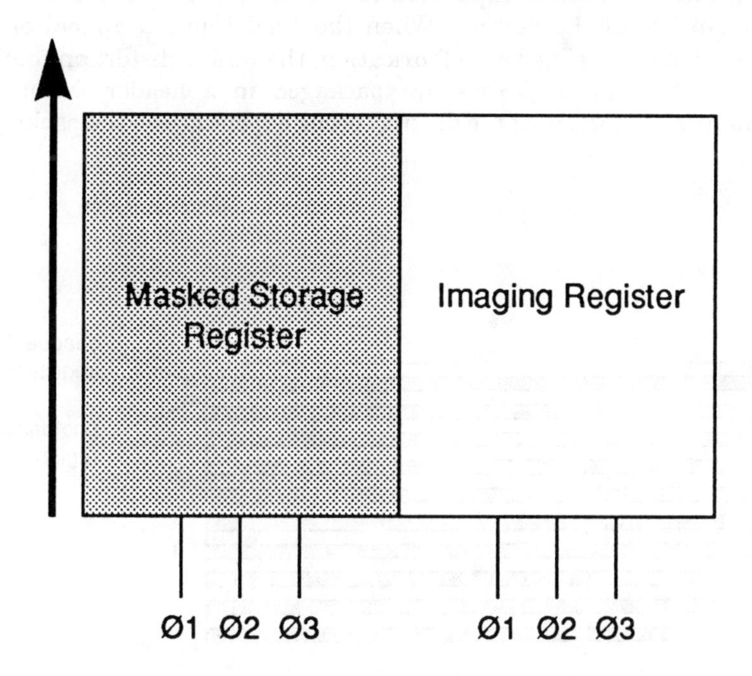

Figure 2.24 Frame-transfer array CCD showing the image and storage arrays.

(CTF, see section 2.3.6.2) are low on one axis due to the gaps between photoactive regions.

2.3.6 Optical Considerations

CCDs used in scientific applications are primarily employed as optical detectors, and as such the optical characteristics of the device should be considered. Frontside-illuminated CCDs are reflective in the visible region of the spectrum, as has been discussed. In addition to their being reflective, the periodic parallel gate and channel-stop structures cause the devices to have a high degree of optical diffraction. This can lead to significant stray light problems in high precision optical systems. Backside-illuminated CCDs do not suffer from these effects, and this may be a consideration in the selection of a device even when higher quantum efficiency is not needed.

2.3.6.1 Optical Flatness

Another optical consideration is that CCDs are not normally precisely flat. This can cause difficulty in achieving precise focus at all points on the device in optically fast systems.

Frontside-illuminated CCDs often have an approximately spherical surface, with the center being high. This arises due to the different coefficients of thermal expansion for the oxides and nitrides on the front surface of the device. When the final thermal anneal of the wafers is made during device fabrication, the wafers distort on cooling. When individual CCD dies are packaged in a header, additional mechanical distortion of the die may occur depending on the packaging method.

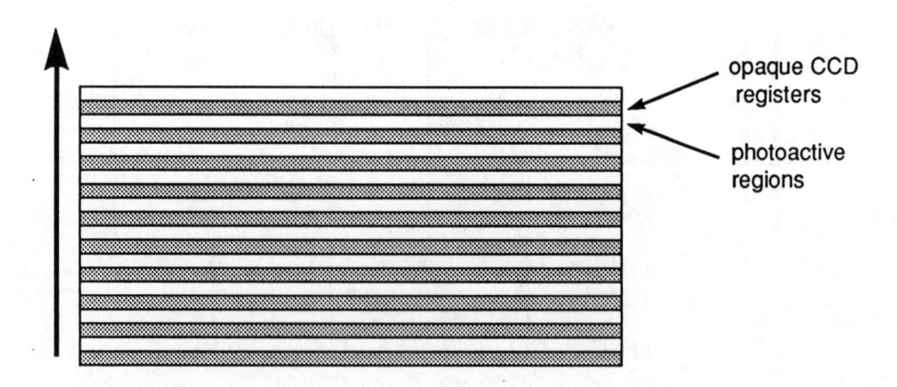

Figure 2.25 Interline-transfer array CCD.

The degree of physical distortion for CCDs varies widely among manufacturers, but as a general rule, a CCD that is 2 cm long in one axis may be expected to have a bow that is 20 to 50 μm high in the center.

Backside-illuminated CCDs suffer even more from physical distortion. In many CCD thinning processes, the CCD is a thin, unsupported membrane at some point during manufacture. The device inevitably becomes highly distorted at this point, which has led to describing the surface as being akin to a potato chip. The distortion is often so great that it can be readily observed by the eye alone. It can make uniform focus across the entire device impossible to achieve except with optically slow systems. Another complication is that the surface profile of the device can change as the device is cooled.

With some thinning processes, such as the one used by Tektronix, the silicon wafer is attached to a rigid glass-ceramic support before it is etched. This approach causes far less distortion than the unsupported approach, but small differences in the coefficient of expansion of the silicon and the substrate cause the same spherically shaped, high-in-the-center profile as is found on frontside devices. The magnitude of the distortion is much greater than for frontside devices, with a bow of 50 to 100 μm for a 2-cm-long device being typical.

2.3.6.2 Resolution and Contrast-Transfer Function

Spatial resolution of CCDs is normally expressed as *contrast-transfer function*, or *CTF*, which is a measure of the detector's response to a pattern of equidistant black and white bars. This is similar to *modulation-transfer function*, or *MTF*, which uses a sinusoidal bar pattern and is commonly used to express the resolution of optical elements.

By definition, the CTF is 100% at zero spatial frequency. As the frequency of the bar pattern approaches the pixel pitch, the contrast between light and dark bars drops. At the Nyquist frequency, where there is exactly one light bar per pixel and one dark bar per pixel, the CTF is typically 70% for visible wavelengths. The CTF drops at longer wavelengths because of the deeper photon penetration, as discussed previously. CTF at Nyquist can drop to 50% or less in the near-IR.

2.3.6.3 Fiber-Optic-Coupled CCDs

Rigid collimated-fiber-optic bundles can be attached to CCDs to give an alternative to lens systems for optical input. This mainly has the advantage of much higher optical efficiency. While practical lens optical systems are in the range of 1% to 10% efficient, fiber-optic systems routinely have efficiencies of over 50%.

The fiber optics used with CCDs employ closely packed core glass of typically 6 to 10 μm diameter surrounded by lower index-of-refraction cladding glass. Fibers of optically absorbing glass (called extramural absorbing, or *EMA*) are dispersed throughout the structure to absorb scattered light and so reduce cross-talk between fibers.

Fiber optics can be made with different magnifications over the range of approximately 0.1 to 10. This is achieved by tapering the bundle from one diameter to another, which has led to the terms fiber-optic taper or minifier (see Chapter 5) for a magnifying or demagnifying element. The numerical aperture of the individual fibers is usually either 0.6 or 1.0 and the optical transmission for a magnification of 1.0 is 50% to 60%. The transmission is relatively independent of the length of the element; most of the light that is lost enters the cladding and is absorbed by the EMA. Transmission of a demagnifying taper is reduced by the square of the magnification factor. This effect occurs because as light travels down a single, tapered fiber it can eventually exceed the critical angle and either exit the fiber or be reflected back out the input.

When a fiber optic is attached to a CCD, the spatial resolution is lowered due to the finite size of the individual fibers. If the fiber size is kept significantly smaller than the pixel size, the loss in resolution can be insignificant. The spacing between the fiber and the CCD is also critical. Because the fibers have a very large numerical aperture, light can exit the fiber at an extreme angle. A small gap between fiber and CCD will allow light to become spread over several pixels and cause a large loss in spatial resolution. It is sometimes difficult to achieve an adequately small gap between fiber and CCD when the CCD surface deviates from planar by more than a few micrometers.

Fiber optics are normally attached to CCDs by either an optical coupling oil or grease, or with an optical cement. With either approach, the different coefficient of thermal expansion of the materials involved can cause failure if the device is cooled to too low a temperature, though some devices have been successfully operated with cryogenic cooling [13].

Fiber optics are typically used in applications such as x-ray imaging where an inorganic phosphor is used to convert high-energy x-rays into visible photons, and the phosphor must be optically coupled to the detector with sufficiently high efficiency to allow single x-ray photon detection [14]. In such applications, the fiber also serves to shield the CCD from potentially damaging x-rays.

2.3.7 Antiblooming

When a pixel becomes saturated with charge and the excess blooms into adjacent pixels, it makes the signal information from nearby pixels unusable. This is a significant problem in applications where extremely high dynamic range scenes are encountered such as in optical astronomy and atomic emission spectrometry [15]. In these situations it is sometimes desirable to use a CCD that has antiblooming capabilities. Several approaches have been developed to reduce or prevent blooming in CCDs for television applications. These different approaches usually have different drawbacks that may make one approach more suitable than another.

2.3.7.1 Horizontal Overflow Drains

The most common antiblooming structure is the horizontal, or lateral, overflow drain. In this approach, an n-type diffusion is placed at every pixel adjacent to the buried channel, as shown in Figure 2.26. The diffusions are interconnected and brought to a contact on the CCD that is biased with a positive potential. A shallow p implant is sometimes placed between the buried channel and the overflow drain to create a potential barrier of controlled height. The overflow drain

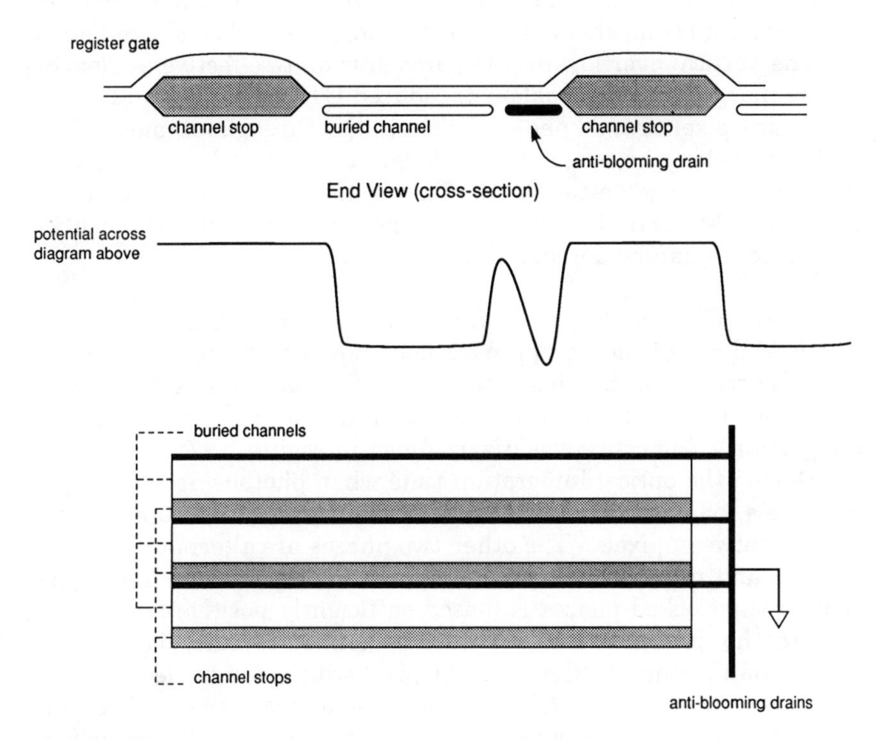

Figure 2.26 Horizontal antiblooming drains.

bias is adjusted to a level where, if the charge capacity of the buried channel is exceeded, the excess charge is collected by the diffusion and conducted away.

The horizontal overflow drain is a very efficient antiblooming structure. It is difficult to exceed the capacity of the diffusion in practical use. The drawback to this approach is that the antiblooming diffusion consumes approximately 30% of the pixel area. This reduces the effective QE of the device by the same amount. Another drawback is that since there is a dead region at every pixel, the CTF at the Nyquist frequency is lower.

2.3.7.2 Vertical Overflow Drains

Vertical antiblooming was developed to avoid the limitations of horizontal drains in television applications. In this approach, the CCD is fabricated using *n*-type substrate and a much thinner *p*-type epi. The epi is sufficiently thin that negatively biased gates deplete the silicon all the way to the substrate (there is no field-free region in the epi) so that there is a potential barrier that extends from the surface to the substrate. When the charge capacity of the pixel is exceeded, the excess charge drifts in the pixel until it approaches the substrate, whereupon it is quickly swept into the substrate and conducted away.

The vertical overflow drain approach is quite effective at clearing excess charge, and the problem of reduced CTF is eliminated because the entire pixel area is photosensitive. QE throughout most of the visible wavelengths is not reduced, but the red and near-IR wavelengths are strongly attenuated. This is not a problem for television use since the near-IR sensitivity is not desired, but for scientific application it can be detrimental.

2.3.7.3 Clocked Recombination Antiblooming (CRAB)

The CRAB antiblooming approach does not require modifications to the CCD structure, but instead the presence of holes at the surface when the surface is pinned is used to annihilate charge during light integration. The concept is diagrammed in Figure 2.27.

During the optical integration time when photons are arriving at the pixels, one clock phase is held constantly negative to act as the barrier between pixels. The other two phases are alternately clocked positive and negative at a rate of approximately 1 kHz. The upper level of the clocked phases is biased sufficiently positive that charge goes to the surface when the buried-channel capacity is exceeded rather than spilling to the adjacent pixel within the buried channel. The lower clock level is sufficiently negative to invert the surface, thus drawing holes from the buried channel into trap sites at the surface. As the two phases are clocked, the charge in the buried channel is moved back and forth, remaining under the gate that is positively

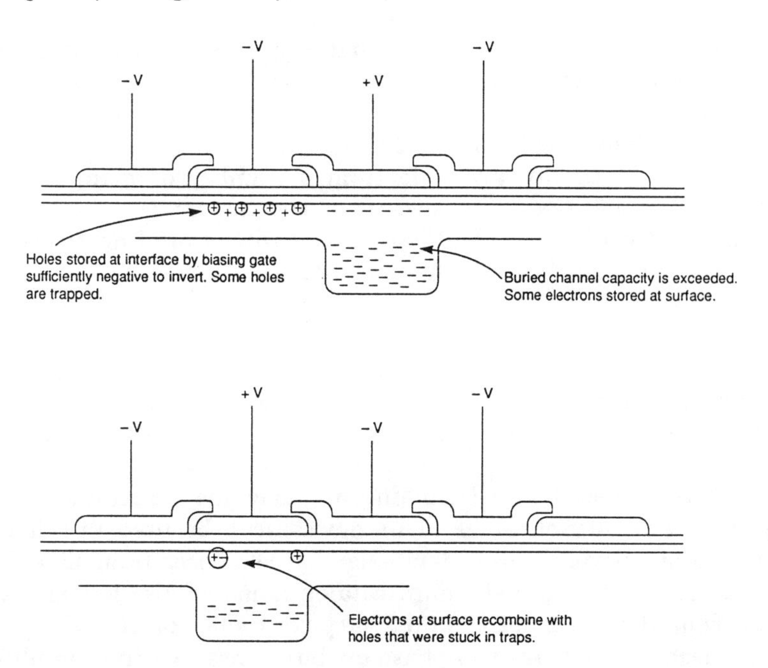

Figure 2.27 The clocked recombination antiblooming approach to antiblooming.

biased. When the buried channel is full, the excess charge moves to the surface where it is annihilated by recombination with holes.

CRAB relies on the fact that both electrons and holes become trapped at surface interface states and remain in the trap up to several milliseconds after the gate has switched potential. Electrons bound in trap sites are annihilated when a gate is switched from positive to negative and holes flood the surface. Likewise, holes bound in trap sites are available to annihilate electrons that are brought to the surface when the gate is switched from negative to positive.

The CRAB approach is not as effective at preventing blooming as are overflow drains, but it can be sufficient to prevent blooming when integration times of several seconds to several minutes are used to collect light from images with moderate light flux, such as is the case in astronomical images. Because the CCD structure is not modified, there is no change in the device QE or CTF. It does require, however, that two different positive gate potentials be used for antiblooming operation and readout. Spurious charge generation can also be a problem with the CRAB approach. This can be reduced to a minimum by keeping the voltage excursion on the clocks to a minimum, limiting the rise time of the clocks, and keeping the clock frequency to a minimum, but the same factors that reduce spurious charge also reduce CRAB effectiveness.

2.3.8 Commercial Devices

Commercial CCDs are available from a number of manufacturers worldwide. In this section several of the major manufacturers of standard scientific CCDs are listed along with a brief description of some of their typical device characteristics.

Dalsa Inc.
Unit C7
550 Parkside Drive
Waterloo, Ontario Canada N2L5V4

Dalsa makes specialty CCDs mainly for machine vision and robotic applications; however, some of its devices can be used in scientific applications. Several linear devices and area devices from 32 × 32 to 1024 × 1024 pixels are standard catalog items. Dalsa has also very recently built some experimental 5000 pixel × 5000 pixel devices. The device architectures are two-phase or four-phase. Output amplifiers are often optimized for high speed operation.

Eastman Kodak Co.
Kodak Research Bldg 81-5
Rochester, New York 14650-2010 USA

Kodak has specialized in devices designed for high-speed, low-noise operation. The distinguishing characteristic of these devices is the very high output-amplifier sensitivity of 10 to 15 mV/e$^-$, which results in a readout noise as low as 10 e$^-$ RMS at 1 mHz pixel rate. The high 1/f-noise corner frequency that is associated with such high sensitivity means that the noise does not diminish with lower readout speeds; in fact, it can increase. The other unique aspect of Kodak devices is their small pixel sizes. Devices with 6.8- or 9-μm pixels are the most common. Two-phase architecture is used exclusively.

EEV
Waterhouse Lane
Chelmsford Essex CM1 2QU England

EEV Inc.
4 Westchester Plaza
Elmsford, New York 20523
USA

EEV was an early innovator in CCD technology and makes devices for television, industrial, and scientific applications. Its scientific devices are primarily intended for low-speed, low-noise applications, though a few devices designed for very high speed operation using many output amplifiers are also available. Pixel formats range from the 385 × 578 of the CCD02 (a device that was a standard for astronomical applications for years due to its low readout noise) to over 1000 × 1000 in the CCD05 series. The CCD02 is also available in a backside-illuminated version. A spectroscopy device is available with 1152 pixels × 298 pixels. Device architecture is three-phase, and 22-μm or 22.5-μm pixels are most commonly used.

EG&G Reticon
345 Portrero Avenue
Sunnyvale, California 94086 USA

While Reticon is familiar to many for its self-scanned photodiode arrays, the compny is a relative newcomer to scientific CCDs. It has made up for lost time by offering devices in formats ranging from 512 pixels × 512 pixels to 2048 pixels × 2048 pixels. A large spectroscopy device of 1200 pixels × 400 pixels is also available. The pixel sizes are generally large (27 μm) except on the 2K-format device, which has 13.5-μm pixels. Output amplifiers are optimized for low-speed, low-noise applications.

Loral Fairchild Imaging Sensors
1801 McCarthy Boulevard
Milpitas, California 95035 USA

The Fairchild operation in Milpitas, California, has made interline transfer devices since 1973, but these were of little interest for scientific use. Loral acquired both Fairchild and the former Ford Aeronutronics operation in Newport Beach, California, and the two operations are now combined as Fairchild Imaging Sensors. Ford began by making CCDs for military applications, but initiated custom-designed scientific devices in 1988. Devices from 512 × 512 to 2048 × 2048 formats are now catalog items. The architecture is exclusively three-phase, and output amplifiers are generally optimized for lower speed, low-noise use.

Tektronix, Inc.
P.O. Box 500
Beaverton, Oregon 97077 USA

Tektronix makes the largest standard commercially available CCD, a 2048 × 2048 pixel device with 24 μm pixels. Tektronix's scientific devices have large pixels, either 24 μm or 27 μm, and low-speed, low-noise amplifiers. The unique capability of Tektronix is its backside-illumination process, which is superior to that of other manufacturers in terms of both sensitivity and device flatness.

Texas Instruments
P.O. Box 225012
Dallas, Texas 75265 USA

Texas Instruments mainly makes television sensors; however, at least one of their devices, the TC213/TC215, is sometimes used for scientific use. This is a 1K $\times$ 1K pixel format device with 12-μm pixels and an output amplifier designed for higher speed operation, similar to the Kodak devices. Texas Instruments uses its patented virtual-phase technology on its commercial devices, which results in higher sensitivity than is available with other frontside devices.

Thomson Componants Militaires et Spatiaux
50, rue J.-P. Timbaud - B.P. 330
F-92402 Courbevoie Cedex France

Thomson Components and Tubes
40G Commerce Way
Totoway, New Jersey 07511 USA

Thomson's product offering is similar to that of its European counterpart, EEV. One difference in the two company's products is that Thomson uses two- and four-phase architectures. Thomson makes a product line geared expressly for scientific use that includes devices of 512 $\times$ 512 and 1024 $\times$ 1024 with 19-μm pixel size. A 2048 $\times$ 2048 device is soon to be available, and a 1024 $\times$ 256 spectroscopy device has been recently released. Output amplifiers have higher sensitivity than other devices designed for low-noise, low-speed operation, but very good noise performance is achieved.

2.4 Charge-Injection Devices

2.4.1 Principles

The principles of operation of CIDs are quite similar to those for CCDs, since both are based on the same MOS technology. There are also many significant differences in the two classes of devices, which result in greatly different operating characteristics and capabilities.

CIDs were developed for commercial use by the General Electric Company. The technology was patented, leading other organizations to pursue the CCD approach to solid-state electronic imaging. General Electric later sold its interests in CID technology to a management group. The resulting company, called CIDTEC (Liverpool, New York) is the only commercial manufacturer of CIDs.

One of the major differences between CIDs and CCDs is that CIDs are made using n-type epitaxy silicon. Holes rather than electrons are the charge carriers collected to form electrical signals. This also means that the polarity of applied gate voltages used to collect or repel charge is reversed. While this reversal of convention is somewhat confusing, it does not significantly alter the fundamental properties of the MOS structures.

The decision to collect holes instead of electrons in CIDs is lost in history. There is no especially good reason to choose one over the other, since the charge mobility is not important as it is in CCDs; only a single charge transfer operation is necessary to read a CID pixel.

A cross-section conceptual diagram of a CID pixel is shown in Figure 2.28. The n-type epitaxy is on top of a p-type substrate. Two gate electrodes are used, one to collect charge and one to sense how much charge is collected. Both gates are initially biased negatively, but the collection gate has approximately twice the applied potential of the sense gate. There is no buried channel, so charge accumulates under the collection gate at the silicon/silicon dioxide interface. To read the charge in the pixel, the potential source for the sense gate is disconnected and the sense gate potential is measured. The potential on the collection gate is then raised to approximately ground, and the charge moves under the sense gate. Again using equation 2.1, this charge movement causes a change in the sense gate potential proportional to the quantity of charge divided by the sense gate capacitance. The sense gate potential is measured and subtracted from the initial measurement as a correlated double sample in order to eliminate KTC noise imposed on the sense gate when the switch connecting it to a potential source is opened. This step is analogous to measuring the output node potential on a CCD output amplifier before the charge is shifted from the serial register onto the output node.

Once the charge in a pixel has been measured, it can be retained in the pixel in its original condition by restoring the potential on the collecting gate to its negative value. The charge can be read again by repeating the process just described. In this fashion, the charge in a pixel can be measured continually as it is being collected. The technique of reading the charge multiple times and averaging the signal is also used to increase the precision of the measurement. The readout process does not alter the quantity of charge in a pixel, and many hundreds or even thousands of reads of a single charge packet are routinely made in scientific CID systems.

If one wishes to clear the charge from a pixel, one needs only to bias both collecting and sense gates near ground potential. When there is no potential well to hold the charge at the surface, the charge will spread throughout the pixel until it encounters the p-type substrate, where it is collected and conducted away (this structure is similar to a vertical overflow antiblooming structure in a CCD). The

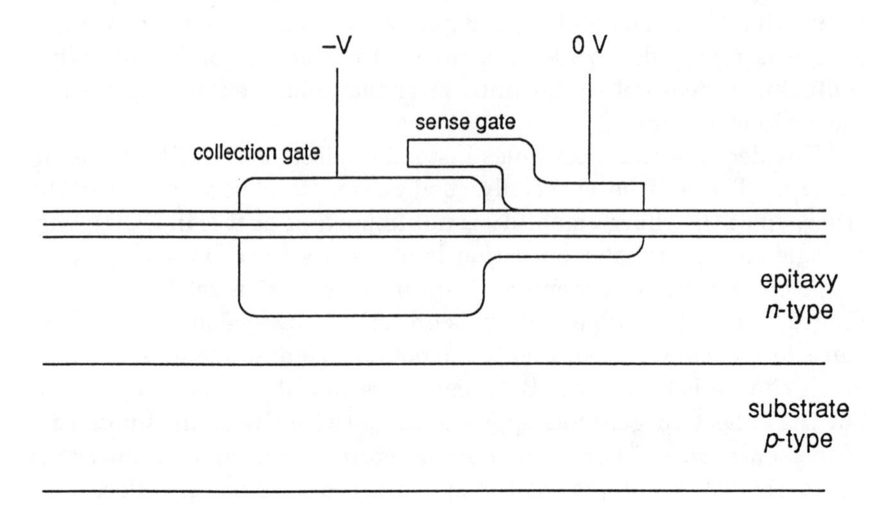

Figure 2.28 CID-element structure in cross section.

original CIDs sensed the quantity of charge in a pixel by measuring the amount of charge that was "injected" into the substrate when both gates were set to ground potential, and this is the basis for the name "charge injection device."

2.4.2 Arrays

A two-dimensional array of CID pixels consists of an orthogonal grid of sense and collection gates on the photosensitive area of a device. The doped polysilicon gates are insulated from the epi with an oxide/nitride dielectric and from each other with oxide in the same manner as in CCD gates. The sense gates are usually "strapped" with a narrow stripe of aluminum in order to decrease resistance. The intersection of a sense and a collection gate defines a pixel. Each pixel is surrounded on all four sides with a highly *n*-doped thick oxide (analogous to CCD channel stops). This structure prevents charge from migrating to adjacent pixels when charge is being injected into the substrate or when a pixel is full.

A pixel in the array is selected by connecting only one of the sense gates to an amplifier and changing the potential on only one collection gate. When the potential is changed on a collection gate, the charge associated with every pixel of that line will move under the sense gate, but only the charge in a single pixel will be sensed.

Note that CIDTEC uses the terms *row* and *column electrodes* for the sense and collection gates, but the function of the row and column has changed over time. The sense gates were the columns at one point, but they are currently the rows. To avoid confusion, the terms *sense* and *collection gates* will continue to be used in this text.

To efficiently choose a single sense and collection gate, shift registers or scanners are placed at the edges of the array. The register, called the select register, serves to close MOS-FET switches that connect the gates to a common bus for connection to external circuitry. On modern CIDs, there is a second deselect register at the opposite end of the array of the select register that connects all unused sense and collection gates to a fixed potential during pixel readout. This eliminates the row-column and column-column crosstalk observed in devices that did not have this feature [16]. The registers operate as digital devices. A bit that turns on the select switches is loaded at the start of the register, then is shifted in one direction using a two-phase clock. Once the bit comes to the end of the register, it is lost and another bit must be loaded. The sense and collection registers operate independently. To address a given pixel, bits must be loaded into both registers, then clocked the appropriate number of times to read the desired x-y location. To access the following adjacent pixel, the appropriate register is incremented by 1. Since the registers cannot be shifted in reverse, it is not possible to read a subregion of the array in raster format without loading in a new bit for every line.

1, The capability to rapidly select pixels to be read using this CID architecture is termed pseudo-random access. Some special CIDs have been built that use address decoders to select the sense and column gate to be used. This allows practically instantaneous selection of any pixel, and is thus termed true random access.

2.4.3 Readout Noise

The main drawback of the CID approach is a low output signal and consequently high readout noise compared with CCDs. The CID output signal is proportional to the charge in a pixel divided by the total capacitance of the sense electrode, part of the select and deselect shift registers, on-chip metallizations that connect the select register bus to the device package pin, and external circuits. This results in a signal of approximately 30 to 50 nV/e^-, about two orders of magnitude less than that of CCDs. This low sensitivity leads to a readout noise higher by approximately the same factor. For example, the readout noise of one common CID, the CID17, is approximately 900 e^-/read at a readout rate of 20 μs/pixel. The readout noise can be reduced by over an order of magnitude by averaging the signal from multiple nondestructive reads, as was discussed before. The noise spectral characteristics are relatively flat over the frequency range of interest, so the effective readout noise decreases as a function of the square root of the number of reads averaged. An order of magnitude reduction in noise is routinely achieved by averaging 100 readouts.

Very recently a new CID has been designed that incorporates a MOS-FET buffer transistor on every sense gate between the gate and the select register. This effectively eliminates the effects of the capacitance of the select register and other on- and off-chip parasitic capacitance. This has reduced the readout noise floor to approximately 250 e^-/readout. This figure can also be reduced by at least an order of magnitude by averaging multiple readouts, which places the effective readout noise of CIDs in the same magnitude as modern CCDs when time permits the use of multiple readouts.

Pattern Noise and Sensitivity Variations. Many different sense gates are used to measure charge in CID array architecture. Slight differences in the capacitance of the different sense gates will give rise to different signals given the same charge. The pixel-to-pixel sensitivity variation normally does not exceed 5% RMS. This effect is usually corrected by a flat-field correction (described in the next chapter and illustrated in other chapters).

Another consequence of the CID architecture is that an electronic bias signal is created by capacitive coupling of the collection gate potential to the sense gate. The coupling capacitance normally varies to some small degree in different pixels, and thus the bias signal is not consistent from pixel to pixel. This effect is called pattern noise and is present to the equivalent of up to several tens of thousands of electrons [17]. Pattern noise is reproducible and can be eliminated by subtracting a measurement of the bias level for each pixel from a signal.

Both bias subtraction and flat-field correction are also routinely used to correct CCD data, but the cause and magnitude of pattern noise and sensitivity variations are different. Since CCDs use only a single point to sense the charge in each pixel, pixel-to-pixel sensitivity variations and pattern noise should not occur and in fact are difficult to observe in modern devices. In practice, some slight correlated pattern noise may be present owing to slight differences in coupling of the CCD clock signals into the output amplifier. Sensitivity variations occur due to slight differences in optical sensitivity due to pixel-to-pixel differences in gate and oxide thickness and the degree of gate overlap (which is also the case with CIDs). Nevertheless, the pattern noise and sensitivity variations for CCDs are typically much lower than for CIDs. Both effects are an issue only if bias subtraction and flat-field correction are not performed.

2.4.4 Quantum Efficiency

The quantum efficiency of CIDs is higher than that of frontside-illuminated CCDs because significant regions of the epitaxy are not covered by polysilicon sense and collection gates, as can be seen in Figure 2.29. The quantum efficiency is similar to that of virtual-phase CCDs in that the sensitivity in the blue visible and near-UV wavelengths is high due to the exposed photosensitive silicon. The quantum efficiency in the far-UV and vacuum-UV can be increased by the application of organic phosphors in the same manner as is used with frontside-illuminated CCDs. The quantum efficiency in the near-IR wavelengths is limited by the epi thickness, which is approximately the same as that used with CCDs. A quantum efficiency curve for a typical CID17 device is shown in Figure 2.30.

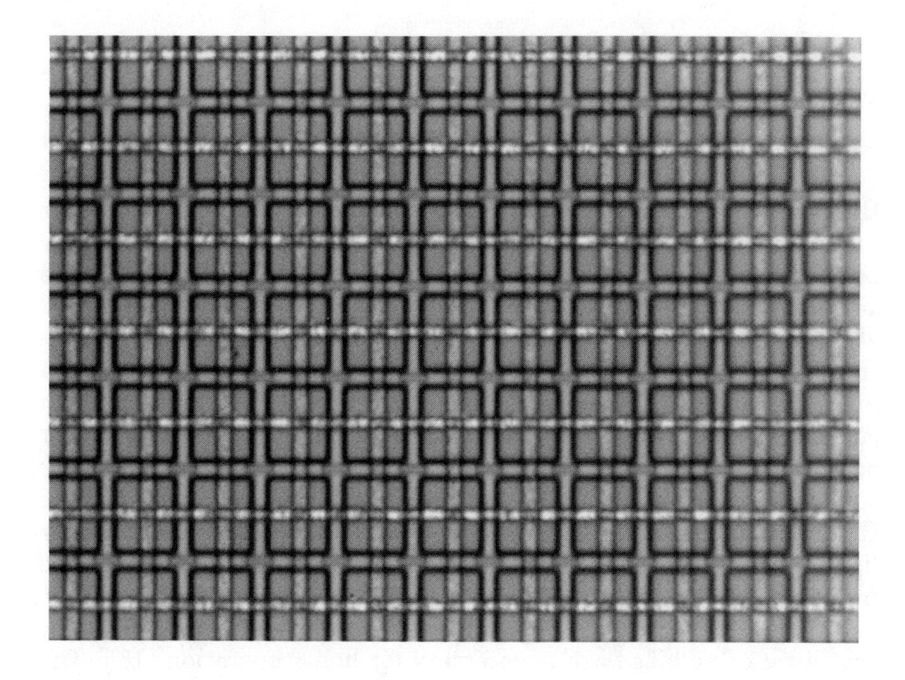

Figure 2.29 Photograph showing polysilicon sense and collection gates in a CID.

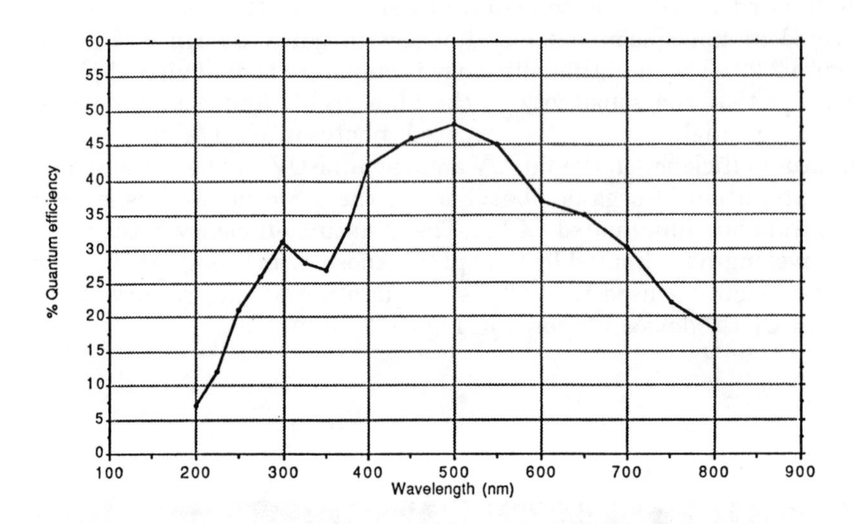

Figure 2.30 Quantum efficiency curve for a CID17.

2.4.5 Charge Capacity, Dark Current, and Linearity

CIDs are surface channel devices, and this leads to several important differences in performance compared with CCDs. The charge capacity per unit pixel area is generally higher because charge is stored at the surface, closer to the gate. A CID can hold approximately 1500 to 2000 $e^-/\mu m^2$ compared with 600 $e^-/\mu m^2$ for a typical three-phase buried-channel CCD. One drawback to surface channel operation is that if a pixel is completely cleared of charge, it must be "primed" with sufficient charge to fill the surface interface states before linear operation is achieved. The charge used to fill the surface states is sometimes called bias charge or "fat zero" and is typically added to the CID immediately after it is cleared by quickly flashing low-intensity light-emitting diodes. The amount of charge necessary to fill the interface states was quite high until recently implemented changes in materials and processing reduced the surface state density to the point that little or no bias flash is necessary for linear operation [18]. The shot noise in the bias charge is removed if a bias frame is collected as a nondestructive readout frame (NDRO) and subtracted from the image frame.

The charge-transfer efficiency of CIDs is not an important parameter, as it is with CCDs. This allows operation at liquid nitrogen temperatures where dark current is virtually undetectable without degradation in other performance characteristics. CIDs are not capable of operation in multiphase pinned mode due to the lack of a buried channel. The dark current is approximately 0.1 to 0.2 nA/cm^2 at 25°C which is the same as CCD dark current when operated in non-MPP mode [19].

2.5 A Comparison of CIDs and CCDs

The very different architectures of CIDs and CCDs lead to complementary properties that are best exploited for different applications. The CCD is now widely recognized as the detector of choice in many applications where light levels are extremely low and consequently high dynamic ranges are experienced. An example of an application in which CCDs have nearly completely replaced other detectors is optical astronomy.

CCDs have a more limited charge-storage capacity per pixel compared with CIDs. This means that the ultimate precision of measurements made in a single readout is less. Another limitation of CCDs is that in order to measure the charge in a given pixel in an arbitrary location in the array, the charge in all pixels between it and the output amplifier must be either read or dumped without reading. This greatly increases the time needed to read a pixel or subgroup of pixels out of the array.

One of the main advantages to CCD detectors today is the wide variety of commercially available devices that have been optimized for particular applications. Likewise, a variety of detector systems employing CCD detectors are readily available.

CIDs in many ways have more diverse capabilities than CCDs. The ability to randomly or pseudo-randomly access pixels and read them non-destructively gives the ability to monitor the signal in various regions of the array while light is being collected. This capability, termed variable integration time, allows one to collect signal until a desired SNR is reached rather than having to collect light for a fixed integration time and later determine the SNR after readout [20]. The inherent antiblooming capability is useful in this mode of operation as it prevents loss of information when regions of the array are intensely illuminated. Although the readout noise of the CID is higher than that of the CCD, the higher pixel charge capacity and the ability to mix nondestructive and destructive readouts provides the capability to measure signals to higher precision and with a wider synthetic dynamic range.

The use of CIDs in scientific applications has been limited in part by the presence of only a single commercial supplier of devices and because slow-scan scientific cameras using CIDs have not been commercially available. As this situation changes, more applications may be identified for which CIDs are better suited than CCDs.

References

1. Sze, S.M. *Physics of Semiconductor Devices.* Wiley, New York, 1981.
2. Beynon, J.D.E.; Lamb, D.R. *Charge-Coupled Devices and Their Applications.* McGraw-Hill, New York, 1980.
3. Janesick, J.R.; Elliott, S.T. *Proceedings of the Astronomical Society of the Pacific Conference,* 1991.
4. Janesick, J.R.; Blouke, M. *Sky and Telescope* **1986**, *77(3)*, 238.
5. Castelli, C.; Wells, A.; McCarthy, K.; Holland, A. *Nucl. Instrum. Meth. Phys. Res. A* **1991**, *A310(1-2)*, 240.
6. McLean, I.S. *Electronic and Computer-Aided Astronomy.* Ellis Horwood, Chichester, U.K., **1989**.
7. Pemberton, J.E.; Sobocinski, R.L.; Sims, G.R. *Appl. Spectrosc.* **1990**, *44*, 328.
8. Sims, G.R.; Schempp, W.B.; McCurnin, T.W. "Contrast Transfer Function Measurements of a Small-Pixel Format MetaChrome II-Coated CCD," application note. Photometrics, Ltd., Tucson, Arizona.
9. Horowitz, P.; Hill, W. *The Art of Electronics.* Cambridge, New York, 1989.
10. McKay, C. *Proc. SPIE,* in press.
11. Chandler, E.C.; Bredthauer, R.; Janesick, J.R.; Westphal, J.A.; Gunn, J.E. *Proc. SPIE* **1990**, *1242*, 27.
12. McCurnin, T.; Schooley, L.; Sims, G. *Proc. SPIE* **1991**, *1448*, 225.
13. Schempp, W.V., Photometrics, Ltd., Tucson, Arizona, personal communication, 1993.
14. Bueno, C.; Barker, M.D.; Condon, P.E.; Betz, R.A. "Solid State Imaging Methodology", U.S. Air Force Report WL-TR-92-4003, March, 1992.
15. Bilhorn, R.B. *Proc. SPIE* **1991**, *1448*, 74.
16. Sims, G.R.; Denton, M.B. *Opt. Eng.* **1987**, *26*, 999.
17. Sims, G.R.; Denton, M.B. *Opt. Eng.* **1987**, *26*, 1008.
18. Pilon, M., Thermo Jarrell-Ash, personal communication, 1993.
19. Carbone, J.; Arnold, F.; Zernowski, J.; Vangorden, S.; Pilon, M. *Proc. SPIE* **1991**, *1447*, 229.
20. Sims, G.R.; Denton, M.B. *Talanta,* **1990**, *37*, 110.

3
Specialized Readout Modes and Spectrometers

Jonathan V. Sweedler

Department of Chemistry, University of Illinois
Urbana, Illinois

Robert B. Bilhorn

Coating Technology Division, Eastman Kodak Company
Rochester, New York

3.1 Introduction

Optical spectroscopic methods of analysis are among the most powerful and versatile analytical techniques used by modern chemists. Spectroscopic measurements based on absorption, fluorescence, Raman, and emission offer tremendous selectivity and sensitivity. These spectroscopic techniques both can give information about a molecule's environment and can help unravel the composition of extremely complex samples. Optical spectroscopy can even detect single molecules in rigorously controlled situations [1–3].

Not surprisingly, there exists an extremely wide range of instrumental configurations greatly dependent on the specific experimental task at hand. The overall goal of a spectrometer/detector system is to measure a spectrum without introducing either distortions or noise into the measurement, so that the measurement is limited by the noise of the incoming photon stream. That goal requires efficient collection of radiation from the source as well as minimum degradation by such factors as stray light and detector noise.

While significant progress has been made in understanding and improving the performance of spectroscopic instrumentation in the last decade, the ability to analyze many samples is still limited by constraints of the spectrometer and detector system and not by fundamental limits inherent in the measurement process. Almost all spectrometer systems today fall far short of the goals of not influencing the measurement process and of having the measurement limited only by the inherent uncertainty in the photon stream.

The overall purpose of this chapter is twofold: (1) to give the reader some direction in selecting the appropriate method for obtaining the desired spectral information and the optimum method for using a charge-transfer device detector, and (2) to list references that should aid the reader in finding additional information. The unique readout modes and features of CTDs that have a significant impact on their spectroscopic performance are emphasized. Specific details about particular applications are described elsewhere in this volume.

3.1.1 Progress in Optical Detection

While the goals of the *ideal* optical system are, for the most part, obvious, most methods of optical detection suffer from a number of shortcomings. Currently, the two extremes in detector formats are the photomultiplier tube (PMT) and photographic film.

3.1.1.1 The Photomultiplier Tube

The PMT is close to the ideal single-channel detector. It offers the ability to detect a very few incident photons (i.e., photon counting) and has unparalleled time resolution [4–6]. A properly designed PMT system offers an absence of read noise, low dark current, and extreme sensitivity. Because of the performance characteristics and the commercial availability of a wide variety of PMT detection systems, PMTs have become synonymous with high-sensitivity single-channel detection.

Despite the overall good performance and tremendous success of PMTs, these detectors have many limitations, the largest of which is their single-channel nature. Almost all analytical measurements involve making multiple intensity determinations, whether to acquire a spectrum at multiple wavelengths or to obtain sample and reference intensities; typically, sequential determinations are used. As will be shown in Section 3.1.3, multichannel methods offer improved performance in most common situations.

In addition to the single-channel nature of PMTs, their performance is limited by the performance of available photocathode materials. The best possible quantum efficiency is on the order of 30%, with typical values in the 5 to 20% range [4–6]. At wavelengths longer than 900 nm, the quantum efficiency typically drops below 1%. In addition, while it is possible to construct photocathodes optimized for response within nearly any UV-visible wavelength region, no one photocathode is yet available that has high quantum efficiency (QE) throughout the ultraviolet to near-infrared wavelength range. Compared with silicon detectors, they have high dark-count rates and are sensitive to light shock.

3.1.1.2 Photographic Film

While the use of photographic film is no longer common for spectroscopy, the method has a number of impressive characteristics: Film can respond at all wavelengths shorter than 1.3 μm, a broad region of the spectrum can be simultaneously covered, and weak features can be observed given sufficiently long observation times [7]. The resolution available from current photographic plates is enormous—much higher than is available from other array detectors. For example, a 50-cm-square photographic plate with a resolution equivalent to 2000 lines/mm offers the staggering spatial information content of greater than 10^{10}. For this reason, film is still commonly used in astronomical surveys, and it will be a long while before array detectors can compete in terms of number of resolution elements.

Unfortunately, photographic plates also have a number of very serious limitations, among which one of the most serious is the development process, making film a non-real-time detector. In addition, film response is relatively nonlinear, the dynamic range is less than that available with electronic detectors, quantitation is more difficult than with electronic or solid state detectors, and film is subject to reciprocity failure.

3.1.1.3 Simultaneous Electronic Detection

The use of two-dimensional array detectors to measure the entire spectrum is not a new idea [8–12]. The overall desire has been to develop an array detector that combines the sensitivity, linearity, and dynamic range of PMTs with the extremely high resolution of film. However, past attempts to use array detectors have met with limited success; the available detectors, whether orthicons, vidicons, intensified vidicons, or photodiode arrays, are not as sensitive detectors as the single-channel photomultiplier tube, and so trade-offs between the multichannel advantages of the detector array and this loss in sensitivity limited the usefulness of these approaches. In addition, these detectors have problems with hysteresis, lag, limited dynamic range, poor linearity, and cross-talk [13–16]. Chapters 8 through 11 in Ken and Marianna Busch's book on multielement detection systems provides additional information about available imaging detectors [16]. CTD's characteristics of high resolution, sensitivity, and large dynamic range have resulted in the use of these detectors in a wide variety of applications.

In addition to CTDs, several other sensitive area detectors are becoming available, including the resistive anode array [17–18]. The resistive anode array combines many of the features of photon-counting PMTs with high spatial resolution; characteristics include no read noise, high time resolution, low dark currents, and high spatial resolution. Unlike CTD detectors, the device can discriminate against cosmic ray events, the importance of which will become apparent in the following sections. However, the QE of resistive anode devices is limited by the poorer QE of photocathode materials. More importantly in wide dynamic range applications, the device has serious problems when observing faint spectral features while intense features are on the focal plane simultaneously. Nevertheless, the resistive anode array detector offers impressive performance, especially in low-light applications requiring time resolution.

3.1.2 CTD Characteristics

In Chapter 2, the operation and characteristics of a variety of CTDs were covered in detail. The following section emphasizes the most important characteristics of the CTD detectors impacting their use in spectroscopy: specifically, array format, noise performance, and spectral response. In addition, many other characteristics can be of paramount concern in a particular application, including freedom from blooming, readout speed, and immunity from interference from high-energy radiation.

3.1.2.1 Array Format

The detector should not limit a spectroscopic system; rather the ideal detector should be available in the correct physical format and size. Although not optimally configured for many current spectroscopic systems, CTDs are available with a wide variety of photoactive areas and numbers of detector elements. The large number of detector elements available with CTDs allows extremely high-resolution spectra to be obtained simultaneously. For example, an echelle spectrometer used in conjunction with a rectangular CTD containing several hundred thousand detector elements is theoretically able to cover the ultraviolet, visible, and near-infrared spectral regions with greater than 0.01-nm resolution [19]. Because of the large variety of available detectors, a CTD often can be selected for a specific application by matching the number of detector elements and device performance to experimental requirements; however, the relatively small size of the individual detector elements in most CTDs is a significant problem that requires ingenuity in optical design [15].

3.1.2.2 Detector Noise

A properly designed photon-counting PMT system is able to detect individual photoelectrons; essentially no noise is associated with the actual readout of this detector. For integrating detectors, the situation is very different; integrating detectors such as CTDs and photodiode arrays (PDAs) have a significant read noise—the noise introduced by the detector and associated electronics in reading out a single-charge packet. In CCDs, the sequential transfer of charge from the photosensitive area to a low-noise amplifier eliminates the multiplexing circuitry necessary in CIDs and PDAs, greatly reducing the capacitance on the amplifier input; therefore, the read noise is extremely low in these devices.

For low-light-level spectroscopy, the dark current, or thermal generation of signal, is an important detector parameter directly affecting the maximum observation time. For silicon array detectors, the majority of the thermally generated charge appears at defects in the bulk silicon and at the surface silicon/silicon oxide interface; consequently, dark current is dependent on the manufacturing process and geometry.

CTDs used in low-light-level applications are cooled to between -30 and -150°C so that the dark current is extremely low; however, CCDs cannot be cooled to arbitrarily low temperatures to further reduce the dark current because the ability to transfer the photogenerated charge decreases as the temperature is reduced, giving an absolute temperature limit of operation for most CCDs of approximately -150°C. In a CID, the charge is transferred between electrodes in a single detector element. Consequently, they can be operated at lower temperatures because the small amount of charge left behind in one transfer is collected in the next transfer, and therefore charge losses are not cumulative as in CCDs. Dark-current levels in the range of 0.01 to <0.0003 e^-/s/element for CCDs and levels below 0.008 e^-/s/element for CIDs are achievable with practical cooling means. Dark currents of these levels are insignificant in many analytical spectroscopic applications. Exposure durations on the order of minutes are required before the dark current in a pixel is measurable and, theoretically, exposures of years are required before the devices saturate. When a high degree of binning is used, however (see Section 3.2.1.2), the dark current from many pixels is combined, and it often becomes necessary to cool the devices as much as possible in order to make very-low-light-level measurements that are free from dark-current shot noise.

3.1.2.3 Spectral Responsivity

In many situations the most important characteristic of an optical radiation detector influencing the SNR of a measurement is the detector quantum efficiency. The intrinsic quantum efficiency of all silicon detectors (PDAs, CTDs, etc.) is high compared with the quantum efficiency of available photoemissive materials in the visible to near-infrared wavelength region. Peak QEs for CTDs can range from 30% to over 90% for some devices. The measured quantum efficiency of silicon detectors varies depending on the structure of the detector; the effects of device characteristics such as gate structure, antireflective coatings, depletion depth, down-convertors, and backside illumination have been described in Chapter 2. When selecting a CCD for a particular wavelength region, especially in the near-IR or UV wavelength regions, it is important to examine actual QE data for the device of interest and not make assumptions about performance on the basis of similar types of CTDs.

The bandgap energy of silicon restricts the use of detectors made of this material to wavelengths shorter than ~1100 nm. Considerable effort has been devoted to extending the longer wavelength range of CTD detectors by making them out of a variety of materials including germanium, indium antimonide, and the silicides of platinum, palladium, and indium [20]. The performance of these nonsilicon CTDs is expected to improve rapidly. Near- and mid-IR spectroscopy will greatly benefit with the advent of high-quality, low-noise multichannel detectors that respond in these regions. Chapter 4 describes the performance of these new detectors in detail. Likewise, a similar level of effort has been expended to improve the responsivity of the devices in the ultraviolet, vacuum ultraviolet, and even x-ray spectral regions.

3.1.3 Signal-to-Noise Ratios of Single-channel, Multichannel, and Multiplexed Techniques

To help readers understand the significance of multichannel detection techniques, a brief overview of the signal-to-noise ratio (SNR) performance of the common spectroscopic methods is presented. This comparison points out the experimental conditions under which certain techniques provide superior SNRs and the reasons for this performance. The following discussion describes the driving force for developing multichannel detection schemes in terms of the higher obtainable SNR.

Currently there are three common approaches to measuring light intensity over more than one wavelength interval. The first is a sequential technique in which a single detector is used with a dispersive optical system to monitor the various wavelength intervals individually. With the second technique, multiple detectors or a detector array is used to monitor multiple wavelengths simultaneously. The third technique involves multiplexing the signal and detecting the multiplexed signal with a single-channel detector (i.e., Michelson interferometry and Hadamard spectroscopy). In these multiplexed techniques, spectral information from multiple wavelength intervals is measured simultaneously with the single-channel detector; the information is encoded so that the intensity-versus-wavelength information can be extracted.

A fourth possible approach is use of a combination of multichannel and multiplexing techniques. In this type of system, the signal is optically multiplexed and the entire interferogram is measured simultaneously with a multichannel detector. Hybrid techniques that combine features of the three basic approaches are also possible and have been exploited. These for the most part combine scanning with detection by multiplexed or multichannel detectors. The relative SNR advantages of these hybrids are readily apparent from extension of the discussion below and so will not be dealt with in detail here.

The relative SNR superiority of multichannel systems over other approaches depends on a number of factors including the type of noise source dominating the intensity measurements and the number of wavelength intervals (N) to be measured. The noise sources tend to be statistically independent and thus add in quadrature; therefore, each type of noise source can be considered independently. The performances of multiplexed and multichannel systems are compared with the performance of single-channel systems under conditions where one type of noise source dominates. The noise sources considered here are categorized as sources that are independent of signal intensity (i.e., detector noise), sources that are proportional to signal intensity (i.e., fluctuation noise), and sources that are proportional to the square root of signal intensity (i.e., photon or dark current shot noise).

Equations 3.1 through 3.5 summarize the SNR that is obtained with a PMT or a CTD in terms of the three types of noise. SC denotes a single-channel sequential measurement; MC denotes a multichannel parallel measurement; MX denotes a multiplexed measurement with a single detector.

$$SNR_{SC,PMT} = \frac{(t/N)R_l}{[t/N(R_d + R_b + R_l) + t^2/N^2(\xi^2 R_l^2 + \xi_b^2 R_b^2)]^{1/2}} \quad (3.1)$$

$$SNR_{SC,CTD} = \frac{(t/N)R_l}{[N_r^2 + t/N(R_b + R_l) + t^2/N^2(\xi^2 R_l^2 + \xi_b^2 R_b^2)]^{1/2}} \quad (3.2)$$

$$SNR_{MC,PMT} = \frac{tR_l}{[t(R_d + R_b + R_l) + t^2(\xi^2 R_l^2 + \xi_b^2 R_b^2)]^{1/2}} \quad (3.3)$$

$$SNR_{MC,CTD} = \frac{tR_l}{[N_r^2 + t(R_b + R_l) + t^2(\xi^2 R_l^2 + \xi_b^2 R_b^2)]^{1/2}} \quad (3.4)$$

$$SNR_{MX,PMT} = \frac{(t/2)R_l}{[tR_d + t/2(\sum_{i=1}^{m} R_{li} + \sum_{i=1}^{m} R_{bi}) + t^2/4(\sum_{i=1}^{m} \xi_{li}^2 R_{li}^2 + \sum_{i=1}^{m} \xi_{bi}^2 R_{bi}^2)]^{1/2}} \tag{3.5}$$

where:

t	=	total measurement time
N	=	number of channels used (features measured)
R_l	=	count rate due to analyte
N_r	=	detector read noise
R_d	=	dark current rate
R_b	=	count rate due to background
ξ_l	=	flicker constant for analyte signal
ξ_b	=	flicker constant for background signal

The signal-to-noise ratio equations used for the comparison of spectroscopic methods are from Bilhorn and colleagues [21] (see also [22]). Multichannel and single-channel systems employing both PMT-type and CTD-type detectors are considered.

The use of an integrating detector in a Michelson interferometer is not considered. Acquisition of a signal that changes in time with an integrating detector is not optimal because the detector read noise is introduced with every sampling of the time-varying signal. If an integrating detector must be employed with a Michelson interferometer, the best SNR can be achieved with a single scan of the mirror and with as few samples as possible. A much more effective way to record an interferogram with an integrating detector is to record the interferogram spatially rather than temporally [23,24]. Rather than a Michelson interferometer, an integrating multichannel detector can be used effectively with a holographic interferometer (in essence, a tilted mirror Michelson interferometer; see Section 3.4.2.1). The advantage of high light throughput is combined with many of the desirable features of multichannel detectors.

In the following sections, cases are examined in which individual noise types are dominant.

3.1.3.1 Detector Noise Dominance

In the case where detector noise dominates (read noise is independent of time), multiplexed detection results in the often-cited Fellgett's advantage of $N^{1/2}$. Detector noise dominance is common with the photoconductive detectors used in the IR. This is one of the several reasons for the success of Fourier Transform infrared (FTIR) over conventional dispersive techniques in the IR. Multichannel detection with an integrating detector results in an advantage of N times the SNR of a single-channel method. This result assumes a detector read-noise-limited situation with the same type of detector and is due to the time-independent read noise. The advantage of a multichannel integrating detector system over a single-channel PMT or photoconductor system depends on the individual detector-noise parameters and detector quantum efficiencies.

The charge-holding capacity of individual pixels must be considered when estimating the best photon-shot-noise-limited SNR that can be obtained with a CTD in a single exposure. Because the charge-storage capacity of the detector elements in modern high-density arrays is typically less than 300,000 e^-, multiple elements may need to be dedicated to each image section or spectral feature to achieve the measurement goals.

3.1.3.2 Photon-Shot-Noise Dominance

One of the most universal noise sources in photon detectors is photon shot noise. Photon shot noise arises from the random (in time) arrival of photons at the detector [25]. This results in an uncertainty when measurements of photon flux are made in a finite period. The uncertainty is approximated by the Poisson distribution, which gives a noise value proportional to the square root of the signal level. In addition, the dark-current shot noise can also be approximated using Poisson statistics. Dark-current shot noise is often the limiting detector noise in PMTs and photoconductors such as the mercury cadmium telluride detector [20].

In photon-shot-noise-limited systems, the nature of the spectrum affects the relative improvement of multiplexed detection over single-channel sequential detection [26]. When broad-band spectra are considered so that the signal strength is relatively constant throughout the spectral range, the predicted performance is comparable to that achievable with single-channel sequential detection. However, when narrow or line spectra are considered, the relative intensity of the line of interest compared with the mean intensity of the whole spectrum determines the magnitude of the advantage or disadvantage [26,27]. Experimenters have observed that the proximity of spectral features of interest to other intense features determines the observed SNR [28,29]. Weak spectral features adjacent to intense spectral features are adversely affected, yielding poorer SNR than is obtainable with a single-channel system.

Multichannel detection systems are unaffected by the nature of the spectrum when photon-shot-noise-limited conditions prevail. Regardless of whether broad-band or narrow-line spectra are being detected, multichannel systems yield a signal-to-noise-ratio improvement of $N^{1/2}$

for N channels over single-detector sequential systems. This result assumes no crosstalk (stray light, ghosts) between the channels in the multichannel system.

3.1.3.3 Fluctuation Noise

Multiplexed systems are adversely affected by fluctuation noises (also known as flicker noises) compared with single-channel sequential systems [30]. For example, random changes in source intensity are indistinguishable from changes in intensity caused by mirror movement. As is the case with photon shot noise, the nature of the effect in a multiplexed system depends on the structure of the spectrum being measured. When intensity is relatively constant over the wavelength range and fluctuation occurs to the same extent at all wavelengths, a multiplex disadvantage of $1/N$ is observed. If the fluctuation occurs mainly in the intensity of bright spectral lines superimposed on a relatively dark background, the noise tends to be localized in the vicinity of the bright spectral line.

A summary of the merits of one technique relative to another when one of the three noise sources is dominant is given in Table 3.1. These results are obtained by examining the ratios of equations 3.1 through 3.5. The results show multichannel systems having no advantage over single-channel systems when fluctuation noise is dominant. Increasing the measurement time per channel has no effect on measurement SNR because noise accumulates at the same rate as signal. However, fluctuation in most analytical sources follows a $1/f$ type of behavior over limited ranges [31]. Because the noise power decreases at higher frequencies, the effects of fluctuation noise can be reduced by decreasing the total measurement time. Consequently, a multichannel system offers an advantage because the point in time at which fluctuation noise exceeds photon shot noise is reached in all channels simultaneously at approximately $1/N$ of the time required by a single-detector sequential system. Parameters that are changing or drifting do not have the opportunity to change as drastically because the measurement time is reduced. Additionally, because all channels are measured simultaneously, all measurements are made under the same set of changing conditions.

To summarize the SNR comparison results, multichannel techniques are expected to produce the highest SNRs in most common analytical situations. While the models used in these calculations are relatively simple and include the effects of only a few types of noise, they emphasize the tremendous gains possible with the multichannel systems described in this book.

A number of factors other than SNR must also be considered when evaluating the overall performance and utility of a particular analytical technique. For example, the absolute wavelength accuracy afforded by

Table 3.1 SNR Comparison of Multichannel and Multiplexed Methods vs. Single-Channel Methods

Type of Dominant Noise Source	$\frac{SNR_{MC,PMT}}{SNR_{SC,PMT}}$ [a]	$\frac{SNR_{MX,CTD}}{SNR_{SC,CTD}}$	$\frac{SNR_{MX,PMT}}{SNR_{SC,PMT}}$
Noise = constant[b]	$N^{1/2}$	N	$N^{1/2}$
Noise ∝ $(\text{signal})^{1/2}$ (photon shot noise)	$N^{1/2}$	$N^{1/2}$	1[c]
Noise ∝ signal (fluctuation noise)	1[d]	1	$1/N^{1/2}$ [e]

[a] MC = multichannel; SC = single channel; MX = multiplexed; N = the number of channels.

[b] Detector noise in photomultiplier tubes and photoconductors results from shot noise in the dark current. The noise has a square root dependence on time. Detector noise in CTDs and PDAs is associated with the read process and is independent of time; thus the different results when detector noise is dominant.

[c] These results are obtained when signal as a function of wavelength is constant. If there is structure in the spectrum, photon shot noise tends to degrade the SNR in the vicinity of intense spectral features. Fluctuation noise also tends to be localized in the vicinity of strong spectral features. In the extreme case of line spectra, evaluation of the SNR figure of merit when fluctuation or photon shot noise is dominant is not possible because it varies from line to line and depends on the complexity of the spectrum.

[d] In some cases, a multichannel advantage may exist in fluctuation noise-limited systems. For the same SNR, measurement time is reduced by a factor of *N*. This shifts the measurement bandwidth to higher frequencies. Fluctuation noise may be less significant at higher frequencies because of its approximate proportionality to $1/f$.

[e] Assumes approximately uniform intensity throughout the spectrum.

Michelson interferometry is required in some applications. Also, stray or scattered light may be higher in a polychromator used with a multichannel detector compared with a scanning double monochromator used with a PMT. Nonetheless, SNR is an important figure of merit, especially when high-precision results are desired, or when decreased analysis time allows increased sample throughput rates and improved productivity.

A variety of advantages associated with multichannel detection with solid-state arrays have thus far been neglected. A significant advantage is the flexibility of wavelength selection offered compared with multichannel detection with a limited number of discrete detectors. This was the primary driving force behind a number of attempts to replace multiple PMTs placed behind laboriously positioned slits with multichannel detectors of far inferior performance (e.g., vidicons) in atomic emission spectroscopy. Modern CTDs offer a number of other advantages including increased reliability as a result of their ruggedness, relative simplicity of operation, and low cost per resolution element.

3.2 Specialized Readout Modes

The majority of CCDs and CIDs are designed for use in imaging applications. Linear arrays are typically used in line-scan applications and two-dimensional arrays are primarily intended for direct imaging. The design (number and size of pixels, register and preamp design) and architecture (number of phases, buried or surface channel) generally reflect the intended readout mode and application. Although a device operated according to its intended readout mode may prove useful in spectroscopy, improvements can generally be made by operating the device in a specialized readout mode. Fortunately, the nature of CCDs and CIDs is such that the flexibility exists for operation in widely varying ways. For example, readout speeds may be greatly reduced to improve noise performance and device clocks may be altered to affect readout of only specific portions of the device. Because the design and architecture of CIDs and CCDs affect their performance in scientific applications, selection of a device that offers both the desired spectral properties (wavelength response, noise, linearity, full-well capacity) and that operates in the desired readout mode is essential. Also, as the number and variety of applications of CCDs and CIDs in science and engineering expand, more and more devices are being designed and produced with operation in one or more of the specialized modes in mind.

3.2.1 CCD Readout

As described in Chapter 2, CCDs measure the photogenerated charge accumulated in pixels by sequentially shifting the charge to an on-chip preamplifier located at the periphery of the device. Depending on the architecture of the device—full frame, frame transfer, or interline transfer—the specifics of how this is done vary. In scientific imaging and spectroscopic applications, however, either full-frame or frame-transfer CCDs operated in full-frame mode are generally used. The main reason is that all of the available area of the device is used for signal acquisition. The drawback is that the illumination must be shut off in some way during readout to prevent image smearing. This generally precludes integrating the signal at the same time that a previous frame is read out.

Full-frame and frame-transfer CCDs also allow considerable flexibility in readout. The two most widely used capabilities are subarray readout and charge binning, which are also possible with interline-transfer CCDs. Taking advantage of these capabilities requires that the CCD-control electronics be able to operate the serial and parallel register independently and, for optimum performance, that the pixel digitization routine also be under separate control.

3.2.1.1 Subarray Readout

A subarray in this context is defined as some contiguous group of pixels, usually rectangular, within the CCD array. Subarray readout refers to reading out only this group of pixels and is the simplest specialized readout mode. The operation consists of eliminating the charge barriers in the serial register while the parallel clocks are advanced to the subarray's parallel origin, thereby dumping the charge that has accumulated in the array up to the origin of the subarray. Next, the barriers are re-established in the serial register and the lines making up the subarray are shifted one at a time into the serial register. Slewing to the serial origin of the subarray is done with the reset gate set to drain away charge. Once the serial origin is reached, readout commences in the normal fashion. At the end of each subarray line, the remaining charge in the serial register must also be dumped so that it is not combined with the next row when it is shifted into the serial register. For a similar reason the remaining charge in the CCD should be cleared after the last row of the subarray is read out. If it were not, it would be combined with the charge from the next exposure. More sophisticated systems allow multiple subarrays to be read from a single exposure by keeping track of which pixels and which rows of pixels to dump. One can also see that the subarray shape need not be rectangular; theoretically a contiguous block of pixels of any shape could be read out as a subarray. However, it may be that the bookkeeping involved becomes so difficult that the disadvantages outweigh the advantages.

Subarray readout is useful for restricting the number of pixels that must be processed by the data system to only those that contain useful information and for increasing readout speed. In imaging applications, a region of interest might be established interactively from a full-frame readout before beginning an involved image acquisition sequence. In spectroscopy, a subarray might be established for the region of the detector corresponding to the illuminated portion of the spectrometer entrance slit.

3.2.1.2 Charge Binning

Charge binning refers to the combining of photogenerated charge from adjacent pixels on the detector prior to readout (see Chapter 2). The main reason for using binning is to improve SNR, dynamic range, and readout speed [32]. Because detector read noise is independent of signal level, combining the charge from two equally illuminated pixels results in doubling the SNR in cases where detector read noise dominates. As only a square-root improvement is achieved with conventional signal averaging, the ability to bin is a significant advantage of the CCD. In photon-shot-noise-limited situations, the expected square-root improvement is also achieved.

Obviously, the trade-off with binning is loss in spatial resolution. Also of significant importance is the fact that dark current is binned along with signal charge. In applications where long exposure times are combined with a high degree of binning, considerably better cooling is required to prevent dark-current shot noise from becoming the limiting noise source.

Binning can be performed in both the serial or parallel directions and in both directions simultaneously. Parallel binning simply involves shifting multiple rows into the serial register before the serial register is read out. Similarly, serial binning involves shifting multiple pixels from the serial register into the summing node of the output amplifier prior to shifting from the summing node onto the actual output diffusion. Modern scientific CCDs are designed with binning in mind and thus the serial register and summing well charge capacities generally exceed those of the parallel register. Care must be exercised, however, to prevent blooming in the serial register or the summing node due to excessive binning, even though the exposure has been adjusted to prevent blooming in the parallel register. The judicious combination of subarray readout and charge binning is the analyst's most powerful tool for optimizing SNR, illumination requirements, integration time, and total measurement time. The existence of these two capabilities has made the CCD the detector of choice for many types of modern research spectroscopy and imaging.

3.2.1.3 Time-delayed Integration (TDI)

The time-delayed integration mode of CCD readout allows the acquisition of long image swaths from a moving scene and is at present not commonly used in spectroscopy. However, as the examples in the next section demonstrate, this mode offers some interesting properties and is expected to become a more common readout mode in these applications. The technique was first developed in the mid-seventies for use in airborne reconnaissance, but since that time TDI has been applied in document scanning and inspection applications [33], spectral imaging [34,35], and analytical spectroscopy [36]. TDI is a technique for creating a scanned image with a two-dimensional imager, much like a linear sensor is used to scan an image. In conventional two-dimensional imaging, a trade-off exists between field of view and spatial resolution because of the number of pixels in the imager. If resolution and field of view requirements exceed what can be achieved in a single exposure, then multiple images must be combined. TDI is a viable alternative to combining still images.

As illustrated in Figure 3.1, TDI involves synchronizing the movement of an image across the sensor with the operation of the CCD parallel clocks so that the photogenerated charge produced by the illumination from the scene moves with the image. Thus, all of the pixels along a column of the CCD are used to integrate charge from a point in the scene. Smearing is avoided by maintaining careful

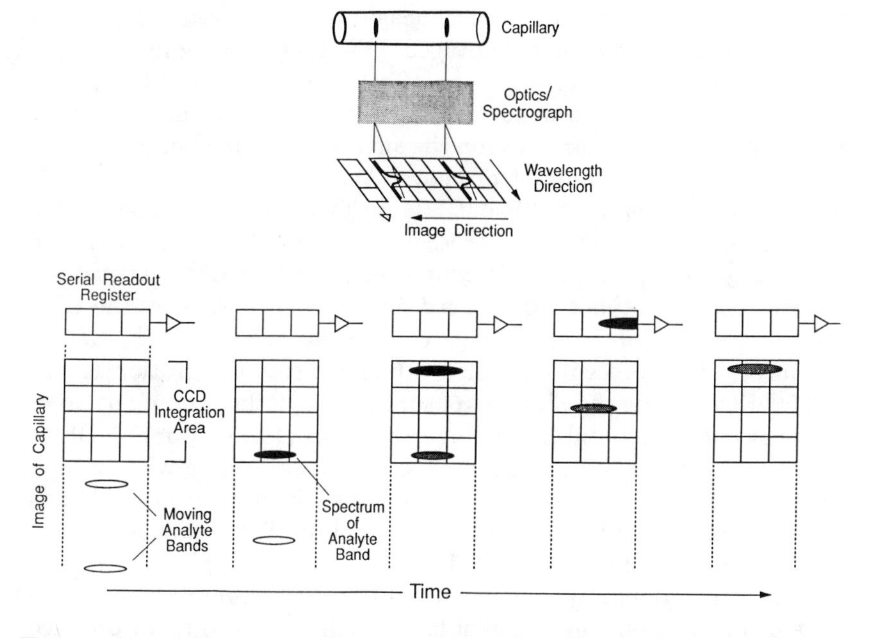

Figure 3.1 Diagram of the time-delayed integration mode of CCD readout. (Reprinted with permission from reference 34; copyright, 1991, American Chemical Society.)

synchronization between the movement of the scene and the shifting of the charge in the CCD, and images of unlimited scan length can be collected. An improvement in sensitivity over line-scan systems is achieved that is directly proportional to the number of lines in the two-dimensional sensor in the scan direction.

3.2.2 CID Readout

As described in Chapter 2, CIDs measure the photogenerated charge accumulated in pixels by shifting the charge back and forth under electrodes contained within each detector element. As opposed to CCDs, the charge never leaves the detector element where it is collected, and hence the readout modes of CIDs are considerably different. Many CCD readout modes that allow charge combination such as binning have not been applied to CIDs. On the other hand, CIDs allow the charge information contained in a detector element to be read using the nondestructive readout (NDRO) mode described in Chapter 2, and they are extremely resistant to charge blooming.

Random Access Integration (RAI). After the exposure of the CID to a scene, the quantity of charge in each detector element is measured. As described in Chapter 2, the measurement of charge is not a destructive process in the CID; after the measurement, the charge remains and can either be remeasured or removed. The NDRO mode has two significant advantages. First, by averaging multiple rereads from a single exposure, the effective read noise of the device is reduced. Second, the dynamic range of the sensor can be greatly extended. The difference between the smallest level of charge measurable and the largest amount of charge that can be stored in a CID detector element is not sufficiently large for many applications—for example, atomic emission spectroscopy. Fortunately, excess charge does not bloom (spill into adjacent elements) in CIDs. Therefore, different exposure times can be used for different regions of the CID, *with the exposure times determined during the exposure of the CID to the source.* This procedure, called random access integration (RAI), uses NDROs during the single exposure to follow the accumulation of charge at both weak and intense spectral lines [19,21,37,38]. NDROs are used continuously during the exposure to determine when a high signal-to-noise measurement is possible at each spectral line. Once a high SNR can be achieved, the precise amount of charge is measured and the exposure time is recorded.

The conventional dynamic range of the CID is combined with the range of integration times available for each spectral line to produce the dynamic range needed for atomic emission spectroscopy. The RAI method is the most efficient means of measuring the intensity of a number of spectral lines because the exposure time is adjusted dynamically according to signal-to-noise requirements based on the intensity of each spectral line.

3.3 Calibration and Background Correction

In many imaging and spectroscopic applications, the goal of the researcher is to obtain absolute intensity information (photometric accuracy) from the detector. In this case, the responsivity variations from individual detector elements, the spectrometer, and the optics must be corrected. In addition, spikes from high-energy events such as cosmic rays and background radiation need to be removed. The following two sections outline methods to detect and remove the high-energy events and to allow calibration of the CTD output.

3.3.1 Cosmic Rays and High-Energy Events

In addition to near-IR to x-ray photons, silicon detectors can respond to cosmic rays and background radiation, with each high-energy event producing a large number of electron-hole pairs. Hence, a single event can produce thousands of electrons and can seriously degrade the sensitivity and the quality of the data unless corrections are made. Exposure times as short as 1 minute often may be sufficient to produce a high-energy-particle event that interferes with the appearance of spectra or images; the frequency depends on a number of factors including materials used in the construction of the CCD camera and the laboratory building, the geographic area, the altitude, the solar cycle, and other variables. High-energy-event removal is easiest when one is observing a scene that consists of low-frequency variations that are many detector elements wide—for example, typical bandwidths in molecular spectroscopy. In many Raman and other molecular spectroscopic experiments, the slit-height dimension contains no high-frequency information; therefore, the high-energy events can be removed by writing routines that look for isolated spikes in a relatively smooth background. Typically, the average value of the surrounding elements replaces the value for the element that contained the high-energy event. In some situations, only a few undetected high-energy events can affect the detection limit. Binning information into large subarrays makes high-energy-event removal more difficult; specifically, the redundant information offered by reading out individual detectors in a subarray individually instead of binning may be useful when long exposures are involved in aiding in the identification and removal of spurious signals produced by high-energy-particle interactions with the CCD. In instances when the focal plane of the detector contains many narrow, high-frequency features (such as looking at a star field or at the focal plane of a well-focused echelle system), discrimination between features that are spurious and those that are not is much more difficult. In such cases, multiple exposures can be taken to identify the cosmic rays and other radiation by looking for inconsistencies in the series of exposures. This method is commonly used in astronomical imaging; typically, the values of each detector element from at least three "identical" exposures are compared, and all spikes are removed.

3.3.2 Bias Subtraction and Flat-Field Correction

Ideally, each detector element in the detector array has the same offset, dark current, and slope (response to light). One of the goals of the bias and flat-field correction is to correct the variations in offset and slope from the average value. To do this, the intercept and slope for each detector element must be determined. This process requires the acquisition of at least two calibration images: a bias image and a flat-field image. The first image corrects for additive effects and the second for multiplicative effects.

The predominant additive effect is an offset or bias in the detector system output for zero signal levels. The offset may not be uniform for all detector elements (fixed pattern response) and so it is removed by subtracting an image taken with no light falling on the sensor and with as short an exposure time as possible to avoid dark current. A second additive effect is dark current. Dark-current accumulation may not be uniform for all detector elements, so a dark exposure is made using the same exposure time as intended for the measurement. Subtraction of this dark frame corrects for the fixed pattern response and offset described above and for the dark current (but not the dark-current shot noise).

Flat-field correction removes the effects of detector responsivity, variations in the illumination pattern from the light source, spatial attenuation due to imperfections in optics, dust particles, and apodization due to optics. These effects are all multiplicative, and so they are removed with a multiplicative correction. The flat-field image is taken using a uniform exposure to an image producing a significant (for example, 25% full scale) response. The user must correct for all nonuniformities, and control for wavelength, aperture, focus, and many additional parameters; improperly applied flat-fielding can be worse than no correction. It is also critical to avoid saturation of any pixels. One standard flat-field process is [33]

$$\text{Corrected image} = \frac{(\text{Raw} - \text{Dark})}{(\text{Flat field} - \text{Dark})} \times (\text{Mean value})$$

Multiplying by the mean value of the dark-subtracted flat-field image ensures that the pixel values of the corrected and raw images are roughly comparable. While to a large extent many deleterious effects can be corrected for, the process involves subtracting or multiplying the data by calibration images and hence can degrade the signal-to-noise ratio of the desired data. If the read noise of the CCD is the dominant source of noise, then subtracting a single bias frame increases the noise by $2^{1/2}$; however, using an average of 25 bias frames, the noise is increased by less than 10%. On the other hand, for high-light-level spectroscopy, the SNR is little affected by the noise in the bias frame. Many linear CCDs consist of a photoactive area and odd and even readout registers. In these devices an odd/even responsivity variation of several percent is common [39]; simple odd/even correction can be determined once for each device and used for as long as the detector or associated electronics are not adjusted.

In many nonimaging spectroscopic applications, flat-fielding is not as important as in photometric or imaging applications because calibration curves are used to determine quantitative information. For example, when using a CCD/spectrograph as a detection system for fluorescence, one determines the concentration of an unknown analyte by comparing values obtained from a known rather than an absolute illumination level. Several excellent descriptions and the actual software routines for sophisticated calibration schemes are available from the Image Reduction and Analysis Facility*; this package is a general-purpose software system for the reduction and analysis of scientific data, developed by the National Optical Astronomy Observatories [40,41].

TDI Flat Fielding. TDI operation of a CCD can also offer advantages over use of a two-dimensional CCD for acquisition of a sequence of still images to be pasted together later. Aside from the obvious advantage of eliminating the pasting step, the TDI acquisition method is less sensitive to nonuniformities in the scene illumination and in the response of the imager. In conventional imaging, nonuniformities in the illumination, variations in efficiency of the optical system as a function of field angle, and pixel-to-pixel variations in the response of the sensor show up in the final image. These artifacts can all be removed through flat-fielding.

In practice, obtaining good flat-field images can be difficult and the operation of correcting many images can be time consuming. Images collected using TDI mode need to be corrected in only one dimension rather than two. Because each point of the scene is imaged onto all pixels, through all field angles, and through all illumination conditions

---

* Operated by the National Optical Astronomy Observatories, P.O. Box 26732, Tucson, AZ 85726.

in the scan direction, there is nothing to correct in the scan direction. The cross-scan correction is the same for all cross-scan rows and so requires only a flat-field vector rather than a flat-field image. The flat-field vector is obtained by averaging many rows from a TDI image of a relatively flat scene. Because of this averaging there is a noise advantage in applying a flat-field correction with a vector. Whereas correcting an image with a flat-field image acquired without averaging (as is normally done) increases the noise by approximately the square root of 2, flat-fielding with a vector introduces essentially no additional shot noise. In applications where TDI imaging is being used to produce very precise low-contrast images, the flat-field vector can be obtained from the sample image itself, eliminating the step of acquiring an additional image for correction purposes. Care must be exercised in using this approach since, if there are any features in the image aligned with the scan direction that persist through a significant portion of the image used for computation of the flat-field vector, the correction will eliminate these features from the image.

3.3.3 Charge-Trapping Effects

Some CCDs can suffer from additional problems that impact their use in spectroscopy. For example, a few early CCDs suffered from charge-trapping effects, in which some of the first charge transferred from the parallel to the serial registers was not measured [42,43]. As these CCDs were commonly used for ultra-low-light-level Raman spectroscopy, this effect could cause detector-induced bandshape distortions, which are a great concern because the significance of the results often depends on subtle changes in bandshape. Harris and co-workers have recently described a quantitative investigation into charge-trapping effects on Raman spectra for the Thomson CCD [42]. For a user of CCDs prone to charge-trapping problems, the output can be seriously distorted at low illumination levels, and various corrective measures need to be implemented. However, many CCDs do not suffer from detectable charge-trapping problems, and so careful selection of CCDs is an important method of avoiding such problems for low-light-level applications.

3.4 Spectroscopy with Charge-Transfer Device Detectors

A considerable difficulty still facing the widespread application of CTD detectors in analytical spectroscopy is the very different geometric requirements placed on optical systems designed for these detectors compared with systems designed for PMTs. The total area available per detector element is considerably smaller than that of a typical

PMT photocathode. Current optical systems designed to be used with PMT detectors are therefore not, in general, optimal for use with CTDs; however, those designed specifically to be used with CTDs often offer the advantages of smaller size, lack of moving parts, higher optical throughput, and lower cost.

Before deciding on a spectrometer and detector system for a particular application, the goals for the system must be known: the required spectral resolution, wavelength coverage, dynamic range, optical efficiency, time resolution, and flexibility. Once these criteria are understood, the appropriate system can be designed. The vast majority of published applications using CCDs have used standard spectrometers and simply replaced the PMT or photographic film with the CCD detector. The following sections describe in greater detail the gains that can be made when the unique readout modes and other special features are combined with both standard and custom spectrometers. The sections are divided into the three categories of systems depending on how the wavelength information is obtained: spectroscopic imaging, one-dimensional spectroscopy, and two-dimensional spectroscopy. Greater detail is provided for those combinations that are less well known or less commonly implemented.

3.4.1 Spectroscopic Imaging

Spectroscopic imaging refers to systems that use the detector array as an imaging detector and obtain wavelength information by using filters or spectrometers compatible with two-dimensional imaging. In fact, this category includes the most common and straightforward implementation of CCDs for scientific applications: using a scientific CCD as a detector for fluorescence or absorption microscopy by replacing the standard 35-mm camera or C-mount video adapter with the scientific CCD detector module. These scientific CCDs are ideally suited for optical microscopy because of their geometric fidelity, large format, and extreme sensitivity. Because of these features, CCDs are starting to dominate digital fluorescence microscopy, in terms of both slow-scan scientific instruments [44–49] and intensifier-video CCD combinations for high-speed applications [50,51]. In most cases a filter is used to select both the emission and the excitation wavelengths. The interested reader is referred to references 44 through 51, since the use of CTDs for microscopy is outside the scope of this chapter.

In addition to applications that require only a two-dimensional image at a single wavelength, many applications require image information at multiple wavelengths. This, of course, becomes three-dimensional data (intensity information for a particular X, Y, and wavelength). To provide this, a variety of implementations have been developed; these include multiple wavelength filters, scanning systems using tunable filters such as the acoustooptical modulator [52], Fabry-

Perot interferometers [53], spectrometers [54,55] and systems that acquire one-dimensional image information. The linear image is then scanned across the object [49]. Combination methods have been developed, including the unique Raman system developed by Morris and colleagues, which consists of a Hadamard transform/imaging system to provide wavelength and imaging capabilities [56,57]. With the use of nearest neighbor deblurring [58], this system can obtain a series of images at different wavelengths and provide depth profiling, for true four-dimensional data. John Sedat and co-workers have developed a three-dimensional, optical sectioning, fluorescence-imaging system using CCD detection [46,58]. Another example of a three-dimensional imaging system is the wavelength-resolved tomographic system developed by Hieftje to image the distribution of various atomic species in plasmas [54,59]. While a wide range of systems have been developed already for wavelength-resolved imaging applications, we expect the variety of instrumental configurations and number of reported applications to expand rapidly in the near future.

3.4.2 One-dimensional Spectroscopy

This category consists of CTD/spectrometer systems that contain wavelength information in one of the axes of the detector. The detector can be either a linear or a two-dimensional array. In the latter case, the second dimension can be used to increase the sensitivity of the measurement by increasing the effective detector-element size (binning), to provide one-dimensional imaging capabilities, to allow multiple simultaneous measurements, or to provide time resolution.

3.4.2.1 Linear Detector Arrays

Conventional spectrographs disperse light across a curved focal plane of 100 to over 1000 mm in length. When using a linear CTD or PDA for spectroscopic measurements that do not require very high resolution, it is necessary to disperse light across a flat detector that is commonly only 10 to 50 mm wide. The use of flat-field concave holographic gratings designed specifically for solid-state array detectors provides a simple, single-element spectrograph with low reciprocal linear dispersion and very high throughput [60–62].

Unfortunately, most linear CTDs have small, approximately square, detector elements typically only 5 to 25 μm on a side [39,63, 64]. On the other hand, linear CCDs require few overlying gate electrodes and hence can have higher quantum efficiencies than two-dimensional arrays [39,63,64]. Because of the small geometries, efficient use of these detectors requires image demagnification to more efficiently measure the light from a spectrograph with tall narrow slits. The image of a tall slit can be compressed with a cylindrical lens [39]. The great advantage of the linear CCD is in the area of cost; such

CCDs can be less than $100 for up to 3456 elements [39], but complete systems optimized for low-light-level spectroscopic applications are just becoming available.

Linear arrays of photodiodes that have been optimized for spectroscopy have high pixel length-to-width aspect ratios (50:1). Similar CID and CCD geometries are also possible. CIDs with interdigitated row and column electrodes such as those found on the "super-pixel" CID [65] would maintain high charge-transfer efficiency at high readout speeds with large pixel areas. High pixel-aspect-ratio linear PDAs with CCD readout are becoming available. For example, Dalsa Inc. offers a 2048-element linear CCD system with 13- by 500-μm detector elements that can be read out at up to 20 MHz. Such linear CCDs can maintain the high quantum efficiency of the square geometry linear CCDs by using ion implants to establish virtual phases (field gradients) so that no electrode structure is required over the photoactive area.

Perhaps the most encouraging trend in this area is the development of pseudolinear arrays by Thomson, Loral, and EEV. As one example, the 2688 × 512 element CCD being developed by the Smithsonian Astrophysical Observatory [66] has a 40 × 8 mm photoactive area and is ideally suited for spectroscopy. As it has been designed with the nondestructive "skipper" readout amplifiers [67], it presumably can have a readout noise approaching a single electron. Such pseudolinear arrays can be used as true linear arrays by binning all the elements in the short dimension; however, for diagnostic tests or added dynamic range, the slit-height-axis elements can be read out individually.

Spatially Encoded Interferometry. Most UV-visible spectrometers available today use dispersion to obtain spectra. On the other hand, interferometric techniques have been used in the UV through near-IR wavelength range; various examples include the Michelson interferometer [29,68] and the common-path holographic interferometer [23,24,69,70]. A Michelson interferometer uses a single-channel detector in a multiplexed mode. The common-path holographic interferometer consists of an interferometer that measures all wavelengths simultaneously with a multichannel detector and hence is a combined multichannel, multiplex system. CTDs offer numerous advantages as detectors for such systems.

Stroke and Funkhouser described the use of photographic emulsion to record a spatial interferogram from a Fourier-transform interferometer [71]. Several investigators have recognized the significant advantages of directly obtaining the interferogram using a multichannel detector such as a PDA or CCD and performing the transform digitally instead of holographically using photographic plates

[23,24,69,70,72]. Because of the unique configuration of the common-path interferometer, the system has many of the important advantages of simultaneous multichannel spectroscopic systems and those inherent in interferometry. Fourier-transform spectroscopy has great advantages for obtaining spectra from extended sources. This throughput (Jacquinot) advantage can yield superior signal-to-noise performance compared with conventional dispersive systems. The multiplex (Fellgett) effect does not normally lead to large gains in the UV and visible regions because of excellent UV-visible detector performance; in some cases, these multiplex effects can lead to serious problems [69,73,74].

For the common-path interferometer, the interferogram is spatially resolved and simultaneously recorded using a CCD detector array. Because every point in the interferogram is measured during the exact same integration period, the system can be used to measure transient phenomena. A Michelson interferometer obviously cannot record interferograms from sources that change substantially during a single mirror scan. A significant advantage is the lack of moving parts in the common-path system, allowing an inexpensive, stable optical system compared with conventional moving-mirror interferometers. The lack of moving parts alleviates the need for a laser for position reference, although a wavelength reference may be desirable for accurate wavenumber calibration. Lastly, the use of an array detector with real-time readout allows extremely simple alignment of a portable, compact system.

The optical block diagram of an experimental system is shown in Figure 3.2a. The triangular common-path interferometer is a source-doubling interferometer with the CCD detector and the virtual sources each located at the conjugate focal points of the focusing mirror. The light from the source is divided into two by the beam splitter, with these beams traveling the same path through the interferometer in opposite directions. The two diverging beams are collimated by the parabolic mirror and form an interference pattern on the CCD detector. The distance between the mirror position shown as a solid and the dashed lines in Figure 3.2a dictates the separation between the two virtual sources and thereby determines the spatial frequency of the observed fringes and hence the resolution of the resulting spectrum.

Because all points of the extended source produce the same fringe pattern at the same position at the detector focal plane, the interferogram is independent of the extent of the source; the system needs neither a slit nor an aperture. The throughput of this system is much greater for an extended source than that of conventional dispersive systems and can even be larger than that obtainable with a Michelson interferometer [75]. The cylindrical lens is used to increase the light-

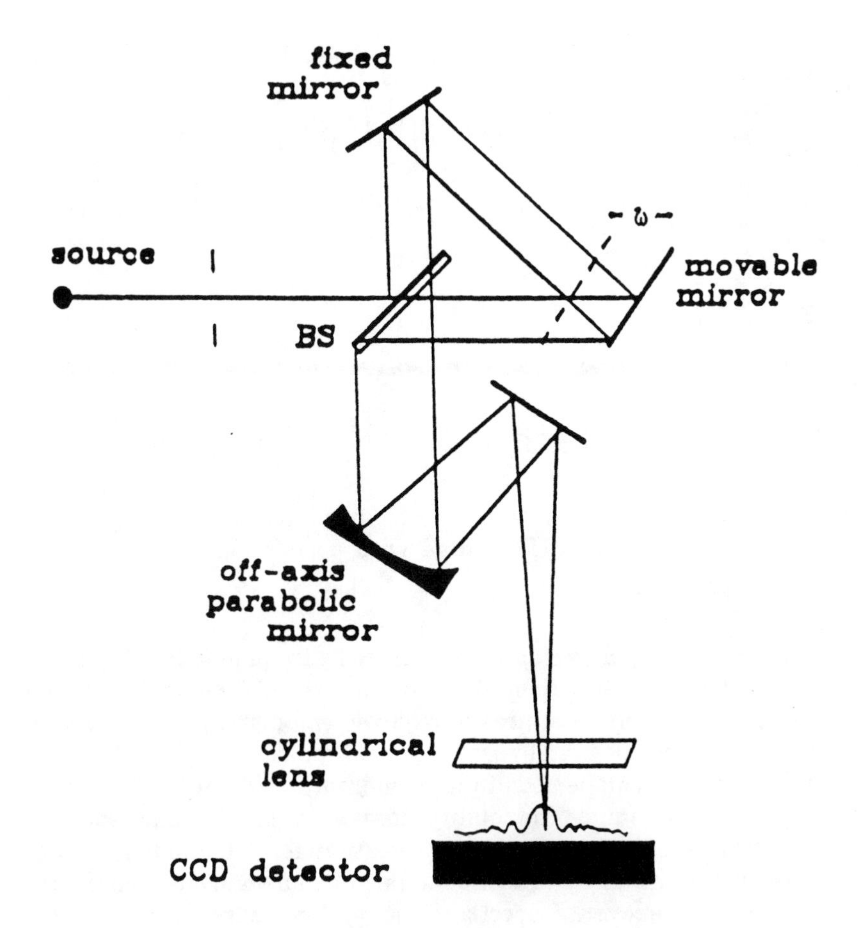

Figure 3.2a Optical diagram of the triangular common-path system. (Reprinted with permission from reference 24.)

gathering ability of the system when using a linear CCD. To recover the spectral information from the interferogram, a complex FFT is applied to the data from the CCD. The resolution of the resulting spectrum is dictated by geometrical considerations and the number of detector elements illuminated by the interferogram. The maximum spatial frequency of any component in the interferogram that can be observed without aliasing has a period of two detector elements.

The interferogram obtained using the common-path system with a mercury pen lamp is illustrated in Figure 3.2b. The mirror displacement has been set to produce low spatial frequency fringes.

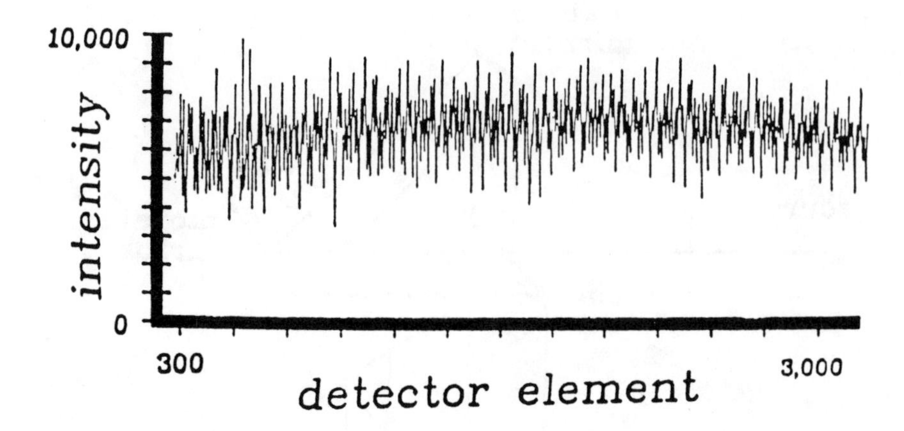

Figure 3.2b Spatial interferogram from a mercury pen lamp source. (Reprinted with permission from reference 24.)

The interferogram shows spectral features throughout the 15,000 to 45,000 cm^{-1} range, approximately between 200 and 600 nm. While the spatial interferometer requires a linear detector array, a two-dimensional detector can be used to produce one-dimensional spectral images of the source and can be used in implementations described below for a dispersive system. The ability to obtain spatial and spectral information simultaneously has application for emission/excitation fluorescence work and diagnostic studies of sources; it has particular applications whenever a spectroscopic system is required to view extended sources.

3.4.2.2 Two-dimensional Detector Arrays

One-dimensional dispersive spectroscopy with a two-dimensional CTD detector allows spectrometers designed to be used with PMTs to be used with CTDs without any modifications to the size of the focal plane image. The whole area of the entrance slit may be used for light collection, the second axis of the CTD being used simply to collect the additional light. The capability of CCDs to bin charge from adjacent pixels is particularly useful in this application since it allows the custom matching of detector-element size to the size of slit images. Rows or columns of pixels that correspond to the slit-height direction can be completely binned together, resulting in a signal proportional to the sum of all the photogenerated charge collected at that wavelength interval but with the read noise of only a single detector element. Subarray readout is used to exclude regions of the CCD that are not illuminated by the slit so that dark current from only the portion of the CCD used for light collection is included with the measurement. As described previously, in low-light-level applications a significant gain in SNR is possible.

Subarray Size and Charge-binning Considerations. Several other considerations should be kept in mind when selecting subarray size and binning factors for use with a two-dimensional CCD in one-dimensional spectroscopy. The primary advantage of binning is improvement in SNR when detector read noise dominates the measurement. Because the read noise levels are so low in modern slow-scan CCD cameras, very little light intensity is required before the measurement becomes dominated by photon shot noise. Once this is the case, binning offers no advantage over summing or averaging in computer memory in terms of SNR, although a small advantage in convenience may be realized. Under these circumstances it may be advantageous to reduce the binning in the slit-height direction and take advantage of the additional information provided by having spectra that represent different positions in the focal plane.

Caution must be exercised when averaging several spectra collected simultaneously using binning in the slit-height direction. If the illumination along the slit is nonuniform, pixel saturation may occur in one of the binned spectra without the problem occurring in the other spectra. Because analog-to-digital-converter saturation is usually set to occur before blooming, quantitative information is lost in the spectrum containing the saturated peaks. If the spectra are averaged, the effect may be subtle enough to go unnoticed, resulting in slightly misshapen peaks. Automated routines, which collect spectra from multiple binning groups, detect high-energy-particle events, and then average the spectra to produce a result, must also either check for saturation or present the spectra to the user before averaging.

Optimizing charge binning and exposure time for a given experiment involves cognizance of the limiting noise sources. For example, binning is not desirable when sufficient light is available to bring pixel intensity values into the photon-shot-noise-dominated range in a measurement time that is practical and convenient. Binning should only be considered when dark-current shot noise or detector read noise begin to dominate a measurement or when measurement time can be shortened without incurring a severe SNR penalty.

Perhaps the most highly idealized case of optimizing binning is that first described by Epperson and Denton [32]. Binning is dynamically varied to achieve a dynamic range wider than that possible with fixed binning parameters. The slit image is oriented so that it is parallel to the serial register. In this way, rows shifted into the serial register contain pixels representing the intensity distribution along the slit height. The first several pixels representing the beginning of the slit image (after reaching the serial origin of a subarray if it is specified) are read out using no binning. On the basis of the signal level measured for these few pixels, a binning factor is chosen for the rest of the row. If signal levels are high, no binning is used; for intermediate signal levels, a binning parameter is chosen that will optimize SNR. If very low signal levels result, the rest of the row is read as a single pixel. The spectrum is reconstructed by averaging the binning groups and then dividing by the binning parameter.

Figure 3.3 shows the spectrum of a low-pressure mercury discharge collected using a single exposure and dynamic binning. Charge from weak spectral features was binned 80 × twofold into a single charge packet, while charge information from intense lines was read without binning. The simple dynamic range of this RCA camera system was approximately 10,000 (50 e^- noise, 500,000 e^- full-well

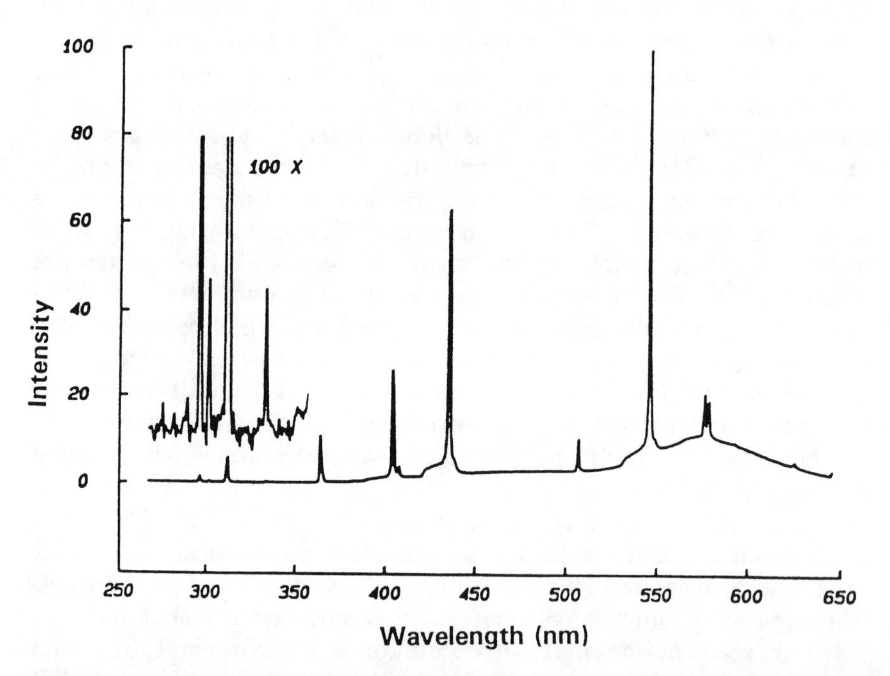

Figure 3.3 Mercury spectrum obtained using variable binning. The inset is the spectral region from 270 to 360 nm expanded 100-fold. (Reprinted with permission from reference 32; copyright, 1989, American Chemical Society.)

capacity), but by using the variable binning method, the dynamic range of the spectrum is >500,000 for a single exposure.

Scanning Multichannel Spectroscopy Using TDI Mode. When a multichannel detector is used with a spectrometer that disperses light in only one dimension, a compromise between resolution and wavelength coverage is often required. As noted previously, the largest commercially available two-dimensional CCD arrays have only several thousand pixels on a side. These devices are often prohibitively expensive, so CCDs having sizes of 512, 576, or 1024 pixels on a side are more commonly used. Depending on the particular spectroscopic technique, several hundred to one or two thousand resolution elements may not be sufficient to cover the desired wavelength range at adequate spectral resolution. For example, in atomic emission spectroscopy, resolution of 0.01 nm over the ultraviolet and visible spectral regions (180 to 800 nm) is desirable. A 620 nm wavelength range covered at 0.01 nm resolution requires 124,000 resolution elements to just satisfy the Nyquist criterion. In Raman spectroscopy, where CCDs are clearly the detector of choice, it is desirable to cover the spectral region comprising shifts of from 0 to 4000 wavenumbers from the excitation frequency at anywhere from 4 to less than 1 wavenumber resolution. Here again, sufficient wavelength coverage and resolution cannot be provided by a readily available CCD.

The traditional approach to dealing with the resolution/wavelength coverage dilemma is to observe smaller spectral regions and to piece the resulting spectra together. Aside from the difficulties associated with maintaining wavelength calibration, problems associated with throughput variations as a function of diffraction angle produce undesirable baseline effects that result in discontinuities at the individual spectral boundaries. Knoll and colleagues have [76] proposed a partial solution to this latter problem in a photodiode-array-based instrument. The authors demonstrate collecting spectra at much smaller wavelength intervals than those necessitated by the size of their array. The spectra are added in the regions of overlap so that the variations in throughput tend to average out. The authors reduce spectral collection time in proportion to the increase in the number of spectra collected so that the overall measurement time is held constant. For example, to cover a wavelength range that is five times that covered in a single exposure with their photodiode array, they could use 50 exposures at 90% spectral overlap and reduce the individual exposure times by a factor of 10 from that which would be conventionally used. The authors correctly acknowledge that additional detector read noise is introduced using this procedure.

The unique readout capabilities of CCDs allow a similar advantage to be realized in a much more elegant fashion using the TDI readout mode [36]. In this case, the previously described TDI mode is used so that the wavelength scan rate is synchronized to the CCD shift rate. The CCD needs to be oriented so that the motion of the spectrum is parallel to the column axis of the CCD and the wavelength scan direction is toward the serial register. One of the largest advantages is the reduction in detector read noise. All of the other advantages of using a two-dimensional CCD in linear spectroscopy are also retained. Subarray readout is still possible in the slit-height direction so that the dark current from areas of the array not used for collecting light can be rejected. Binning in the slit-height direction is also possible so very high sensitivity can be achieved. Dynamic binning can also be used, provided that the readout is completed before the next wavelength interval is ready to be shifted.

TDI operation of a CCD for analytical spectroscopy offers the advantages of multichannel detection combined with the flexibility of single-channel scanning systems. The two-dimensional CCD format allows the same multichannel advantage to be realized as in the stationary case but the synchronization of the spectrometer scan with the shifting of photogenerated charge eliminates the restrictions placed on wavelength coverage and resolution by the finite size of the CCD. Collection of a spectrum using a CCD in TDI mode also avoids the problems associated with variation in throughput as a function of diffraction angle; each element of the spectrum is continuously integrated as it passes through all of the angles that are intercepted by the CCD, thereby averaging out any effects associated with angle. Each resolution element of the spectrum is collected under identical geometrical and temporal conditions just as in a scanning system using an exit slit and a photomultiplier tube.

Another more practical advantage of TDI mode used with a CCD for one-dimensional spectroscopy is that it is compatible with the spectrometers that most analytical spectroscopists are currently using. Rather than incurring the expense of designing and building a spectrometer matched to the two-dimensional format of a CCD and accepting the trade-offs in performance that will invariably be required, all that is required is a relatively simple interface and computer software additions. Scanning monochromators commonly use a stepper motor on a sine drive. This arrangement provides a constant wavelength scan rate with a constant step frequency and is ideally suited for adaptation to the scanning multichannel concept. The pulses from the stepper motor are simply used to synchronize the transfer of charge in the CCD, after being passed through a suitable divider circuit. The advantage of slaving the camera to the wavelength drive is that any scan speed can be selected and the readout will be automatically adjusted. Scan speed then becomes an additional variable that can be used to control sensitivity (in addition to binning).

Figure 3.4 shows a region of the emission spectrum of mercury observed from a low-pressure mercury discharge lamp collected using a CCD in the TDI mode. The spectrum was collected using a CCD with 576 elements in the dimension in which the light is dispersed. The monochromator uses a 2400-groove/mm grating and a 1-m focal length to give approximately 0.008 nm/pixel and a stationary wavelength coverage of 4.8 nm; the spectrum in this figure covers 70 nm. Figure 3.5a shows the Hg 313-nm doublet and the Hg 312.6-nm line as measured with the CCD and grating stationary. With a 10-μm entrance slit, the 23-μm detector-element size limits the resolution. Figure 3.5b shows an enlargement of the same portion of the Hg emission spectrum shown in Figure 3.4; the observed factor-of-two loss in resolution is due to the stepwise transfer of charge in the CCD as opposed to the continuous movement of the spectrum across the spectrometer focal plane.

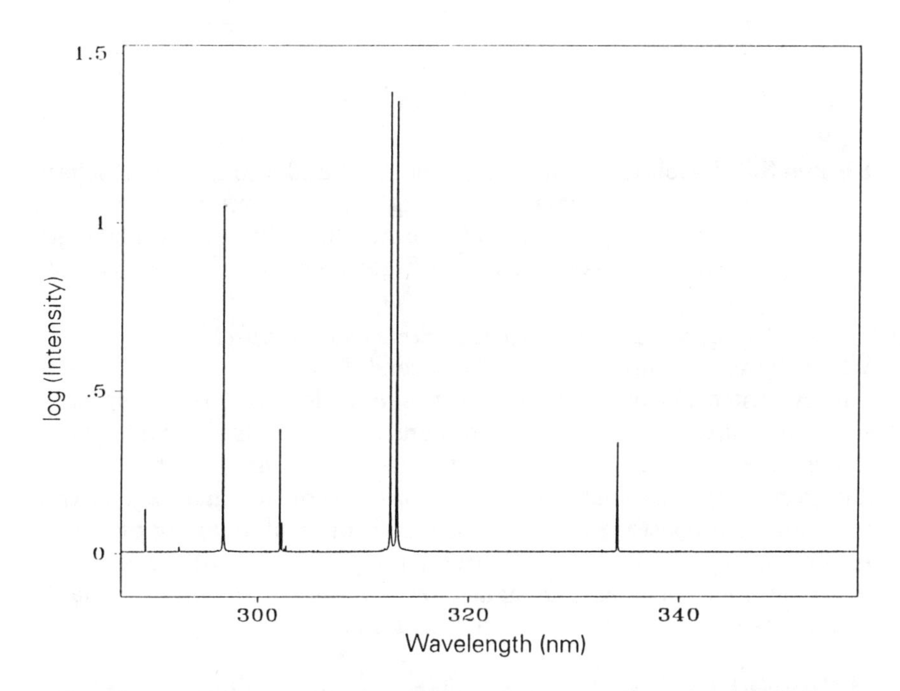

Figure 3.4 Spectrum of a low-pressure mercury discharge lamp collected using a CCD in the TDI mode. It consists of 8396 points; the CCD has 576 pixels in the wavelength dimension. The stationary wavelength coverage is 4.8 nm.

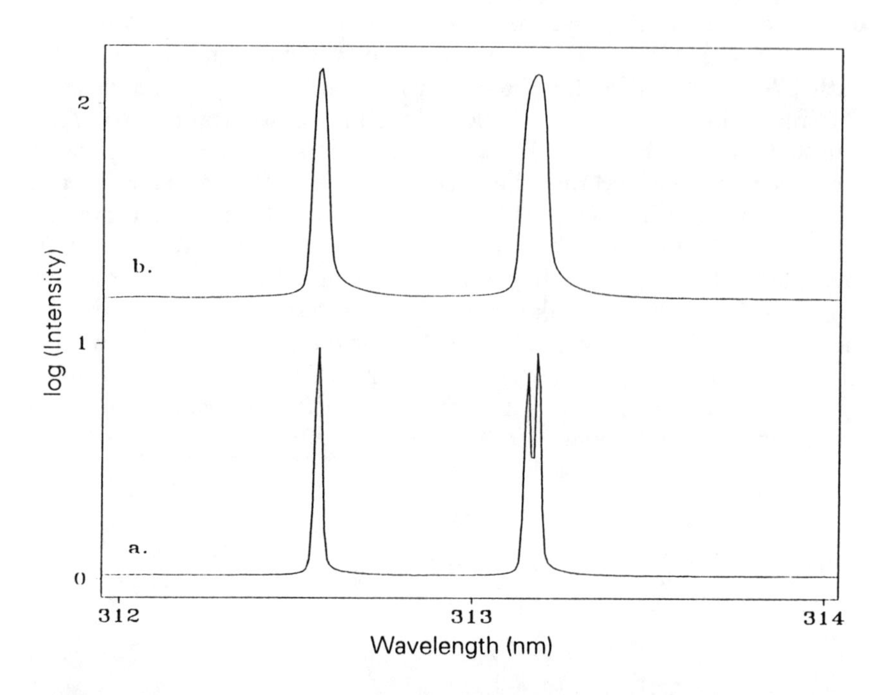

Figure 3.5 Resolution comparison for fixed and scanning multichannel spectrometer showing three mercury lines. (a) Stationary grating, 252 points. (b) The same wavelength coverage expanded from Figure 3.4.

3.4.2.3 Spatially Resolved One-dimensional Spectroscopy
While the most common use of the second dimension of the CTD is to improve the sensitivity, dynamic range, and flexibility of the spectroscopic measurement, the second dimension can be used for additional information such as in imaging and time-resolved applications. When designing a system that uses the slit dimension for imaging, *careful* selection of a spectrograph is required to prevent imaging problems introduced by the grating; otherwise image quality will be reduced. Spectrographs specifically designed for imaging are becoming available and are recommended in these applications [60–62].

Multipoint and Multisample Spectroscopy. One of the most obvious uses for the second dimension offered by the two-dimensional array is multiple samples. As an example, fiber optics can be used to bring light from various points in a manufacturing process to the entrance slit of an imaging spectrograph to do process control. Thus, a single spectrograph/detector can monitor tens to hundreds of

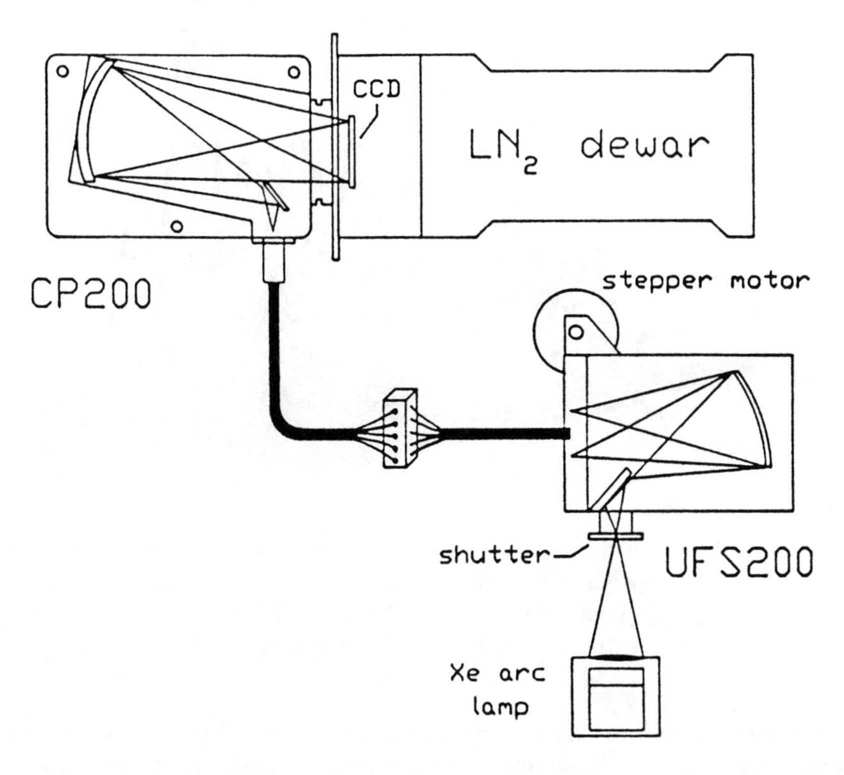

Figure 3.6a Instrumental geometry for multidimensional fluorescence experiments. (Reprinted with permission from reference 61; copyright, 1989, by International Scientific Communications, Inc.)

locations/samples, and, as all are using the same instrumentation, cost and calibration problems are reduced. Piccard and Vo-Dinh [77] demonstrate one implementation of such a system consisting of a fiber array at the entrance slit of a spectrograph/CCD. Another example involves the fluorescence system shown in Figure 3.6a, which includes fiber-optic bundles and two spectrographs, one for emission and one for excitation [61]. This type of multidimensional fluorescence system offers an improved method for obtaining simultaneous, highly sensitive emission/excitation matrices [78]. The fluorescence emission/excitation matrix (EEM) from a mixture of perylene and anthracene collected by the four optical fiber bundles is shown in Figure 3.6b.

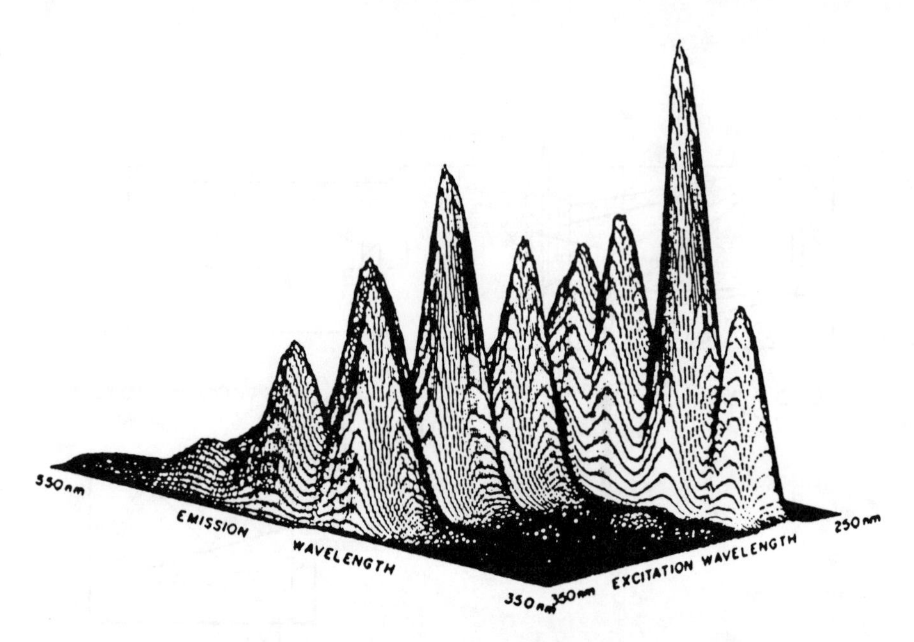

Figure 3.6b Fluorescence EEM for a perylene/anthracene mixture with four fiber-optic bundles. (Reprinted with permission from reference 61, copyright (1989) by International Scientific Communications, Inc.)

TDI for One-dimensional Spectral Imaging. Another analytical application for TDI is observing analyte bands in a separation experiment during the separation. The method can be used for observing bands separating during two-dimensional separations (TLC or slab electrophoresis) and capillary (linear) separations.

Figure 3.7 shows a block diagram for one such system, where the optics form a 250 μm × 6 mm image of a 63 μm × 2 cm section of the separation capillary onto the slit of an imaging spectrograph. On the detector focal plane, one dimension of the CCD contains wavelength information, while the second is the capillary image. During a separation (whether chromatographic or electrophoretic), the analyte bands move down the capillary, and hence the fluorescence image moves across the CCD.

If the CCD readout rate is synchronized with the analyte band movement, then the effective integration time is the entire time the band remains in the observation zone. Unlike nonimaging systems, this increase in observation zone does not correspond to a decreased spatial resolution (i.e., separation efficiency). Figure 3.8 shows a TDI-electropherogram taken of a 3×10^{-18} mole injection of two fluorophores (sulforhodamine and fluorescein) into a multichannel capillary-

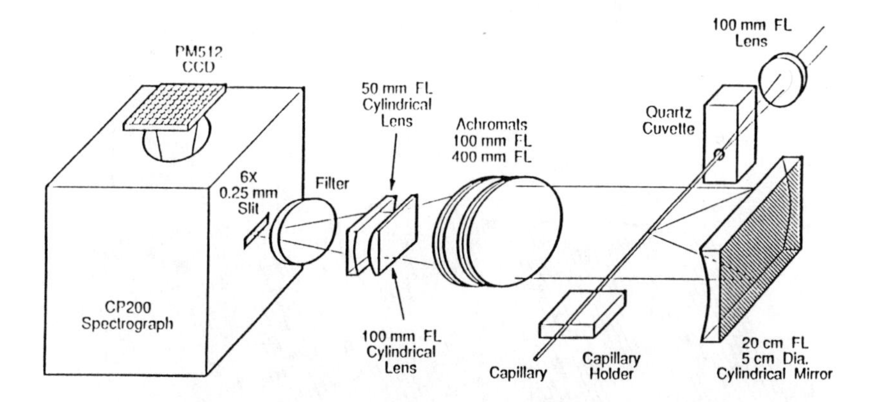

Figure 3.7 An optical block diagram of capillary-electrophoresis/laser-induced-fluorescence system. (Reprinted with permission from reference 34; copyright, 1991 American Chemical Society.)

electrophoresis/laser-induced-fluorescence (CE/LIF) instrument [34,35]. By using the TDI readout mode, the spatial (capillary) information is lost, and the output of the system is an infinite series (i.e., as many as obtained during the separation) of spectra, each of which corresponds to a several hundred micron wide "tracked" analyte (or solvent) band migrating though the entire several centimeter observation window. (The increase in the Raman and Rayleigh lines as the run progresses is caused by the accumulation of signal for a longer time for each succeeding spectrum.) An important advantage of this approach over previous fluorescence detection methods in capillary electrophoresis is the acquisition of multiwavelength fluorescence information; this allows multiple fluorophores to be used for such applications as DNA sequencing.

3.4.2.4 Time-resolved Spectroscopy: The Fast Spectral Framing Mode

Fast spectral framing is a variation on frame-transfer operation of a CCD, in which a portion of the CCD is not used for imaging but instead is used to store photogenerated charge from a previous image. In conventional frame-transfer operation, half of the CCD is masked and is used to store the image from the unmasked half during pixel readout. Shifting the image from the image register to the storage register happens rather quickly relative to the time required to read

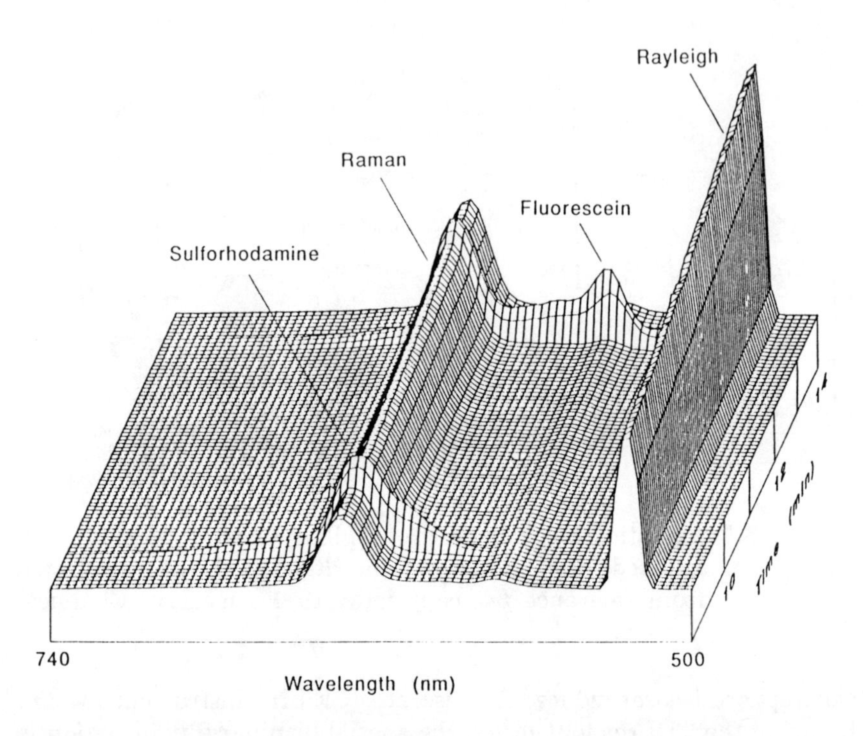

Figure 3.8 TDI electropherogram of 3 amol of sulforhodamine and fluorescein injected into a CE capillary. (Reprinted with permission from reference 34; copyright, 1991, American Chemical Society.)

out the storage register, so the shift is normally done without closing a shutter or blocking the light in some other way. During the time that the storage register is read out, integration of the next frame is taking place in the image register.

In a slow-scan camera, the time required to read out pixels is much greater than in a conventional video rate camera, so the notion of capturing motion or achieving time resolution on a scale shorter than several seconds is not often considered. If, however, a portion of the CCD is masked and not used for imaging, several frames can be taken in rapid succession by using the masked portion of the CCD as an analog storage register. For example, if five-sixths of a CCD were masked, five images equal in area to one-sixth of the total CCD area can be taken in rapid succession. Once the five images are stored, they need to be read out at the much slower pixel-readout rate. Time resolution sufficient to follow the shattering of a light bulb has been demonstrated [79].

The fast spectral framing technique can be extended to images having a height of only a single row of pixels. This format is useful for spectroscopy and is the use from which the name is derived [15,79]. Here a large number of spectra can be recorded at high speed before the "analog memory" is full and the CCD must be read out. In this technique all of the array is masked off except for the last row (the row at the end of the array *opposite* the serial register). As an alternative, a spectrometer that has good imaging properties can be used so that by illuminating only a small part of the entrance slit, as with a fiber optic, only a small portion of the CCD is illuminated. Figure 3.9 shows a series of spectra of a xenon flash lamp discharge collected at 6-μs intervals using the fast spectral framing technique with a mask positioned over most of the CCD. (For clarity, not all of the collected spectra are shown.) While in this case the mask was manually positioned over relatively few rows of the CCD, it is possible to have exactly one (or as many as desired) rows masked off with an opaque aluminum coating during the fabrication of the CCD. Thomson CSF has recently announced the commercial availability of such a device [63].

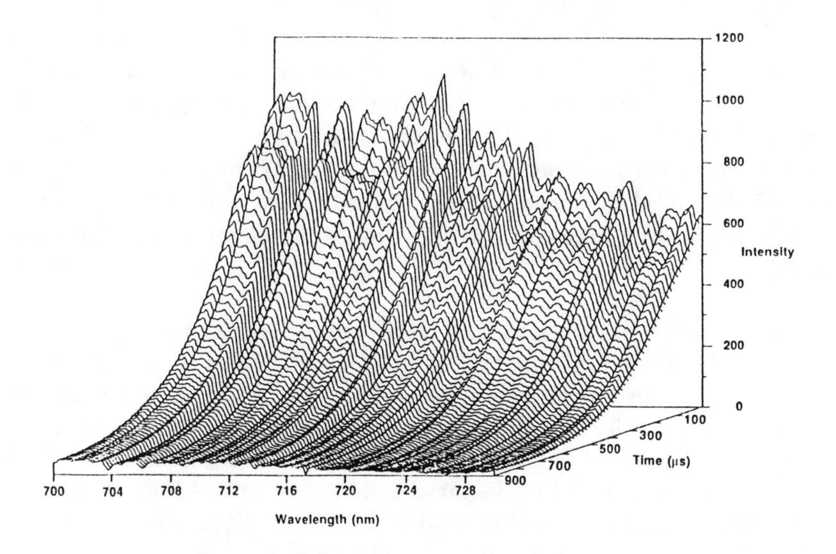

Figure 3.9 Time-resolved spectra of a xenon flash lamp acquired using a CCD operated in the fast-spectral-framing mode. A 320-point spectrum was obtained every 6 μs. (Reprinted with permission from reference 79.)

An advantage of the fast spectral framing technique for recording spectra of changing sources is that the shift time between individual spectra can be freely varied. For example, an exponentially increasing delay time can be used to follow an exponentially decreasing signal. In this scheme, spectra are collected rapidly near the beginning of the event so that it is well characterized, but spectra can also be collected throughout the decay without running out of space on the array for their storage.

Fast spectral framing offers a number of advantages over conventional rapid scanning spectrometer systems that use rotating mirrors or gratings. Most important, the multichannel advantage allows improved SNR. The number of parallel measurement channels equals the number of pixels in a row of the CCD so that the advantage, assuming all other parameters are constant, can be as great as a factor of 64 (square root of 4096 pixels). The spectral throughput of a spectrometer system based on fast spectral framing can also be higher. Conventional rapid-scan spectrometers use small mirrors or gratings that are rotated or swept at high speeds by galvanometer movements or the like. Maintaining low-mass moving parts is essential so it is often necessary to sacrifice the high spectral throughput that is provided by large optics. Because all components are fixed in the fast spectral framing technique, the high-efficiency, commercially available *f*/2 spectrometers are ideally suited in this application.

An additional advantage of the fast spectral framing technique is higher speed operation than that provided by conventional rotating-mirror rapid-scan spectrometers. The maximum collection rate with the CCD-based instrument is limited by the time required to shift the charge in the array by one row and can be faster than 1 μs. Trade-offs exist between shift speed, charge transfer efficiency, and operating temperature, but times on the order of a few microseconds are easily obtainable.

3.4.3 Two-dimensional Dispersive Spectroscopy

The two-dimensional format of many CCDs and CIDs can be efficiently used for high-resolution spectroscopy when different spectral sorting systems are used for the different axes of the detector array. The most commonly employed system is the echelle spectrometer [80]. This spectrometer employs two dispersive elements, one of high dispersion (the echelle grating) and the other of low dispersion, oriented so that their directions of dispersion are at right angles to each other [80].

Figure 3.10 shows an optical block diagram of an echelle system employing a prism as the second dispersive element. The echelle grating disperses light into a number of superimposed orders and the prism separates these orders. As shown in Figure 3.11, the result is

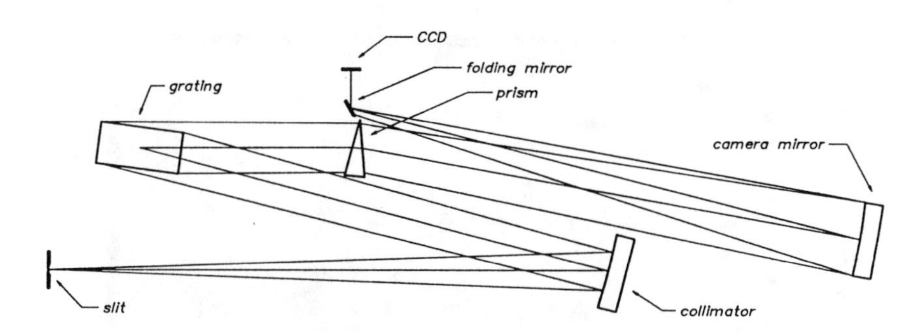

Figure 3.10 Echelle system optical diagram showing the echelle grating, the prism cross-dispersing element, and the CTD detector.

a focal plane image that looks like a series of linear spectra arranged like text on a page, each line containing a contiguous piece of the whole spectrum [81]. While the same considerations discussed above for producing image sizes compatible with the format of the detectors for spectrometers that disperse the light in one direction apply for echelle spectrometers, the design of the system is constrained by the requirements of producing an output compatible with both dimensions of the array. In addition to systems that use dispersion in both axes, systems that use spatial interferometry in one axis and dispersion in the second axis, and dual interferometric systems, are also possible.

3.4.3.1 Echelle Spectrometers for Atomic Spectroscopy

To cover the UV to near-IR range with the 0.01-nm resolution required for atomic spectroscopy, several hundred thousand resolution elements are needed. As linear arrays of this size are both impractical and not available, the echelle arrangement is one of the few alternatives for obtaining the required resolution with a rectangular detector array. Not surprisingly, most applications involving echelle systems have involved atomic spectroscopy. Several echelle spectrometers have been developed using CID detectors [19,38,82,83] and CCD detectors [81,84–87]. Because these systems are covered in greater detail in a later chapter, only a brief overview is included here.

The spectrometers used are similar in principle to echelle spectrometers designed for use with photographic film or arrays of PMTs; however, the image size is greatly reduced in order to produce an echellogram compatible in size with silicon detectors. The flexibility afforded by these systems in terms of wavelength selection is an important feature compared with conventional scanning or

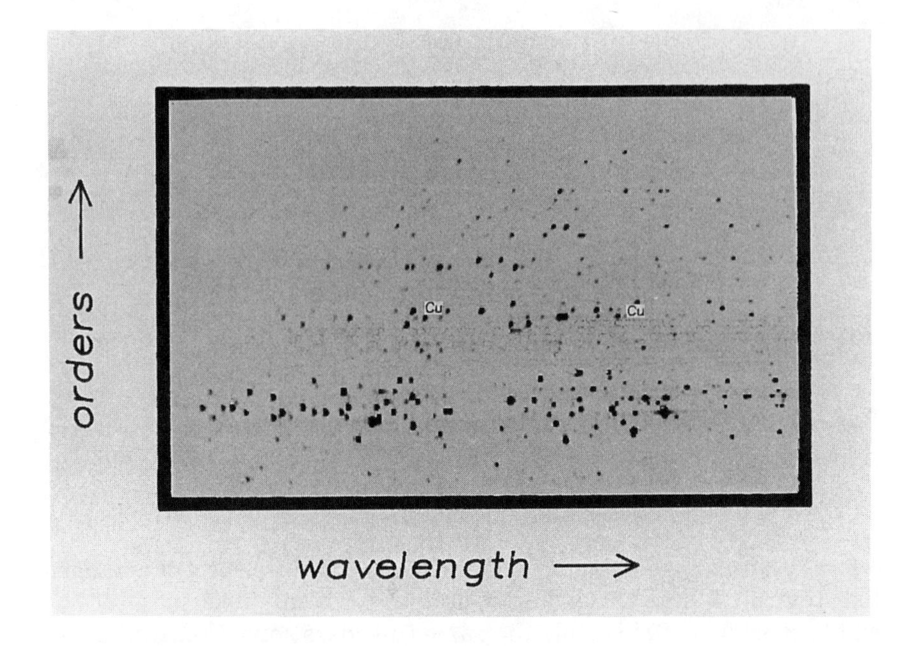

Figure 3.11 Background-subtracted CID echellogram of a 30 s integration of mixture of Cu, Fe, and Ca. (Reprinted from reference 81 with permission from Pergamon Press Ltd, Headington Hill Hall, Oxford OX3 0BW, UK.)

polychromator systems. The choice of spectral lines to be used for an analysis is custom tailored to the components of the sample being measured. The most intense spectral lines of an element are selected for components present at very low levels, and less intense lines resulting from nonresonance transitions are used for elements at high concentrations in order to avoid problems with self-absorption. Multiple spectral lines are used for each element to enhance precision, and known spectral interferences are avoided in the cases where other spectral lines are available [88]. Background correction procedures are applied to every spectral line using the data from adjacent detector elements.

The number of detector elements employed in existing CID/echelle spectrometer systems is rather low compared with what is available with modern CCD detectors. The use of a large CCD with an echelle spectrometer allows the attainment of much higher resolution at the same or increased wavelength coverage. Calculations show that large CCDs coupled with an echelle spectrometer are capable of achieving the approximately 0.001-nm resolution over the ultraviolet and visible wavelength range that is necessary for continuum source atomic absorption spectroscopy. However, when considering a nonuniform order spacing, Nyquist sampling criterion, and other effects that reduce the effective use of detector elements, further improvements are still needed. Besides offering simultaneous multielement capabilities as in emission spectroscopy with a CID, the flexibility of choosing spectral lines is available. The combined use of microsampling furnace techniques that are available for use in atomic absorption spectroscopy and a simultaneous multielement spectrometer holds great promise as a powerful analytical tool for trace element analysis when sample size is limited.

3.4.3.2 An Echelle Spectrometer for Raman

An echelle spectrograph using CCD detection can make an excellent Raman instrument, but the design requirements are different from those described for atomic spectroscopy. In particular, the spectral resolution does not need to be as high, and the light collection efficiency is of paramount importance. A recent report by Pelletier describes a Raman echelle/CCD system using transmission optics and characterizes the spectrographic performance [89]. While the light-collection efficiency and sensitivity of the echelle system are not as high as for linear Raman/CCD systems [90], the simultaneous wavelength coverage is an important advantage in many applications.

3.4.3.3 Crossed Interferometric Dispersive Spectrometer (XIDS)

One of the greatest disadvantages of all forms of multiplexed spectroscopy is the multiplex or Fellgett's disadvantage [26,27,91]. This disadvantage is a result of the distribution of spectral noise from intense features to other spectral features in the spectrum. Unfortunately, no method has been developed that removes this problem with the exception of filtering the light before the light is detected. A number of arrangements using filtering have been used to increase the performance of conventional Michelson instruments including the use of a premonochromator for atomic emission analysis [73], the use of sophisticated filters to remove the Rayleigh light in Raman [92,93], and the blocking of particular elements for Hadamard transform Raman spectroscopy [94].

A unique method for reducing the multiplex disadvantages inherent in interferometry is to combine a cross-dispersive element with the common-path interferometer and use a two-dimensional array detector. Thus, the common-path interferometer is the wavelength selector in one direction and a prism or grating is the cross disperser in the orthogonal direction. For example, using a 100×1000 element CCD array detector, 100 separate interferograms result; each interferogram covers a slightly different wavelength range. As an example, detector row 90 may have light in the wavelength range 601 to 610 nm falling on it, while row 91 has light of 611 to 620 nm. This optical arrangement is similar in principle to the echelle arrangement, where two dispersive elements are placed so that orthogonal dispersion takes place and a "rectangular" focal plane spectrum results. In the crossed interferometric dispersive spectrometer (XIDS) arrangement, the interferometer replaces the echelle grating, with both systems containing a low-resolution dispersive element. The resulting system has the large advantage that any noise present in an optical signal is constrained to the rows where this noisy signal falls.

A simplified three-dimensional representation of the XIDS system is shown in Figure 3.12 [72]. In this figure, a cross-dispersing element

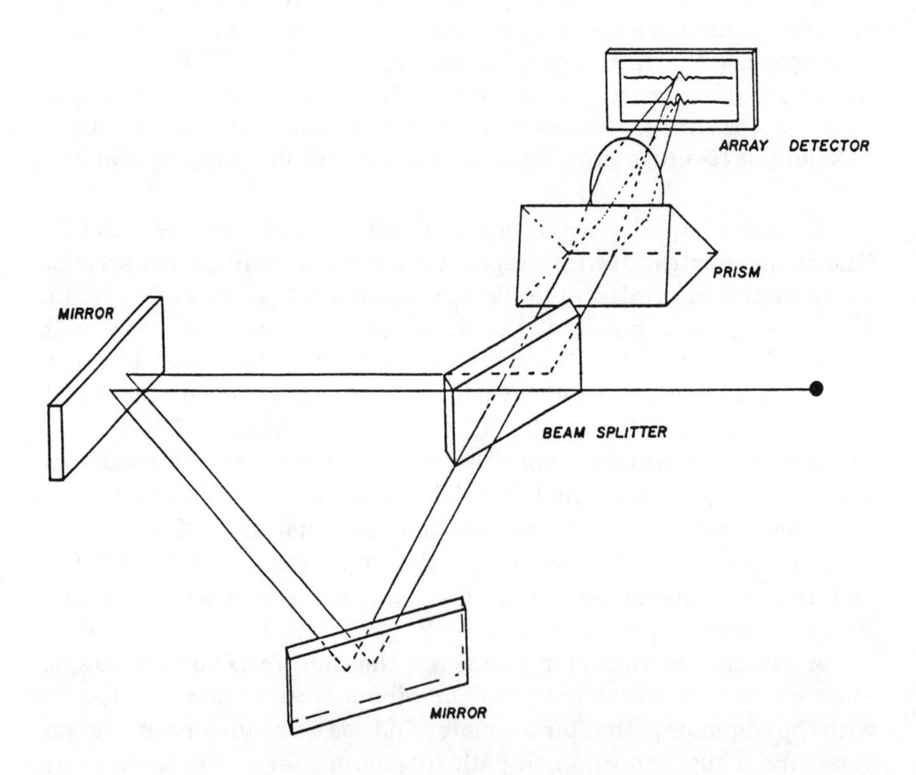

Figure 3.12 Optical arrangement of the crossed interferometer dispersive spectrometer (XIDS). (Reprinted with permission from reference 72.)

has been added to the common-path instrument shown in Figure 3.2. The use of a cross disperser creates a rectangular focal-plane image consisting of multiple interferograms, each one with a slightly different wavelength range.

Figure 3.13a shows the unprocessed output of the instrument, and Figure 3.13b shows the resulting spectra using a mercury pen lamp source. The movable mirror position has been adjusted to give very few fringes so that the 254-nm fringes remain visible in the figure. The changing fringe spacing in this figure is clearly visible, demonstrating that both dimensions contain wavelength information. The interferometric data shown were obtained using only a 190-detector-element subsection of the CCD. The resolution can be greatly increased by using pseudolinear arrays and the appropriate cross-dispersive optics. For example, CCDs with > 4000 detector elements in the interferometer axis are available and allow large improvements in resolution.

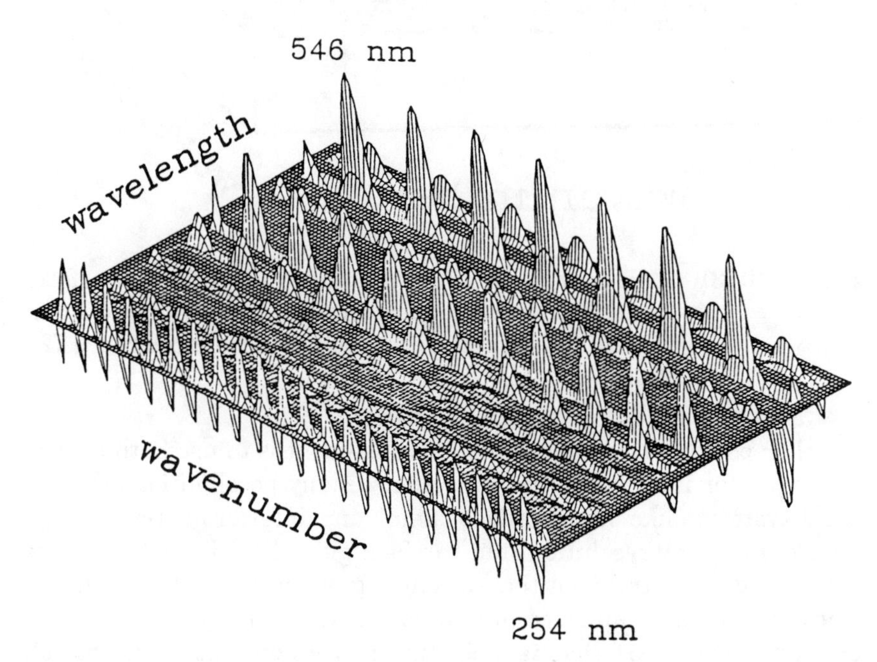

Figure 3.13a Mercury lamp spectrum using a crossed interferometric dispersive spectrometer: unprocessed spectrum with the average value of each row subtracted. (Reprinted with permission from reference 72.)

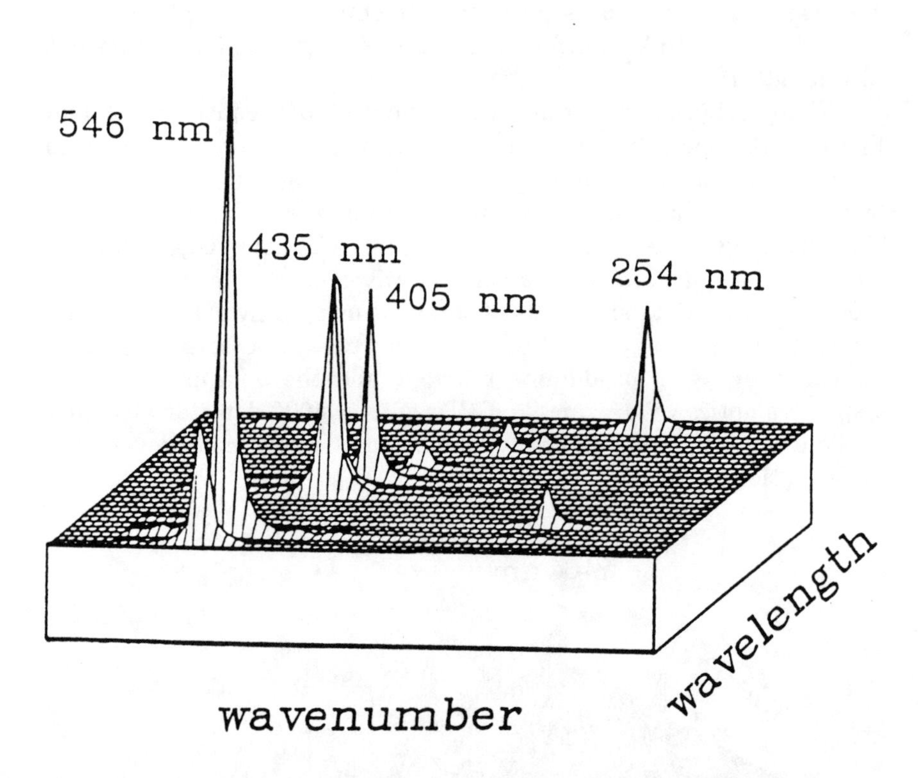

Figure 3.13b Mercury lamp spectrum using a crossed interferometric dispersive spectrometer: transformed spectrum. (Reprinted with permission from reference 72.)

3.4.3.4 Dual Interferometric Systems

Another possible two-dimensional arrangement is to use both axes of the array for interferometric detection. Many possibilities exist for such systems; like an echelle system, one interferometric system should have high resolution (but can be highly aliased), and the other is required to separate out the overlapping information. Callis and coworkers [95] have published preliminary results on a system that uses crossed Fizeau interferometers [96] (i.e., Fabry-Perot interferometers), one of which is operated in a high-resolution mode and produces many overlapping orders, and the other is operated in a low-resolution mode and acts similar to the order-sorting cross-disperser in an echelle. Figure 3.14 shows a block diagram of their prototype instrument, a fiber-optic-based flame-emission analyzer. One of the interferometers is a low-resolution wedge-type interferometer, and the high-resolution information is obtained using the Fabry-Perot interferometer. This

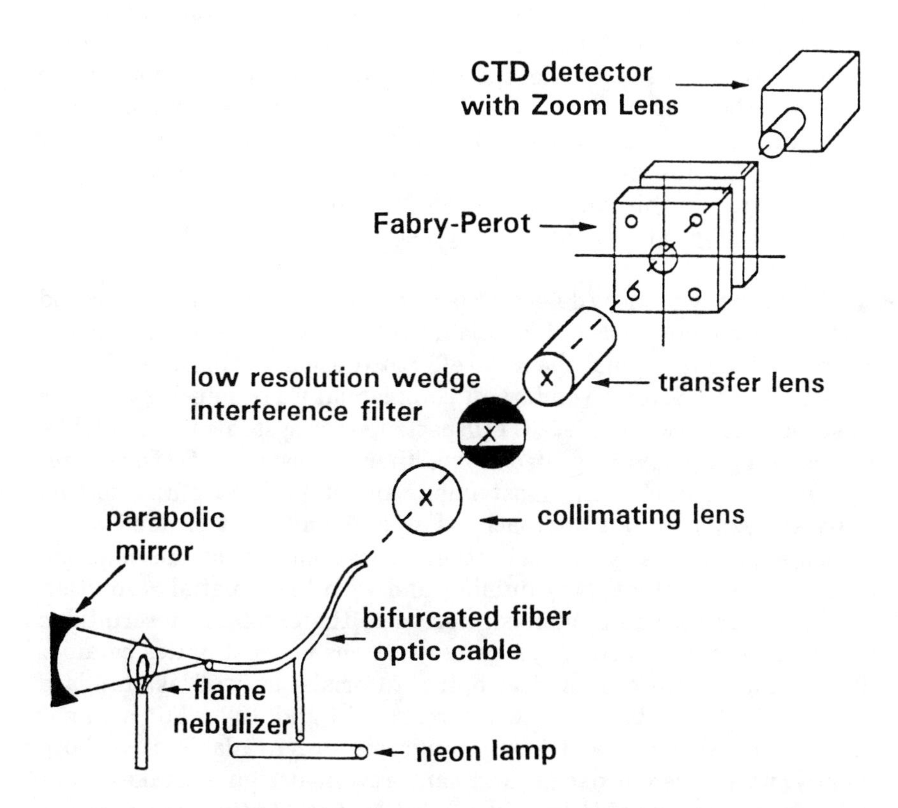

Figure 3.14 Block diagram of a two-dimensional crossed-Fizeau interferometer system designed as a fiber-optic-based flame-emission analyzer. (Reprinted with permission from reference 95.)

approach appears to have moderate resolution, high throughput, and a compact, stable design. Further work is still required to determine the practicality of such systems.

3.5 Designing a Custom System

This chapter is intended to serve as an introduction to the use of CTDs for spectroscopic detection. A large variety of CTD readout modes and CTD/spectrometers have been described. Many of these custom systems offer greatly improved sensitivity, spatial resolution, or time resolution compared with single-channel or previous array systems. Much greater detail on the application of these systems is provided in later chapters. One of the goals of this chapter has been to illustrate the wide variety of configurations possible, demonstrate the gains possible by several unconventional readout modes, and illustrate methods that combine these readout schemes with different spectrometers. While many of the spectroscopic systems developed using CTDs have been applied to a specific area, most hardware implementations can be used in many other applications. It is hoped that, by examining the descriptions of the large variety of approaches to using CTDs covered in this chapter, the reader can gain an overview of the different configurations and readout methods that have been developed to date.

While many of these detector/spectrometer systems offer improved performance compared with available alternatives, many researchers are not willing to expend the large effort to design, optimize, and build such systems. Several vendors of commercial CTD detector systems exist, but very few complete CTD/spectrometer systems are available. In designing (or trying to decide whether to design) a CTD/spectrometer system, one of the most important steps is carefully setting goals and choosing an approach. For a chemist not trained as an optical engineer, designing collection optics that collect the emission from a source both at low *f*-number and with high spatial resolution can be difficult. The reader is referred to the excellent descriptions that can be found in the appropriate sections of most optics catalogs [97–99] for simple geometrical optics tutorials, as well as standard optics texts [100–102], and to references 16 and 103. For a minor investment, several ray-tracing programs are available that both greatly aid in system design and can prevent design mistakes. For example, *BEAM 4* [104] has been used by both authors to design a custom high-resolution echelle system and several spatially resolved time-delayed integration systems.

In the last year, multiple analytical instrumentation companies have announced systems using these detectors. As this trend continues, the researcher will not need to design and build a custom system except in unusual applications. Instead, the researcher will need to select the appropriate system for each particular application. Here again, a knowledge of detector performance will help in the selection process.

References

1. Mathies, R.A.; Peck, K.; Stryer, L. *Anal. Chem.* **1990**, *62*, 1786.
2. Soper, S.A.; Shera, E.B.; Martin, J.C.; Jett, J.H.; Hahn, J.H.; Nutto, H.L.; Keller, R.A. *Anal. Chem.* **1991**, *63*, 432.
3. Whitten, W.B.; Ramsey, J.M.; Arnold, S.; Bronk, B.V. *Anal. Chem.* **1991**, *63*, 1027.
4. Engstrom, R.W. *Photomultiplier Handbook*. RCA Solid State Division, Lancaster, Pa., 1980.

5. Tachino, M.; Suzuki, S.A.; Sakai, S.; Kume, H.; Taguchi, T. , 1991 Pittsburgh Conference, Chicago, paper 72.
6. Hammamatsu Product Literature. Hammamatsu, Bridgewater, N.J., 1991.
7. Harrison, G.R.; Lord, R.C.; Loofbourow, J.R. *Practical Spectroscopy.* Prentice-Hall, Englewood Cliffs, N.J., 1980.
8. Margoshes, M. *Optical Spectra* **1970**, *4*, 26.
9. Margoshes, M. *Spectrochim. Acta* **1970**, *25B*, 113.
10. Mitchell, D.B.; Jackson, K.W.; Aldous, K.M. *Anal. Chem.* **1973**, *45*, 1215A.
11. Wood, D.W.; Dorgis, A.B.; Nash, D.L. *Appl. Spectrosc.* **1975**, *39*, 310.
12. Boumans, P.W.J.M. *Proc. Anal. Div. Chem. Soc.* **1977**, *14*, 143.
13. Talmi, Y. *Anal. Chem.* **1975**, *47*, 658A.
14. Talmi, Y. *Anal. Chem.* **1975**, *47*, 700A.
15. Bilhorn, R.B.; Epperson, P.M.; Sweedler, J.V.; Denton, M.B. *Appl. Spectrosc.* **1987**, *41*, 1125.
16. Busch, K.W.; Busch, M.A. *Multielement Detection Systems for Spectrochemical Analysis.* Wiley, New York, 1990.
17. Veirs, D.K.; Chia, V.K.; Rosenblatt, G.M. *Appl. Optics* **1987**, *17*, 3530.
18. Acker, W.P.; Yip, B.; Leach, D.H.; Chang, R.K. *J. Appl. Phys.* **1988**, *64*, 2263.
19. Bilhorn, R.B.; Denton, M.B. *Appl. Spectrosc.* **1990**, *44*, 1538.
20. Dereniak, E.L.; Crowe, D.G. *Optical Radiation Detectors.* Wiley, New York, 1984.
21. Bilhorn, R.B.; Sweedler, J.V.; Epperson, P.M.; Denton, M.B. *Appl. Spectrosc.* **1987**, *41*, 1117.
22. Winefordner, J.D.; Avni, R.; Chester, T.L.; Fitzgerald, J.J.; Hart, L.P.; Johnson, D.J.; Plankey, F. *Spectrochim. Acta* **1976**, *31B*, 1.
23. Okamoto, T.; Kawata, S.; Minami, S. *Appl. Optics* **1984**, *23*, 267.
24. Sweedler, J.V.; Denton, M.B. *Appl. Spectrosc.* **1989**, *43*, 1378.
25. Bevington, P.R. *Data Reduction and Error Analysis for the Physical Sciences.* McGraw-Hill, New York, 1969.
26. Hirschfeld, T. *Appl. Spectrosc.* **1976**, *30*, 68.
27. Luc, P.; Gerstenkorn, S. *Appl. Optics* **1978**, *17*, 1327.
28. Marra, S.; Horlick, G. *Appl. Spectrosc.* **1986**, *40*, 804.
29. Thorne, A.J. *Anal. Atomic Spectrosc.* **1987**, *2*, 227.
30. Hirschfeld, T. *Appl. Spectrosc.* **1976**, *30*, 234.
31. Alkemade, C.Th.J.; Hooymayers, H.P.; Lijnse, P.L.; Vierbergen, T.J.M.J. *Spectrochim. Acta* **1972**, *27B*, 149.
32. Epperson, P.M.; Denton, M.B. *Anal. Chem.* **1989**, *61*, 1513.
33. *PMIS Image Processing Software User's Manual*, version 1.5; *CC200 Operating Manual.* Photometrics Ltd., Tucson.
34. Sweedler, J.V.; Shear, J.B.; Fishman, H.A.; Zare, R.N.; Scheller, R.H. *Anal. Chem.* **1991**, *63*, 496.
35. Sweedler, J.V.; Shear, J.B.; Fishman, H.A.; Zare, R.N.; Scheller, R.H. *Proc. SPIE* **1992**, 1439, 37.
36. Bilhorn, R.B., U.S. patent 05 173 748, "Scanning Multichannel Spectrometry Using a Charge-Coupled Device in the Time-Delayed Integration Mode," December, 1992.
37. Sweedler, J.V.; Bilhorn, R.B.; Epperson, P.M.; Sims, G.R.; Denton, M.B. *Anal. Chem.* **1988**, *60*, 282A.
38. Bilhorn, R.B.; Denton, M.B. *Appl. Spectrosc.* **1989**, *43*, 1.
39. Sweedler, J.V.; Jalkian, R.D.; Denton, M.D. *Appl. Spectrosc.* **1989**, *41*, 953.
40. Massey, P. *A User's Guide to CCD Reduction with IRAF.* National Optical Astronomy Observatories, Tucson.
41. Tody, D. *The IRAF Data Reduction and Analysis System.* National Optical Astronomy Observatories, Tucson.
42. Lacy, W.B.; Rowlen, K.L.; Harris, J.M. *Appl. Spectrosc.* **1991**, *45*, 1598.
43. Pemberton, J.E.; Sobocinski, R.L.; Sims, G.R. *Appl. Spectrosc.* **1990**, *44*, 328.

44. Jovin, T.M.; Arndt-Jovin, D.J. *Proc. SPIE,* **1992**, *1439*, 109.
45. Arndt-Jovin, D.J.; Rober-Nicoud, M.; Kaufman, S.J.; Jovin, T.M. *Science (Washington D.C.)* **1985**, *236*, 247.
46. Hiraoka, Y.; Sedat, J.W.; Agard, D.A. *Science (Washington D.C.)* **1987**, *238*, 36.
47. Jovin, T.M.; Arndt-Jovin, D.J. *Ann. Rev. Biophys. Biophys. Chem.* **1989**, *18*, 271.
48. Herman, B.; Jacobson, K. Eds. *Optical Microscopy for Biology*. Wiley, New York, 1990.
49. Brakenhoff, G.J.; van der Voort, H.T.M.; Visscher, K. *Proc. SPIE* **1992**, *1439*, 121.
50. Spring, K.R. *Scanning Microsc.* **1991**, *5*, 63.
51. Tsay, T.-T.; Inman, R.; Wray, B.E.; Herman, B.; Jacobson, K. In *Optical Microscopy for Biology*. Wiley, New York, 1990, pp. 219–33.
52. Treado, P.K.J.; Levin, I.W.; Lewis, E.N. *Appl. Spectrosc.* **1992**, *46*, 553.
53. Walton, D.; Vanderwal, J.J.; Zhao, P. *Appl. Spectrosc.* **1992**, *46*, 373.
54. Monnig, C.A.; Gebhart, B.D.; Marshall, K.A.; Hieftje, G.M. *Spectrochim. Acta* **1990**, *45B*, 261.
55. Puppula, G.J.; Mul, F.F.M.; Otto, C.; Greve J.; Robert-Nicoud, M.; Arndt-Jovin, D.J.; Jovin, T.M. *Nature (London)* **1990**, *347*, 301.
56. Govil, A.; Pallister, D.M.; Chen, Li-Heng; Morris, M.D. *Appl. Spectrosc.* **1991**, *45*, 1604.
57. Morris, M.D.; Govil, A.; Liu, K.-L.; Sheng, R. *Proc. SPIE* **1992**, *1439*, 95.
58. Agard, D.A.; Hiraoka, Y.; Shaw, P.; Sedat. J.W. In *Fluorescence Microscopy of Living Cells in Culture, Part B.*; D.L. Taylor; Y.L. Yang, Eds. Academic, San Diego, 353–377, 1989.
59. Monnig, C.A.; Gebhart, B.D.; Hieftje, G.M. *Appl. Spectrosc.* **1989**, *43*, 577.
60. Kolczynski, J.D.; Pomeroy, R.S; Jalkian, R.D.; Denton, M.B. *Appl. Spectrosc.* **1989**, *43*, 887.
61. Jalkian, R.D.; Pomeroy, R.S; Kolczynski, J.D.; Denton, M.B.; Lerner, J.M.; Grayzel, R. *Amer. Lab. (Fairfield, Conn.)* **1989**, *21*, 80.
62. Chromex Product literature, 2705-B Pan American NE, Albuquerque, NM 87107.
63. Yip, W.T.; Tse, R.S. *Rev. Sci. Instrum.* **1992**, *63*, 3777.
64. Hoffman, G.G.; Menzebach, H.-U.; Oelichmann, B.; Schrader, B. *Appl. Spectrosc.* **1992**, *46*, 568.
65. Sweedler, J.V.; Denton, M.B.; Sims, G.R.; Aikens, R.S. *Opt. Eng.* **1987**, *26*, 1020.
66. Geary, J. *Proc. SPIE* **1992**, *1439*, 159.
67. Chandler, E.C.; Bredthauer, R.A.; Janesick, J.R.; Westphal, J.A.; Gunn, J.E. *Proc. SPIE* **1990**, *1242*, 123.
68. Farres, L.M. *Anal. Chem.* **1986**, *58*, 1023A.
69. Okamoto, T.; Kawata, S.; Minami, S. *Appl. Optics* **1985**, *24*, 4221.
70. Barnes, T. *Appl. Optics* **1985**, *24*, 3702.
71. Stroke, G.W.; Funkhouser, A.T. *Phys. Lett.* **1965**, *16*, 272.
72. Sweedler, J.V.; Jalkian, R.D.; Sims, G.R.; Denton, M.B. *Appl. Spectrosc.* **1990**, *44*, 15.
73. Stubley, E.A.; Horlick, G. *Appl. Spectrosc.* **1985**, *39*, 81.
74. Epstein, M.S.; Winefordner, J.D. *Prog. Anal. Atom. Spectrosc.* **1984**, *7*, 69.
75. Vanasse, G.; Sakai, H. In *Progress in Optics*, Wolf, E., Ed. North-Holland, New York, 1976, Vol. 6.
76. Knoll, P.; Singer, R.; Kiefer, W. *Appl. Spectrosc.* **1990**, *44*, 776.
77. Piccard, R.; Vo-Dinh, T. *Rev. Sci. Instrum.* **1991**, *62*, 584.
78. Warner, I.M.; Patonay, G.; Thomas, M.P. *Anal. Chem.* **1985**, *57*, 463A.
79. Aikens, R.S.; Epperson, P.M.; Denton, M.B. *Proc. SPIE* **1984**, *501*, 49.
80. Harrison, G.R.J. *Opt. Soc. Am.* **1949**, *39*, 522.
81. Sweedler, J.V.; Jalkian, R.D.; Pomeroy, R.S.; Denton, M.B. *Spectrochim. Acta* **1989**, *44B*, 683.

82. Pilon, M.J.; Denton, M.B.; Schleicher, R.G.; Moran, P.M.; Smith, S.B. *Appl. Spectrosc.* **1990**, *44*, 1613.
83. Sims, G.R.; Denton M.B. *Talanta* **1990**, *37*, 110.
84. Krupa, R.J.; Leominster, R.; Owen, C. U.S. Patent 4 995 721, "Two-dimensional Spectrometer," February 26, 1991.
85. Scheeline, A.; Bye, C.A.; Miller, D.L.; Rynders, S.W.; Owen, R.C. *Appl. Spectrosc.* **1991**, *45*, 334.
86. Barnard, T.W.; Crockett, M.I.; Ivaldi, J.C.; Lundberg, P.L. *Anal. Chem.* **1993**, *65*, 1225.
87. Barnard, T.W.; Crockett, M.I.; Ivaldi, J.C.; Lundberg, P.L.; Ziegler, E.M.; Yates, D.A.; Levine, P.A.; Saver, D.J. *Anal. Chem.* **1993**, *65*, 1231.
88. Pomeroy, R.S; Sweedler, J.V.; Denton, M.B. *Talanta* **1990**, *37*, 115.
89. Pelletier, M.J. *Appl. Spectrosc.* **1990**, *44*, 1699.
90. Newman, C.D.; Bret, G.G.; McCreery, R.L. *Appl. Spectrosc.* **1992**, *46*, 262.
91. Voightman, E.; Winefordner, J. *Appl. Spectrosc.* **1987**, *41*, 1182.
92. Hirschfeld, T.; Chase, B. *Appl. Spectrosc.* **1986**, *40*, 133.
93. Chase, B. *J. Am. Chem. Soc.* **1986**, *108*, 7485.
94. Tilotta, D.C.; Freeman, R.D.; Fateley, W.G. *Appl. Spectrosc.* **1987**, *41*, 1280.
95. Aldridge, P.K.; Lindahl, E.G.; Callis, J.B. *Phil. Trans. R. Soc. Lond. A*, **1990**, *333*, 29.
96. Fizeau, H. *Ann. Chim. Phys.*, **1862**, *3*, 429.
97. "Fundamental Optics," in *Optics Guide 5*. Melles Griot, Irvine, Calif., 1992.
98. Newport Research Corporation, Irvine, Calif., 1992.
99. Oriel Corporation, Stamford, Conn., 1992.
100. Jenkins, F.A.; White, H.E. *Fundamentals of Optics*. McGraw-Hill, New York, 1957.
101. O'Shea, D.C. *Elements of Modern Optical Design*. Wiley-Interscience, New York, 1985.
102. Levi, L. *Applied Optics — A Guide to System Design*. Wiley, New York, 1980, Vol. 2.
103. Lerner, J.M.; Thevenon, A. "The Optics of Spectroscopy — A Tutorial v2.0," Instruments SA, Inc., Edison, N.J., 1988.
104. Beam 4 Optical Ray Tracer, Stellar Software, Berkeley, Calif., 1989.

4
Arrays for Detection Beyond One Micron

Bruce Chase

Corporate Center for Analytical Science
Dupont Experimental Station
Wilmington, Delaware

4.1 Introduction

The introduction of silicon-based CCD detectors has had a dramatic impact on many different types of optical spectroscopies. The extremely low read noise, low dark noise, and high dynamic range have produced results that are improved by orders of magnitude over results obtained with single channel techniques. The detection limits in fluorescence spectroscopy have been lowered to the point where single molecule detection is now within the realm of feasibility [1]. With these detectors, the improved performance in Raman spectroscopy now permits dilute solution measurements in a process environment as well as routine application in the laboratory [2].

One of the greatest limitations for these detectors lies in their inherent physical properties, specifically the bandgap of silicon. If one wishes to work at wavelengths much longer than 1 μ, different detector materials must be used. These nonsilicon detectors have already been employed in astronomy with substantial success, and many detector arrays have been designed specifically for space flight operation. There have been several reviews of such arrays in SPIE publications [3–5]. While the performance characteristics of nonsilicon CTD arrays are impressive, the devices themselves are not yet readily available commercially. To be generally useful for analytical spectroscopic experiments, such detectors both must be commercially available (beyond the prototype stage) and must have a large number of elements. Some very good spectroscopy can be accomplished with a limited number of pixels, but for routine applications at least a hundred or more resolution elements are needed. Many of the nonsilicon array detectors have been used in the prototype stage by astronomers and other exploratory groups but have not been produced commercially in numbers sufficient for evaluation by analytical spectroscopy laboratories.

4.1.1 The Fluorescence Problem in Raman Measurement

In Chapter 7, Richard McCreery discusses the use of multichannel detectors for Raman spectroscopy. Their impact has clearly been significant in terms of sensitivity and measurement time. As discussed by McCreery, the Raman effect is quite weak and is often overwhelmed by interfering effects such as fluorescence arising either from impurities or from the material of interest. It was recognized very early in Raman spectroscopy that the best way to overcome fluorescence interference is to reduce the energy of the incoming photons below the threshold for excitation of the interfering luminescence. It was this concept that drove the development of Fourier transform Raman spectroscopy [6]; the scattering from a Nd:YAG laser is detected using an interferometer operating in the near IR. The success of this approach prompted several groups to look at the use of slightly shorter wavelength lasers, operating in the red or short wave IR (SWIR) coupled with CCD detection [7]. As shown by McCreery [8] and Schulte [9], this approach can be quite useful and is currently being pursued for use in process monitoring. Unfortunately, some samples still fluoresce at these wavelengths. In addition, the long wavelength limit of silicon-based CCD detectors eliminates the use of a laser operating at 1 μ, because the Raman scattering will occur beyond the 1.1-μ cutoff of silicon. With the introduction of new nonsilicon focal plane detectors operating in the near IR, there is now an alternative to FT-Raman spectroscopy using Nd:YAG excitation. The successful use of these detectors for Raman spectroscopy will ultimately depend on both array sensitivity and cost.

4.1.2 The Anti-Stokes Solution

Before discussing some of these new detectors which operate beyond 1.1 μ, there is one application where silicon-based CCD elements can be used for Raman spectroscopy with Nd:YAG excitation. The conventional Raman spectrum occurs on the long wavelength side of the laser line. If a Nd:YAG laser is used, operating at 1.064 μ (9395 cm^{-1}), the Stokes side of the Raman spectrum falls between 1.0 and 1.6 μ (9395 cm^{-1} to 6000 cm^{-1}). This region is not accessible to a silicon-based CCD detector. However, the anti-Stokes scattering lies to higher energy and results from Raman scattering of molecules in excited vibrational states. (See Chapter 7, Figure 7.1.) Normally this region of the Raman spectrum is not used, because the information is redundant with the Stokes side information, and the intensities are dependent on sample temperature and are usually quite weak. The high sensitivity of the CCD detector can compensate for the weak anti-Stokes scattering, and quite reasonable results can be obtained. This approach was first discussed by Barbillat and co-workers [10]. Figure

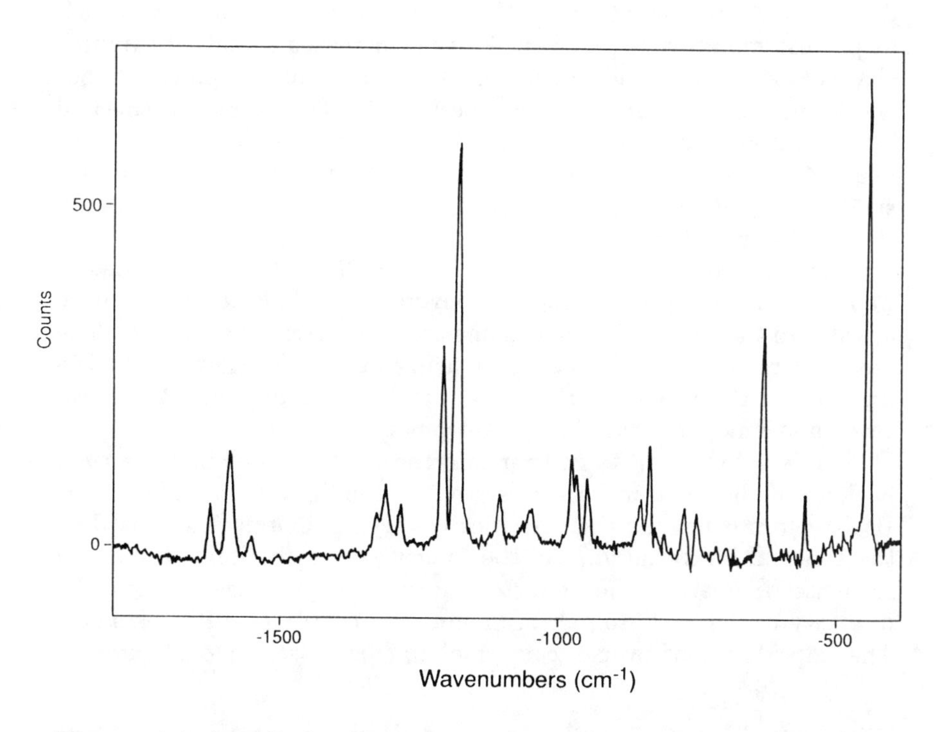

Figure 4.1 Anti-Stokes Raman spectrum of bis-methyl styryl benzene obtained with CCD detection and Nd:YAG excitation at 1.064 μ.

4.1 shows the anti-Stokes spectrum of a dye molecule excited with 1.06-μ radiation and detected with a CCD camera. This approach also has the advantage of totally avoiding any fluorescence that may still be present at these long wavelengths since the fluorescence will always occur at lower energies than the incident laser radiation. The limitations are sensitivity, caused by the temperature dependence of the anti-Stokes scattering, and inadequate spectral bandwidth, caused by the limited dispersion of the ruled gratings in the near IR. These problems limit the anti-Stokes approach as a routine tool for Raman spectroscopy, but the approach may well have application for limited spectral range studies.

4.2 New IR Detectors

After developing and using FT-Raman techniques, we became interested in the new developments in multichannel detection with sensitivity beyond 1 μ. Initially our interest was concentrated on detection of Raman scattering. There are, however, applications such as process monitoring that benefit from increased sensitivity (multichannel advantage) and are not hurt by the reduced spectral bandwidth encountered in near-IR operation. This reduced spectral bandwidth results from the increase in dispersion which accompanies near-IR operation. These new detectors could well impact other areas, such as near-IR spectroscopy, both in laboratory measurements and in the process-monitoring environment.

Many nonsilicon multichannel detectors, including those listed in Table 4.1, have a spectral response beyond 1 μ. While several of these are still in the prototype stage, some are commercially available. The next several years will see an increasing number of these prototype units reach the production stage as the demands of near-IR spectroscopy and imaging drive their development.

As detectors operate further into the IR, the problem of thermal background becomes more severe. In fact, the limiting noise factor for IR detectors is usually the statistically varying photon flux caused by the thermal background. As the long wavelength limit of spectral response increases, the detector noise also increases, since the intensity of the background flux increases according to Planck's law. The importance of proper cold shielding (see Section 4.2.3) and of

Table 4.1 Near-Infrared Multichannel Detectors

Type	Spectral Range (μm)	Size (elements)	Availability
Ge	0.8–1.6	256 × 1	Commercial
InGaAs	0.8–1.6	256 × 1	Commercial
PtSi	1–5	1024 × 1 512 × 512	Commercial
IrSi	1–8	512 × 1	Prototype
InGaAsP	0.8–2.6		Prototype
InSb	1.0–5.5	64 × 64	Commercial
Doped-Si BIB	1–28	16 × 16	Prototype
Hybrid MCT/Si	2–12	256 × 256	Prototype

limiting the observed wavelengths to the spectral region of interest cannot be overstated. The case reaches an extreme for elements that are active well into the IR. The doped-Si BIB prototype detector based on a blocked impurity band structure requires operation at 10 K and very effective cold shielding before it can be used. If the temperature of the detector fluctuates, there is a large change in the thermal background, which becomes indistinguishable from a signal of interest. The problems associated with cryogenic IR detectors have been reviewed by Huppi in an excellent article [11].

As mentioned previously, some of these array detectors are now commercially available. In the 1 to 1.6 μ region, both germanium and indium gallium arsenide arrays can be purchased. Arrays of platinum silicide are available with coverage out to 5 μ. Indium antimonide arrays are also on the market, though with relatively small numbers of pixels in a linear configuration. All of the other arrays that might be useful for spectroscopy are still under development at this time.

4.2.1 Germanium (Ge)

The single element germanium detector has been quite useful for a wide variety of spectroscopic experiments in the region 1.0 to 1.7 μ. It was one of the first detectors used in FT-Raman experiments [12], and a special high-purity, germanium detector biased at high voltage is the current detector of choice for an FT-Raman spectrometer. Within the past three years, multichannel arrays constructed of germanium have become readily available. These arrays are photodiode devices that are normally read out with silicon shift registers, similar to the conventional silicon-photodiode arrays. Since each element is actually wire bonded and individually biased and monitored, there is obviously a limit to the number and density of elements. Lead density can become overwhelming, and the number of bad or inactive pixels goes up dramatically as the array size is increased. Currently, linear arrays of up to 256 elements are available, with 512 a possibility at significantly higher cost. The typical size of an array element is 25 μ by 2.5 mm with an aspect ratio that is well suited to the slit image produced by conventional monochromators. The quantum efficiency (QE) of germanium in this region is quite high, averaging around 0.6. The major loss comes with reflection at the front surface of the detector, since the refractive index is high (4.0). Future products may offer improved antireflection coatings to minimize this problem and increase QE.

The readout and biasing scheme used for these detectors results in a bias offset that can vary significantly from element to element across the array, changing in a periodic fashion as shown in Figure 4.2, a Raman spectrum of anthracene, taken with a single monochromator and a Nd:YAG laser. The total measurement time was 60 ms. The

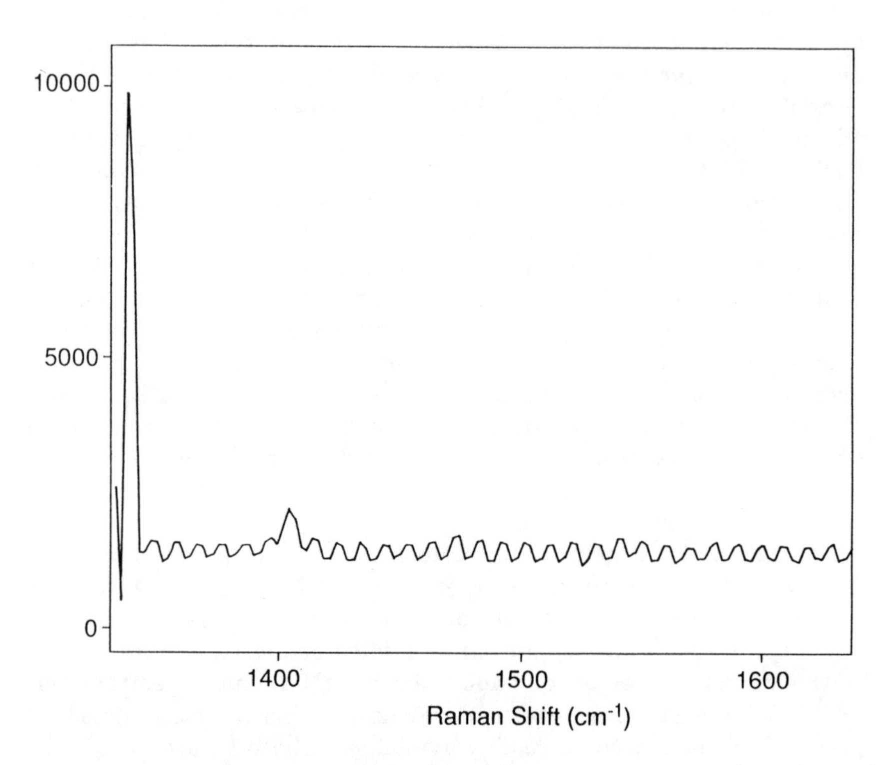

Figure 4.2 Uncompensated Raman spectrum of anthracene obtained with germanium diode array detection and Nd:YAG excitation.

periodic intensity fluctuation is due to the variation in biasing voltage across the array. By acquiring another scan under the same conditions but with no illumination, the bias variation from pixel to pixel can be measured. Since the bias signal is independent of the light level falling on the detector, subtracting the two scans will eliminate this periodic signal [13], as shown in Figure 4.3. The features arising from Raman scattering of anthracene are now observable. By allowing for increased integration time, the signal-to-noise ratio of this spectrum can be increased.

With any integrating detector it is critically important to know the relative levels of read noise and detector noise to ensure the proper operation of the array. By measuring the noise at varying integration times, with no light falling on the detector, the effect of dark noise and read noise can be determined. With CCD detectors, the terms *detector noise* and *read noise* are often used interchangeably, since the dark

Table 4.2 Noise Characteristics of a 128-Element Ge Array

Exposure (s)	Scans	Observation Time (s)	Noise (peak-peak)	Noise (rms)
0.067	1	0.067	97	17.4
0.267	1	0.267	96	10.7
1.067	1	1.067	93	18.6
4.27	1	4.27	93	19.0
17.07	1	17.07	89	20.6
68.26	1	68.26	340	41.8
0.067	4	0.267	253	35.9
0.067	16	1.067	359	72.4
0.067	64	4.27	713	135.8
0.067	256	17.07	1330	291.0
0.067	1024	68.26	3218	624.8
0.067	4096	273	4821	1059.7

counts are often negligible; in this case, the detector noise is the read noise. For near-IR detectors, the read noise, noise induced by the readout process, and the detector noise, noise due to the random statistical fluctuations in background radiant flux and dark current, can both be significant. If read noise is the dominant noise source, the observed noise will be independent of integration time, even though the absolute signal will increase due to integration of dark noise. If detector noise is dominant, then the observed noise will increase with integration time. For the particular germanium detector discussed here, the variation in noise as a function of integration time is shown in Table 4.2. Clearly, the array detector is read-noise dominated up to integration times of 70 seconds. Within the read-noise limit, the maximum integration should obviously be used. Shorter integration times with multiple reads will only increase the final noise figure while not changing the observed signal at all.

The other aspect limiting integration time will be the buildup of dark counts. For CCD detectors, this buildup will lead to saturation

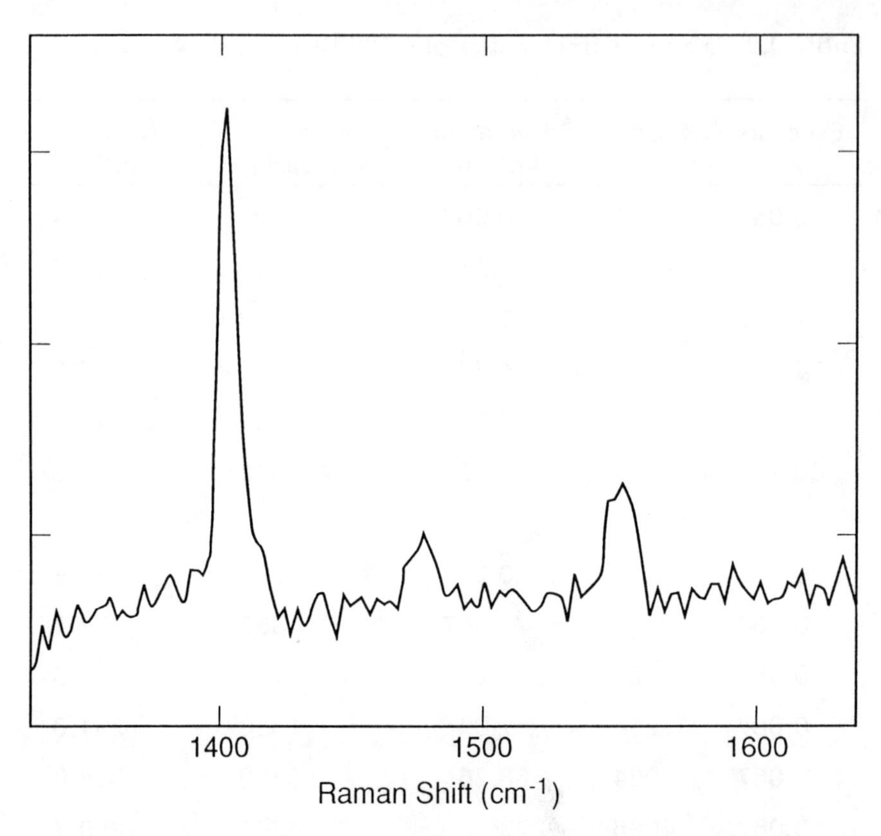

Figure 4.3 Spectrum shown in Figure 4.2 after compensation for bias variation. (Reprinted with permission from reference 13.)

only after hours of integration for cooled detectors. The situation for IR arrays can be quite different due to the much higher thermal background. For a germanium array with a 1.7-μ cutoff, the build up in dark background is not too severe: at the point where detector noise competes with read noise, there is still a significant dynamic range in the detector. As will be seen later, for longer wavelength detectors this situation does not hold.

Once the optimum integration time has been established, one is faced with "block averaging" as the only way to further increase SNR. Multiple scans are taken at a given integration time and are averaged to produce a final result. Improvement in SNR should go as the square root of total measurement time in the averaging process. As previously described, noise can be readily measured by taking two scans under identical conditions and subtracting them. This removes contributions from the biasing voltages and other sources, leaving one with a noise figure 1.4 times higher than the true noise in a single

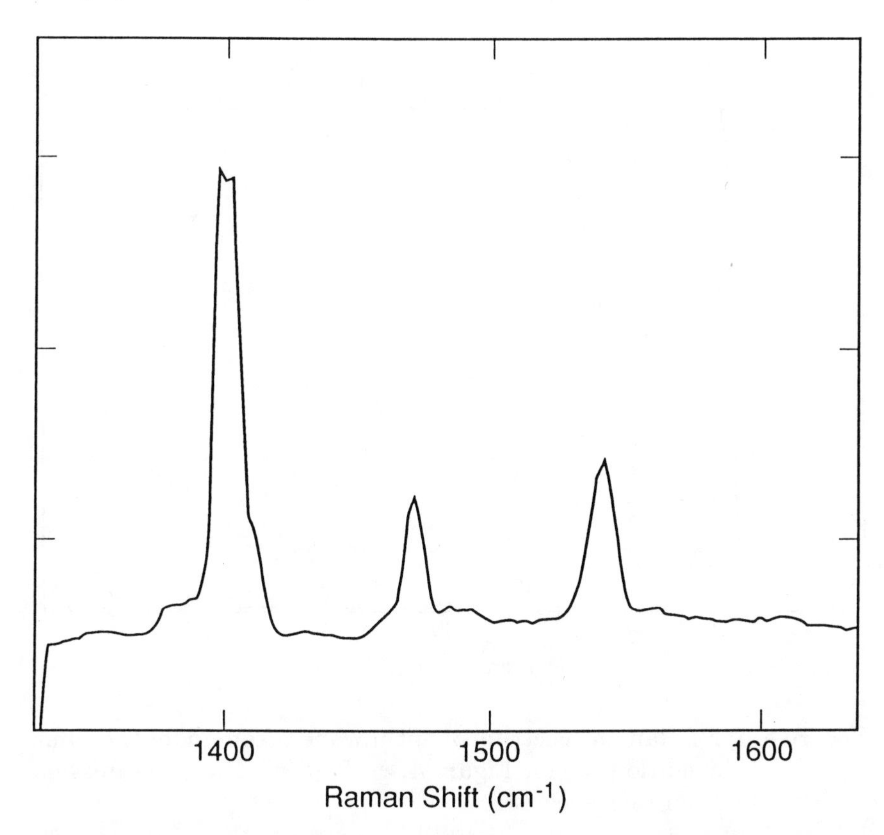

Figure 4.4 Raman spectrum of anthracene obtained with germanium array detection. Total measurement time 120 s. (Reprinted with permission from reference 13.)

exposure. Figure 4.4 demonstrates the effect of varying total measurement time through block averaging on SNR; for this spectrum, total measurement time was 120 s, taken in the form of an average of ten 12-s integrations. Up to a 256-s measurement time, the square-root improvement is observed. At that point, however, there is a contribution from a coherent noise source, that is, a noise source that cannot be reduced by scan averaging. In this detector the coherent signal is related to a $1/f$ noise source at 0.004 hertz [13]. Presumably, further work on the noise characteristics of the readout electronics will reduce this problem and allow longer integration times.

The germanium array can be used for Raman spectroscopy and produces results comparable to those of FT-Raman. Figure 4.4 shows a Raman spectrum of anthracene taken with a germanium-diode-array spectrometer [13]. A single monochromator, JY HR320, was used with dielectric longpass filters to remove the Rayleigh line. For comparison

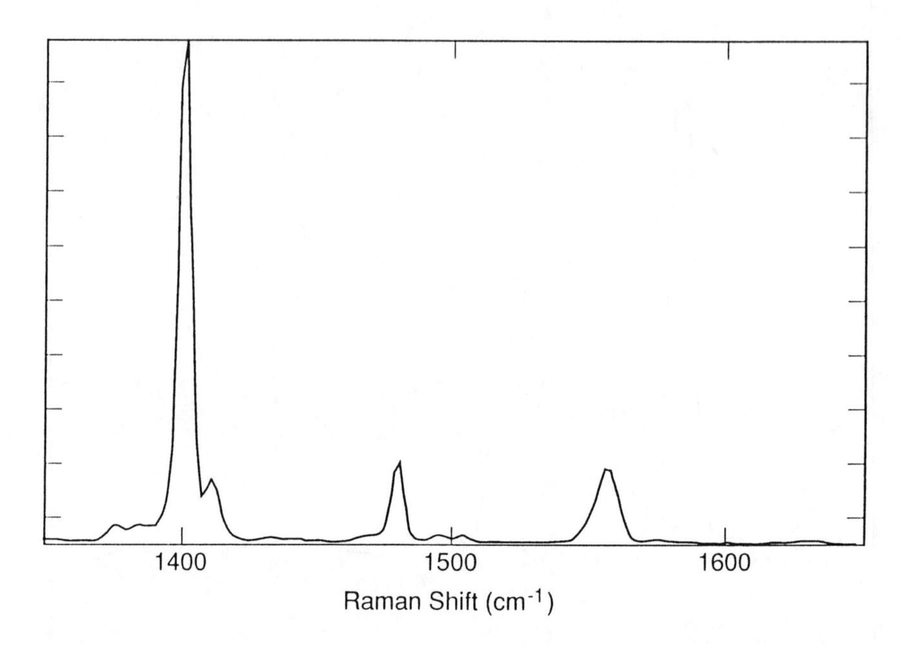

Figure 4.5 FT-Raman spectrum of anthracene taken under the same conditions as for Figure 4.4. (Reprinted with permission from reference 13.)

purposes, an FT-Raman spectrum was taken of anthracene under the same conditions, that is, comparable resolution, laser power, and measurement time, as shown in Figure 4.5. The calculated SNR (rms) of the spectrum in Figure 4.4 was within a factor of two of the spectrum in Figure 4.5. With further improvements in the optical system, I expect the array instrument to outperform the FT-Raman spectrometer for measurement of a small portion of the Raman spectrum. However, the problem of dispersion in the near IR makes it difficult to acquire a large spectral bandwidth across the detector array. High-resolution operation is easy; high spectral bandwidth is difficult.

There is also a problem of pixel-to-pixel reproducibility in response. Figure 4.6 shows a Raman spectrum of toluene taken at 1.06 μ that has a slight background signal from fluorescence of the glass container. What appears to be noise in the spectrum does not change with increasing measurement time. In fact, the structure in the spectrum remains constant. These features are due to slight variations in detector-element response. The obvious remedy to this problem is to use a flat-field correction as described in Chapter 3; this

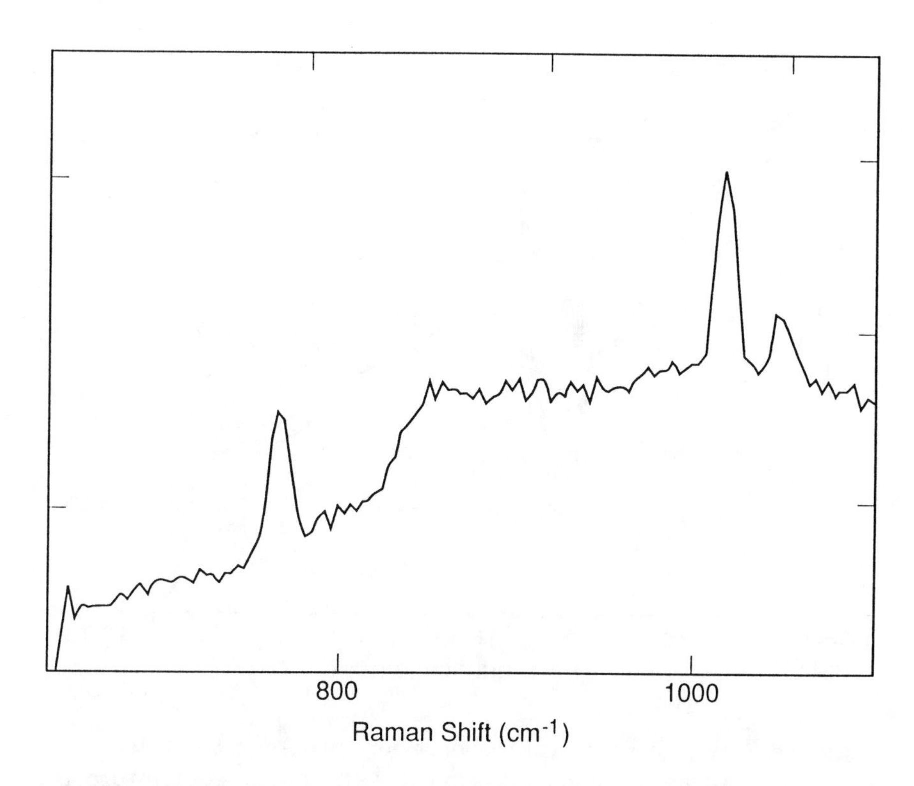

Figure 4.6 Raman spectrum of toluene taken with germanium diode array detection. (Reprinted with permission from reference 13.)

may involve recording a spectrum of a known continuum source and ratioing all measurements to this instrument-response curve.

4.2.2 Indium Gallium Arsenide (InGaAs)

This detector element has been developed as a potential replacement for germanium detectors. The performance characteristics are at least as good as germanium, and there is the potential for longer wavelength operation out to 2.6 μ [14]. As a single-element detector it has produced a lower noise-equivalent power (NEP) and equivalent quantum efficiency. This material has been fabricated into arrays of up to 256 elements and would be expected to rival the germanium arrays in performance [15]. Slight improvements in spectral range of the ordinary InGaAs detector can be achieved by operating at higher temperatures. Figure 4.7 shows the relative spectral response for InGaAs detectors at room temperature, with thermoelectric cooling,

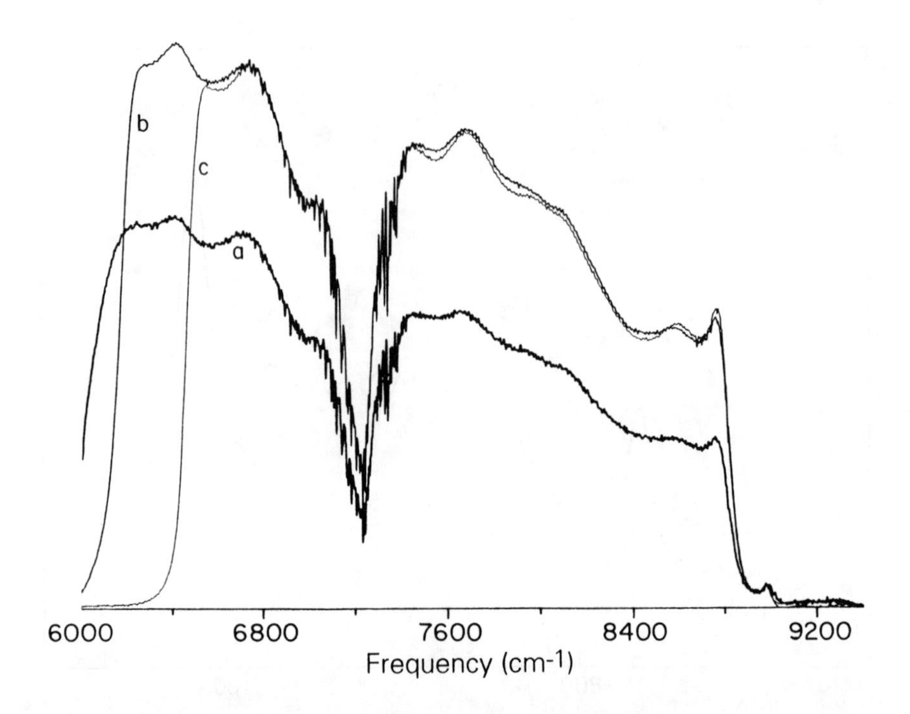

Figure 4.7 Relative spectral response for InGaAs at three temperatures. Raman shift is in cm^{-1}. (a) Room temperature; (b) thermoelectric cooling; (c) 77 K.

and at liquid nitrogen temperature. The price for increased spectral response is degraded dark noise performance. Although there have been no reports to date on the utility of InGaAs arrays for spectroscopy, the performance is anticipated to be very similar to that of Ge arrays. The deciding features will be size and cost.

4.2.3 Platinum Silicide (PtSi)

Although both germanium and indium gallium arsenide arrays have useful electrooptical characteristics, they are limited in their spectral response to wavelengths less than 1.6 μ. As previously mentioned, there is significant interest in operating further into the IR, and one possible material is platinum silicide. This detector element operates out to 5 μ [16,17]. More important, the manufacturing process makes use of prevailing silicon technology, which eliminates the drawback of individual detector wiring as found for the previous two array architectures.

The use of platinum silicide as a detector element is based on the concept of a Schottky barrier diode [18]. Photons are absorbed at the metal interface exciting a hot electron gas, which creates high-energy carriers. These carriers can drift or be propelled by bias fields to the silicon layer where they can be captured. This entire process is often referred to as "internal photoemission." The process is fast and independent of carrier lifetime and diffusion length. Clearly, one of the major benefits of this type of detector is the compatibility with standard silicon-integrated-circuit processing. In practice, the top layer is usually *p*-type silicon, followed by the PtSi layer, a dielectric (SiO_2), and finally an aluminum reflector. The reflector causes the photons (and hot electrons) to multipass the detector, and this arrangement can be designed to act as a slightly resonant cavity. As photons are absorbed, charge resulting from the transiting electrons builds up in the silicon layer and can be read out by a MOSFET switch, which performs charge reset and sensing. This device is quite similar to a standard silicon photodiode array with the mechanism of photon absorption being the important difference. Since the absorbing layer is much thinner than the silicon layer in a photodiode array, the sensitivity of this Schottky barrier detector is significantly lower. A typical quantum efficiency curve is shown in Figure 4.8.

As mentioned previously for IR detectors, the limiting noise figure is set by the background radiant flux from objects in the field of view of the detector. Since PtSi operates out to 5 μ and the background photon flux will roughly follow Planck's law, we expect the limiting noise level to be strongly dependent on the degree to which long wavelength photons are detected. If the spectral region of interest is limited, the noise and dark level can be reduced by optical filtering. To ensure that the filter does not simply reradiate absorbed photons, it must be held at the same temperature as the detector, a process termed cold filtering or cold shielding. Figure 4.9 shows the effect of operation with and without a 1.78-μ cold filter. Cold filtering involves the use of optical filters maintained at low temperature in order to limit the radiation from the filter itself in regions where the filters absorbs. In these regions the transmittance of the filter is zero, so the re-emission can be quite high. If the filter is at room temperature there is significant photon flux from the filter in this spectral region. By lowering the temperature of the filter, this flux is reduced, if not eliminated. Both the background DC level and the associated noise are reduced. Of course, in this condition the detector cannot be used beyond 1.6 μ, and if that range corresponds to the spectral region of interest, one would be better off using either Ge or InGaAs array detectors as described in the previous section.

With or without cold filtering, the PtSi detector will record a background signal in the absence of external illumination. The signal

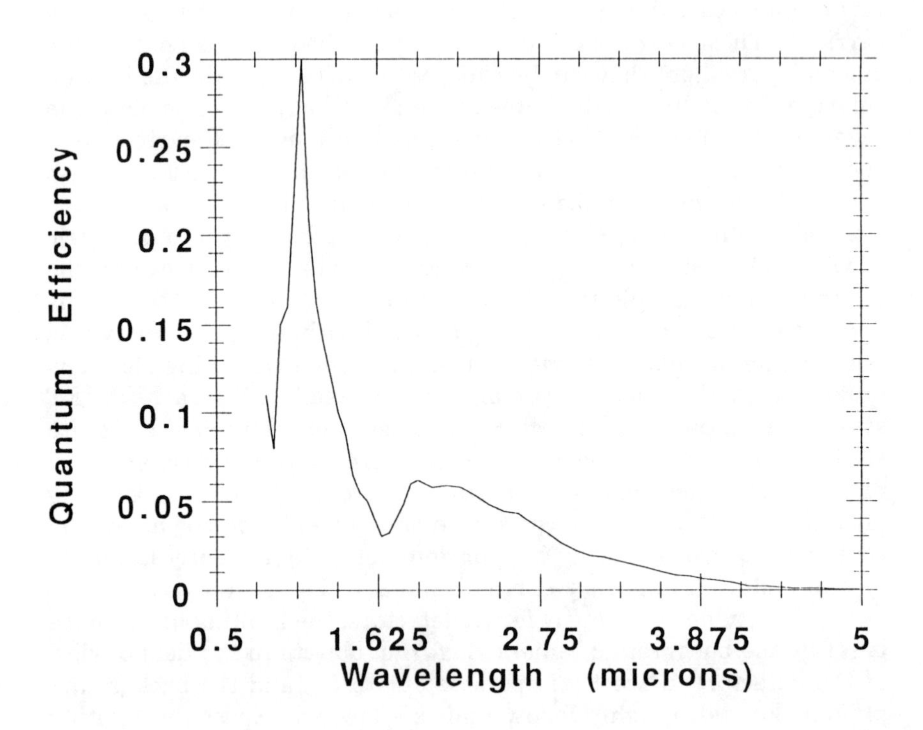

Figure 4.8 Quantum efficiency for a PtSi element.

must be subtracted from the signals of interest to achieve photometric accuracy. For example, to use this detector for the measurement of near-IR transmission spectra, we mounted it on a 0.25-meter *f*/4 single monochromator. A small tungsten source, operating at 1 watt, was used for illumination. Figure 4.10 shows the single-beam spectrum obtained without illumination (curve a), the single-beam spectrum of the source (curve b), and the spectra of the source with two different narrow bandpass filters inserted in the beam (curves c, d). The reference spectrum is subtracted from both the background and sample spectra and the results are then ratioed to produce a transmittance spectrum of the filters, shown in Figure 4.11. A 100% transmittance line is included in this figure. Figure 4.12a shows a spectrum of a polystyrene film obtained in the same manner; Figure 4.12b shows a spectrum of the same sample obtained on a conventional near-IR spectrophotometer.

These results demonstrate the potential for this array detector in near-IR spectroscopy. With the high-intensity sources available in this

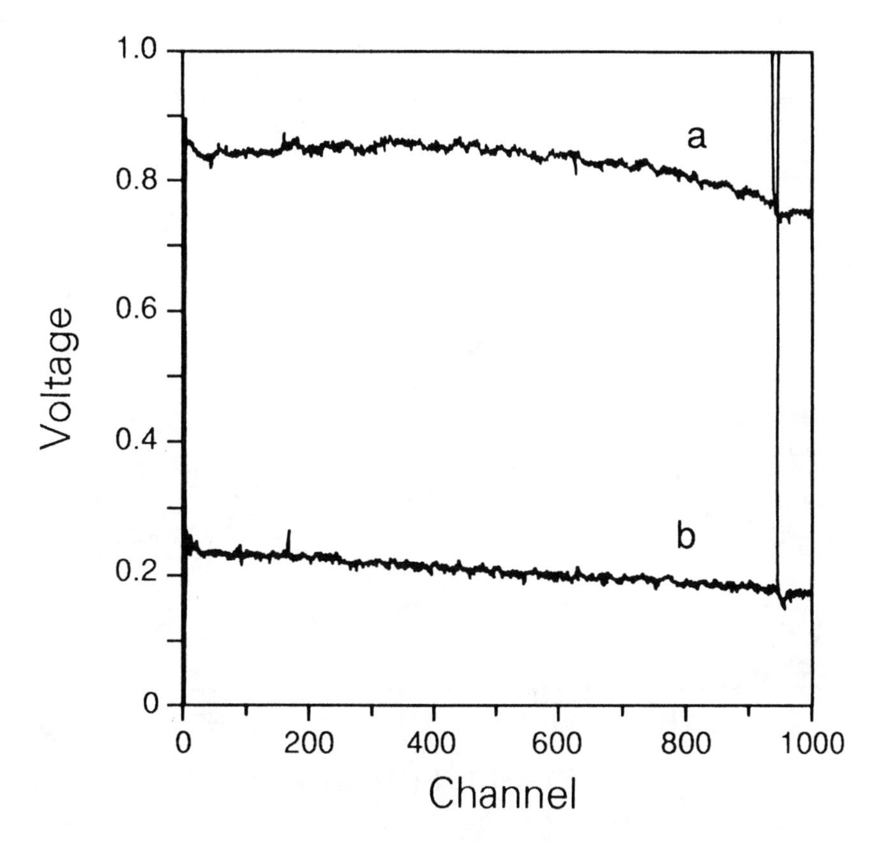

Figure 4.9 Single-beam spectrum obtained with a PtSi diode array. (a) No cold filtering; (b) cold filtered at 1.78 μ.

region, a low quantum efficiency is less of a problem than is the case for visible arrays. The attractive features of a focal plane array detector such as minimal moving parts and excellent wavelength stability could contribute greatly to near-IR spectroscopy. Larger linear arrays and two-dimensional arrays have also been produced using PtSi [19]. Additionally, there are other silicide detectors under development, such as iridium silicide. These devices have the same quantum efficiency, but the spectral range can be extended even further into the IR.

4.2.4 Indium Antimonide (InSb)

The active region for InSb covers from 0.8 μ to 5 μ, with quantum efficiencies approaching 0.70. This is significantly higher than values for PtSi. However, the detectors are not monolithic and the problems

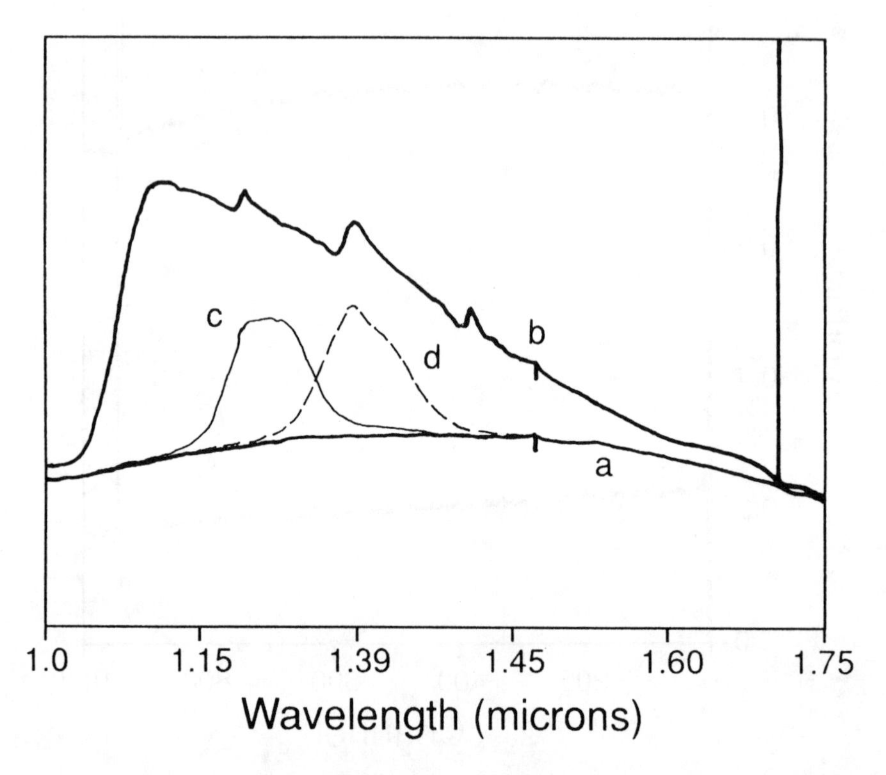

Figure 4.10 Spectra obtained with a PtSi array. (a) Single beam, no illumination; (b) single beam with illumination; (c,d) single beam of narrow bandpass filters centered at 1.22 and 1.29 μ, respectively.

associated with individual detector readout limit the array size. These detectors are currently available in both linear and square arrays. The maximum size is 256 elements for the linear array and 256 × 256 for the square array. The linear arrays are usually read out in a capacitive discharge mode (CDM) where each element is coupled to an FET switch. The detector elements are essentially used as photon-sensitive capacitors; the photon flux is read out by monitoring the *p-n* junction capacitance after a period of integration. As with PtSi, the integration time is severely limited due to the high photon flux arising from a 300-K background. The manufacturing process for the detectors is quite reproducible, and element-to-element variation in responsivity can be under 2%. Signal nonlinearity can also be held to under 1%.

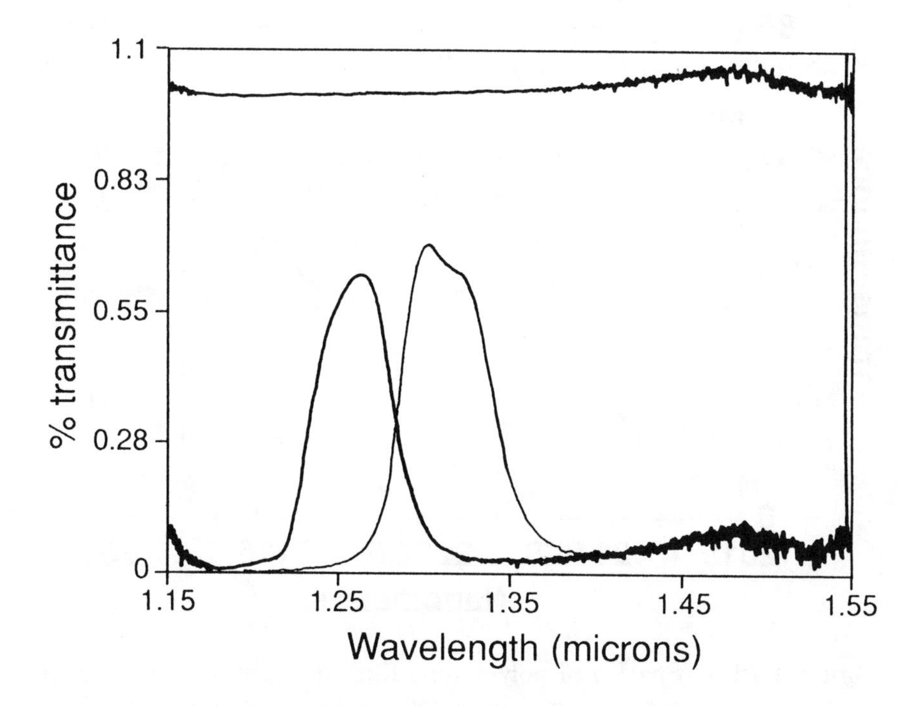

Figure 4.11 Transmittance spectra of the narrow bandpass filters and a 100% line for a PtSi array detector.

Richardson and colleagues used a 32-element InSb array coupled to a low resolution grating monochromator (100 grooves/mm) to obtain IR spectra in the 4- to 5-μ region [20]. Each element was individually wire bonded and the 32 outputs were fed to two 16-channel multiplexers. Since the response time of these detectors is quite rapid (microseconds), the outputs can be gated, and the detector may be used in a time-resolved mode. Preliminary results indicate that for the restricted spectral range (approximately 0.40 μ, limited by the dispersion), an SNR of 400 can be obtained in approximately 1 ms at a resolution of 16 cm^{-1}. This is comparable to rapid scanning FT-IR instruments. Clearly, the major limitations are the limited number of channels and the restricted spectral range.

Limited preliminary work has been done with two-dimensional InSb arrays for spectroscopic imaging [21]. These are hybrid arrays that are usually manufactured by bump-bonding the detector onto a silicon chip. Bump bonding is the process in which an array of detector elements each has a spot of solder (or some alloy) contacted

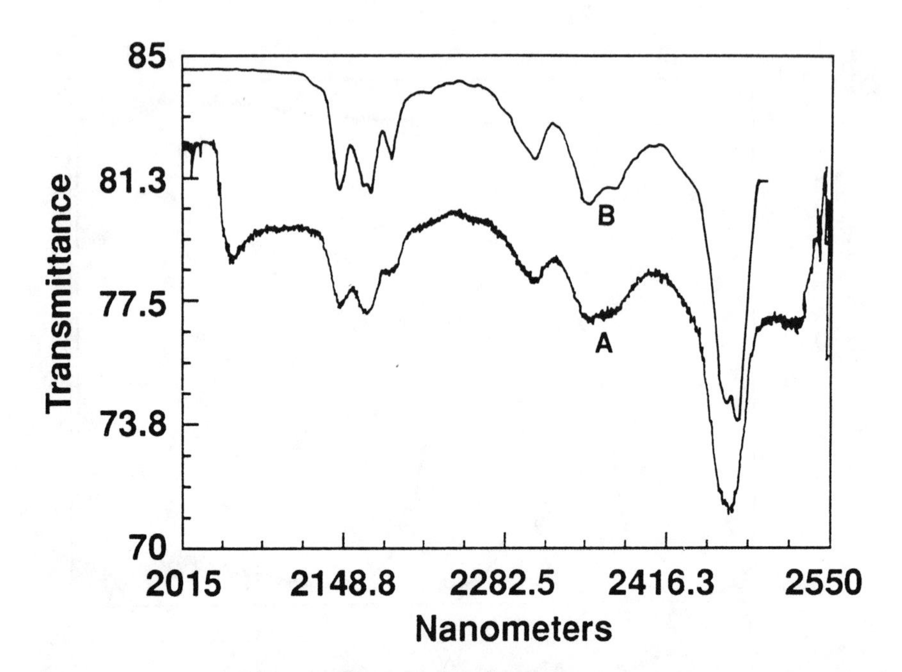

Figure 4.12 Spectra of polystyrene film. (a) Obtained with PtSi array; (b) obtained with conventional NIR spectrometer.

to it, producing an array of "bumps." Then an array (similar dimensions) of detector elements is bonded to the bumps, producing a three-layer structure comprising detector elements, conductive connections (solder), and readout elements (CCD). This allows direct connection of each detector element to a direct-injection, unity-gain amplifier that serves as the integration element. Preliminary work done by Ira Levin and Neil Lewis at NIH has shown that these types of array detectors can be used for microscopic imaging [22]. Figure 4.13 shows a schematic of their spectroscopic microscope. The wavelength selection is accomplished with an acoustooptic tunable filter (AOTF). They have done considerable work with this microscope in the visible region of the spectrum using CCD detection. The utility for both transmission microscopy [23] and Raman microscopy [24] has been demonstrated. They have now extended this work to the IR region using an InSb two-dimensional array detector, 128 × 128 pixels.

Figure 4.14 shows the NIR images of 0.4-mm path-length microcapillaries containing hexadecane (top) and water (bottom).

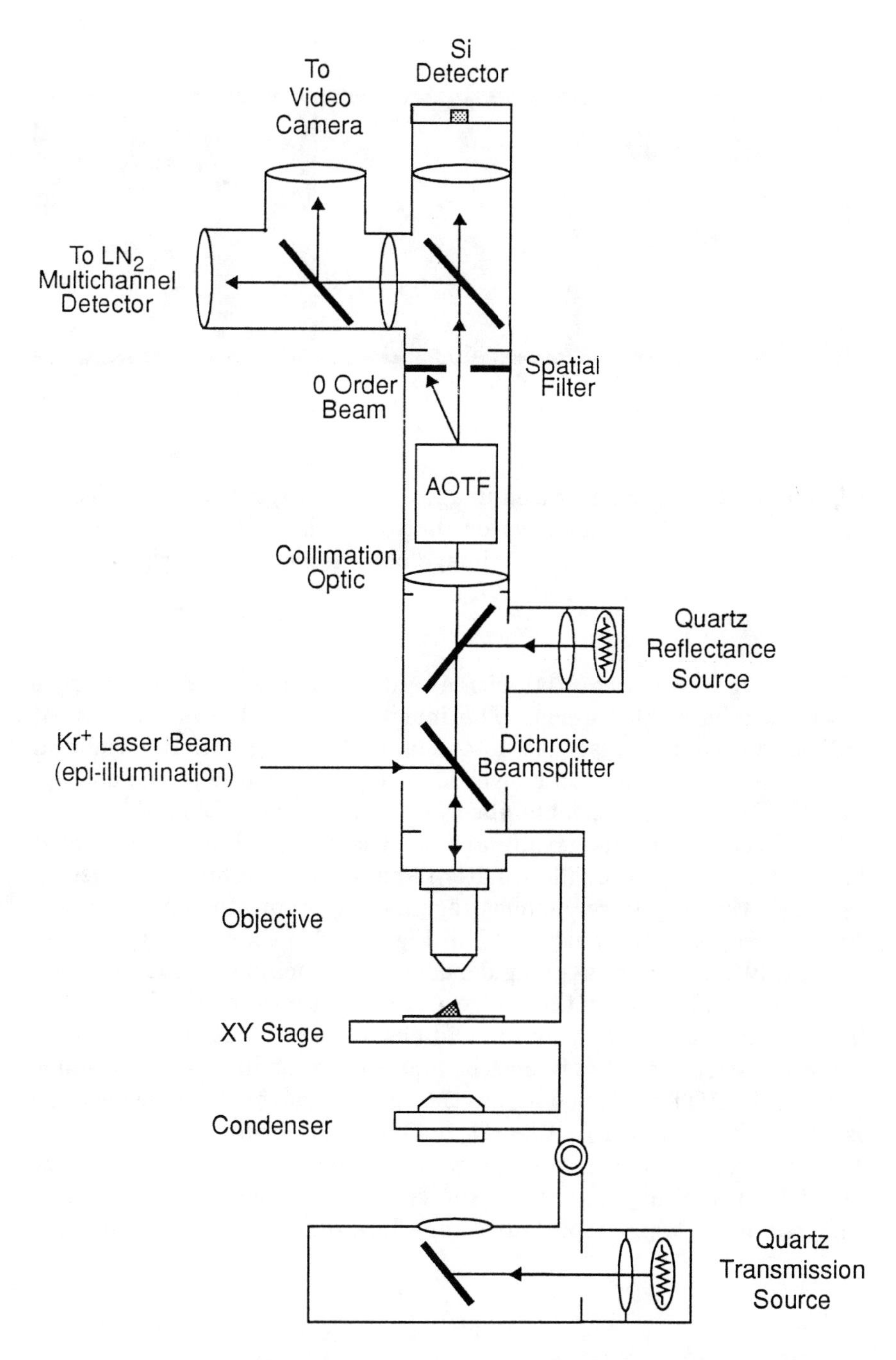

Figure 4.13 Schematic of spectroscopic microscope used by Lewis et al. (Reprinted with permission from reference 23.)

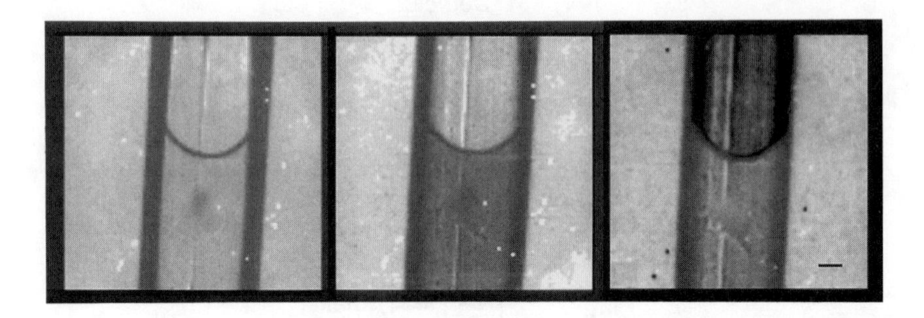

Figure 4.14 Near-infrared images of microcapillaries, hexadecane (top), and water (bottom). Left image obtained at 1.30 μ; center image obtained at 1.44 μ; right image obtained at 1.73 μ.

These images were acquired with an integration time of 31 ms and a block average of 50 frames. The image on the left was acquired at 1.30 μ, the center image at 1.44 μ, and the image on the right at 1.73 μ. At 1.44 μ the water layer is visibly darker due to the absorption by OH. No appreciable absorption is seen at 1.30 μ. At 1.73 μ, where CH absorbs, the hexadecane layer is visible. Individual images have been corrected for illumination and detector nonuniformities by ratioing the capillary images against appropriate backgrounds. Another application is exhibited in Figure 4.15, showing NIR images of freshly cut onion skin epidermis. The image on the left was acquired at 1.30 μ while the image on the right was acquired at 1.44 μ. The contrast is supplied by the OH absorbance. NIR spectra can be extracted from the set of images by monitoring the intensity of a single pixel as the AOTF is tuned across the spectral range. Such a spectrum is shown in Figure 4.16*. Clearly, the potential for high-quality, near-IR microscopy has been demonstrated by the Levin and Lewis group at NIH. The ability to extract spectroscopic information from these images offers a tremendous advantage for a wide range of experiments.

---

* I am grateful to Dr. Neil Lewis and Dr. Ira Levin for providing these data before publication.

Figure 4.15 NIR images of onion skin epidermis. Left image obtained at 1300 nm; right image obtained at 1440 nm.

4.3 Conclusions

Many array detector materials and architectures for the near-IR and IR regions are commercially available. Their performances are suitable for both conventional absorption spectroscopy and, in some cases, Raman spectroscopy. While such data are not available for many of these new detector arrays, this chapter has demonstrated the spectroscopic performances of several array materials. Improvements in the read noise characteristics are to be expected, as are the uniformity and available sizes of these arrays. Many of the detectors listed in Table 4.1 will move from the prototype to the production stage within the next several years.

Near-IR detector array performance is one of the most rapidly evolving areas of array detectors today, and as new designs become widely available, their impact in scientific application will become significant. Mid-IR array detectors comprising bump-bonded mercury cadmium telluride (MCT) onto silicon-CCD chips will provide a real alternative to FT-IR instruments for IR spectroscopy [25]. The potential development of blocked impurity band detectors [26] and superconducting array detectors [27] promises a very exciting future in the area of IR array detectors for both imaging and spectroscopy.

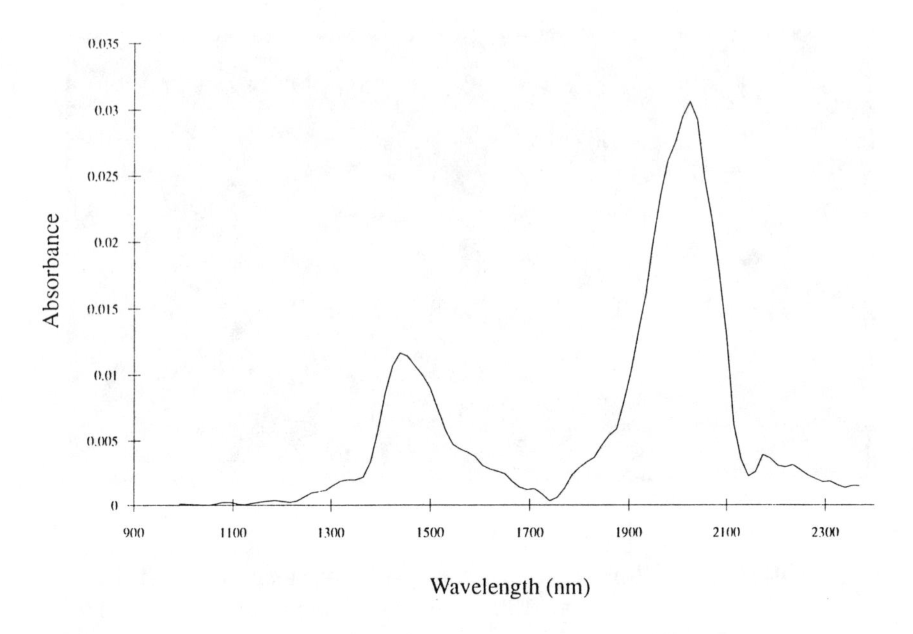

Figure 4.16 Spectrum of onion skin epidermis obtained by monitoring a single pixel as the acoustooptic tunable filter is scanned.

References

1. Denton, M.B., University of Arizona, private communication, 1990.
2. Newman, C.; Bret, G.; McCreery, R. *Appl. Spectrosc.* **1992**, *46*, 262.
3. Bluzer, N.; McKee, R.C.; Shiskowski, R.; Colquitt, L. *Proc. SPIE* **1985**, *570*, 137.
4. Dann, R.J.; Dennis, P.N. *Proc. SPIE* **1986**, *590*, 413.
5. Ballingall, R.A. *Proc. SPIE* **1990**, *1320*, 70.
6. Chase, B. *Anal. Chem.* **1987**, *59*, 881A.
7. Pemberton, J.E.; Sobocinski, R.L. *J. Am. Chem. Soc.* **1989**, *111*, 432.
8. Wang, Y.; McCreery, R.L. *Anal. Chem.* **1989**, *61*, 2647.
9. Schulte, A. *Appl. Spectrosc.* **1992**, *46*, 891.
10. Barbillat, J.; DaSilva, E.; Roussel B. *J. Raman Spectrosc.* **1991**, *22*, 383.
11. Huppi, E.R. In *Spectrometric Techniques, Vol. I.* Vanasse, G.A., Ed. Academic Press, New York, **1977**.
12. Hirschfeld, T.; Chase, B. *Appl. Spectrosc.* **1986**, *40*, 133.
13. Chase, B.; Talmi, Y. *Appl. Spectrosc.*, **1991**, *45*, 929.
14. Olsen, G.H.; Joshi, A.M.; Ban, V.S. *Proc. SPIE* **1991**, *1540*, 596.
15. Olsen, G.; Joshi, A.; Lange, M.; Woodruff, K.; Mykietyn, E.; Gay, D.; Erikson, G.; Ackley, D.; Ban, V. *Proc. SPIE* **1990**, *1341*, 432.

16. Mooney, J.M. *Proc. SPIE* **1985**, *570*, 157.
17. Cizdziel, P.J. *Proc. SPIE* **1991**, *1488*, 6.
18. Theden, U.; Green, J. M.A.; Storey, W.; Kurianski, J.M. *Proc. SPIE* **1990**, *1308*, 58.
19. Fowler, A.M.; Gatley, I.; Merrill, K.M.; Herring, J. *Proc. SPIE* **1990**, *1341*, 52.
20. Richardson, H.G.; Pabst, V.W.; Butcher J.A. *Appl. Spectrosc.*, **1990**, *44*, 822.
21. Blessinger, M.A.; et al. *Proc. SPIE* **1990**, *1308*, 194.
22. Lewis, E.N.; Treado, P.J.; Levin, I.W., private communication.
23. Treado, P.J.; Levin, I.W.; Lewis, E.N. *Appl. Spectrosc.*, **1992**, *46*, 553.
24. Treado, P.J.; Levin, I.W.; Lewis, E.N. *Appl. Spectrosc.*, **1992**, *46*, 1211.
25. Kozlowski L.J.; et. al., *Proc. SPIE* **1990**, *1308*, 2028.
26. Stetson, S.B.; Reynolds, D.B.; Stapelbroek, M.G.; Stermer, R.L. *Proc. SPIE* **1986**, *686*, 48.
27. Quelle, F.W. *Proc. SPIE* **1991**, *1449*, 157.

5
Intensified Array Detectors

Yair Talmi

Princeton Instruments Inc.
Trenton, New Jersey

5.1 Introduction

Photodiode arrays (PDAs) and image-intensified photodiode arrays (IPDAs) for use as spectrometric detectors were introduced as complete systems approximately 15 years ago. Their superiority over traditional single-channel detectors has been established for a variety of applications including Raman, luminescence, transient spectroscopy, colorimetry, flame and plasma emission, and many others.

In the last 5 years the newer charge-coupled device (CCD) technology has rapidly gathered momentum and has actually proven superior to the IPDA for various applications [1–4].

The principle advantages of CCD arrays over IPDAs are:

- Their ability to provide two-dimensional information, for example, spectral versus spatial or two-dimensional formats.
- Their extremely low noise, as low as 4 electrons rms, at the present time.
- Their high spectral response in the VIS-near-IR region.

Their main disadvantages have been:

- Absence of "spectroscopic" formats, that is, with the appropriate aspect ratio of length to width; however, CCDs with spectroscopic formats (e.g., 1024 × 256, 1100 × 330, and 1752 × 532 pixel) have been introduced recently.
- Poor and/or unstable UV-response.
- Relatively smaller well capacity, which can be a slight disadvantage for absorbance spectroscopy.
- Susceptibility to signal blooming and cross talk across the array matrix.
- Relatively higher price (at the time of writing).

For most continuous wave (CW) applications where relatively long signal-acquisition times are possible, the CCD offers significant signal-to-noise ratio (SNR) advantages. Particularly successful has been the application of CCD detectors in Raman and luminescence studies.

While the relative merits of CCD detectors for CW applications have been detailed in numerous papers, they are summarized herein for the convenience of the reader in comparing continuous wave with pulsed applications.

For transient spectroscopy in general and pulse laser spectroscopy in particular, the CCD is often an inadequate detector for two main reasons: First, it does not have sufficient SNR performance to allow detection of very weak signals, that is, down to a single photoelectron level, especially when signal-integration time is ultra-short. Second, CCD detectors cannot be gated, that is, rapidly turned on and off. Gating, as will be shown later, is essential when temporal information (nanosecond to millisecond scale) is needed or when discrimination of very weak pulse signals superimposed on very intense continuous backgrounds is measured. The intensified charge-coupled device (ICCD) detector was developed to address these requirements.

This tutorial-like chapter introduces the concepts of ICCD detectors, their major components, and their principle of operation. It identifies and explains the instrumental parameters most influential for achieving high performance. Following that, a simplified analysis of SNR performance including a comparison between CCD and ICCD detectors is presented. If the user understands this analysis, a detailed description of specific applications will be unnecessary.

Nevertheless, a few examples will be discussed, albeit without great detail. The readers may realize the advantages of ICCD detectors for their specific applications. This treatise is kept as simple as possible, without the burden of complex mathematics for which most readers have no time or patience or the need to digest. For those who may regard this approach as over simplified, apologies are extended a priori.

5.1.1 General Principles of Operation

Commercially available ICCDs consist of two major parts: a proximity-focused image intensifier, usually with a microchannel plate (MCP) electron multiplier that acts as a photon multiplier, and a CCD array that acts as a storage/readout device for the amplified photon image. A photon image incident on the image intensifier is amplified and transferred with minimum distortion to the CCD array via a relay lens or optical fiber image couplers acting both as windows and as image transfer devices. The image intensifier can operate either in the CW mode, where it operates continuously, or in the gate mode, where it operates only for short, predetermined periods. As will be discussed below, the gate mode is definitely the preferable one for ICCDs.

5.1.2 CCD Arrays in ICCDs

The operation and characteristics of CCD arrays have been described in detail in Chapter 2. Here it is sufficient to know that the CCD is an area array of individual pixels that act as individual photon detectors. Incoming photon signals are converted to proportional electron signals. The signals generated in the individual pixels are stored within the pixels in potential wells created with an elaborate set of overlaying electrodes (gates). This complex gate structure is arranged in a multiple-phase configuration that allows the photogenerated electrons to be transferred through the array toward a shift register. This shift register then shifts the charge information sequentially toward a preamplifier that accurately measures the charge values. The analog signals are digitized and reconstructed as spectra or images.

5.2 Image Intensifiers—Principles of Operation

The ICCD is represented by a diagram in Figure 5.1. The image intensifier consists of three major components:

- Photocathode: Transduces the photon image to an electron image.
- Microchannel plate (MCP): Amplifies the electron image.
- Phosphor: Converts the amplified electron image to a corresponding photon image.

The photocathode is usually coated on the inside wall of a quartz window or an optical fiber window. The MCP, as shown in Figure 5.2, is a glass disc consisting of millions of individual hollow microchannels, typically 12 μm in diameter, each acting as an electron multiplier. The geometric registration between the input and output ends of this disc (plate) is very precise to ensure a nondistorted transfer of the amplified image. The phosphor is deposited on the inside wall of an optical fiber window. The amplified electron image emerging from the MCP output is accelerated and focused onto the phosphor, further generating a corresponding amplified photon image. This phosphor-generated image is then transferred through the intensifier's output fiber window to another fiber window cemented to the CCD array. This image then generates a corresponding electron image in the CCD array.

In summary, a photon image incident on the window of the intensifier is transformed into a corresponding but greatly amplified electron image in the CCD. The end result is that even the faintest image is sufficiently amplified to overcome the inherent noise of the CCD. This is the reason for the exceptional inherent "sensitivity" of the ICCD.

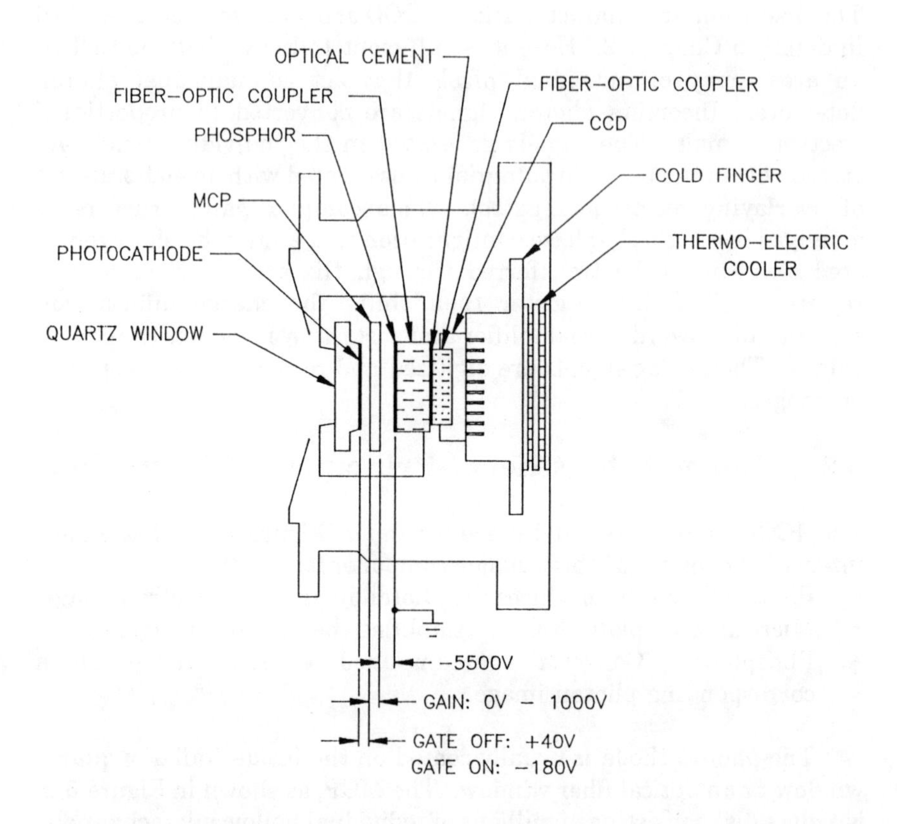

Figure 5.1 Principles of operation of the intensified charge-coupled device.

Typical image intensifiers are 18 mm in diameter, well suited for operation with standard small format CCD arrays, for example, 512 × 512 pixel or 576 × 384 pixel. Larger intensifiers are available and their relative merits will be discussed later.

The photograph in Figure 5.3 shows a 25-mm image intensifier coupled via an optical-fiber reducer (minifier) to a 576 × 384 CCD array. The image intensifier is powered by a network of electrodes (Figure 5.1), which creates a continuous potential gradient across it. Typically, the image intensifier is operated with the phosphor at ground voltage and the photocathode at the highest voltages. This arrangement is preferred because it poses minimum risk due to arcing from the intensifier to the static-sensitive CCD; however, the reverse potential arrangement is also usable. Typical operating voltages in CW mode are as follows:

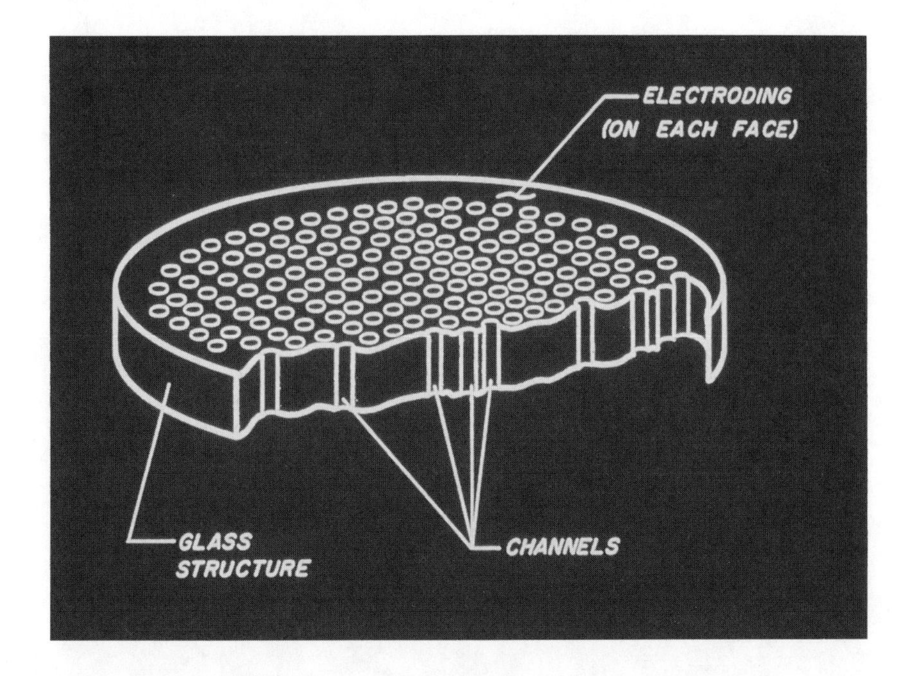

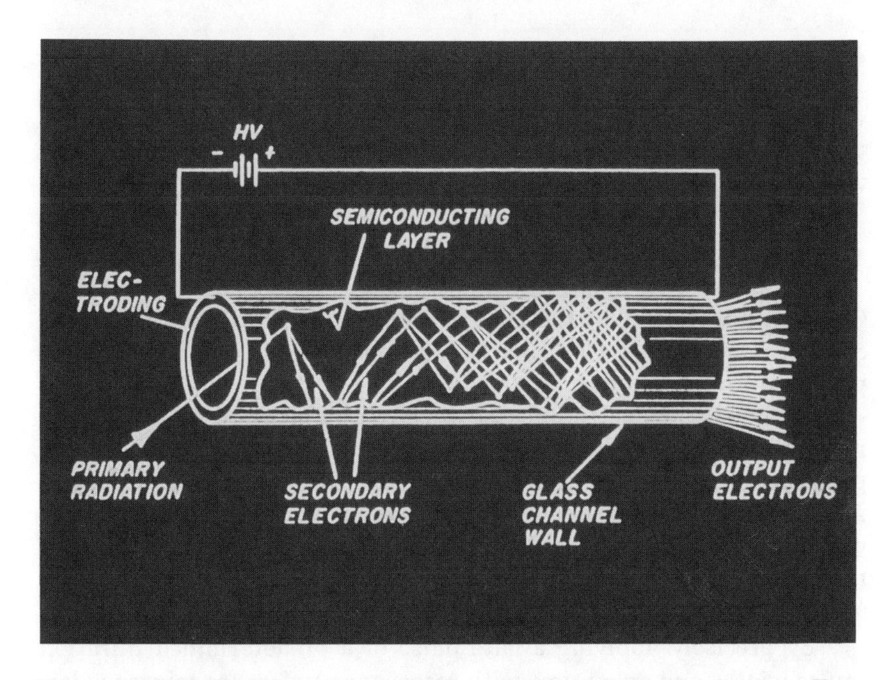

Figure 5.2 Principles of operation of microchannel plates. Top: Cutaway diagram of a microchannel plate. Bottom: Electron multiplication in a single fiber.

Figure 5.3 A minifier mounted on a multichannel detector.

- Phosphor: Ground potential (0 volts)
- MCP output: −5 kV
- MCP input: −5.7 kV to −6 kV (0.7–1.0 kV more negative than the MCP output)
- Photocathode: −5.9 kV to −6.2 kV (0.2 kV more negative than the MCP input)

Clearly, once emitted by the photocathode or MCP walls, electrons are continuously accelerated by the potential gradient toward the phosphor.

In the gate mode the photocathode is maintained 40 to 60 volts more positive than the MCP-input voltage, thereby preventing the photoelectrons emitted from the photocathode from further propagating toward the phosphor. When an external pulse of −200 volts or more is applied to the photocathode, it is temporarily rendered more negative (+40 −200 = −160 volts), and consequently the intensifier amplifies light during the duration of this gate pulse.

By precisely applying a gate pulse of a predetermined duration (pulse width) and delay time with respect to a reference trigger pulse, the image intensifier and therefore the ICCD can be activated at the exact time required to capture an image or a spectrum, for example, a precise time after fluorescence decay has commenced. Typically the reference trigger pulse has a fixed timing with respect to the time evolution of the spectrum being measured.

5.3 Components of an ICCD

5.3.1 Windows

The front window of the intensifier can be made of various materials to provide the required transmission characteristics. Quartz is used for the 180 to 900-nm spectral range. Special proprietary glasses are used for extended NIR, to 930 nm. Chemical interaction between the photocathode and the window can also determine the overall spectral response. MgF_2 is used for operation down to 120 nm. LiF is less useful because of its hygroscopic characteristics. Occasionally, sapphire is also used instead of quartz. When the ICCD has to be interfaced to another device, for example, an optical fiber output window of a streak camera, the window of choice will be an optical-fiber coupler.

5.3.2 Photocathodes

Photocathodes are identical to those used in photomultiplier tubes. Most common are S-20 and S-25. Usually very thin bi-alkali photocathodes are optimal for UV and near vacuum UV (VUV), whereas S-25 photocathodes (Na, K, Sb, and Cs) are more suited for UV-NIR and VIS-NIR. NIR extension requires thicker photocathodes. Generation III intensifiers use GaAs photocathodes, which operate in the 550 to 910 nm range.

The electrical conductivity of the photocathode determines its gate performance. Generally, faster and more precise gating is achieved when the photocathode resistance and capacitance are reduced (see gating section). Because resistance decreases with thickness, NIR optimized photocathodes are generally more appropriate for fast gating. Methods for creating photocathodes with optimized stoichiometry and physical characteristics are often art rather than science, and are usually very proprietary in nature.

5.3.3 Microchannel Plates

The MCP is a glass plate (a few millimeters thick) that contains millions of individual hollow fibers, each of which acts as an individual electron multiplier. Many parameters affect the overall performance of MCPs.

Noise levels of MCPs are mostly affected by:

- Radioactivity in the components.
- Outgassing and ion-feedback, that is, creation of spurious ions during the electron multiplication process.

- Field emission, that is, electron emission due to microcracks and broken debris. This usually induces "hot spots" — regions of intense emission.
- Noise in the measurement chain.
- Noise due to cosmic rays, that is, high-energy particles from space.
- Resistivity of the microchannel.

The overall sensitivity of the MCP depends also on the bias angle of the channel with respect to the photocathode, the geometric shape of the channel inlet, that is, straight or funneled, the diameter of the channels, and the packing methodology (the two affecting the active-to-dead-space ratio).

Another important parameter is the methodology used to scrub (i.e., bombard) the MCP with high-energy electrons. Scrubbing affects the gain, stability, and ion-feedback characteristics of the intensifier.

Microchannels can produce pulses with extremely short time of flight, that is, the sum of intercollision transit times from the point of avalanche initiation at the channel entrance until the avalanche pulse leaves the channel exit is as low as 20 ps. These times depend on the ratio of length to diameter of the channels and the voltage across them.

The recovery time, the time it takes for the microchannels to fully recover and be ready for detection of the next pulse, must also be considered. This time is on the order of a few milliseconds or more. In principle, this recovery time sets the limit on the maximum signal flux that the MCP can handle before it saturates. From experience it is possible to obtain a 16-bit (ADC) dynamic range (65,000 to 1) in CW mode but usually not when detecting ultra-short pulses.

Typically, individual channels produce electron gains of 500 to 1000. Also, the output electron distribution of the avalanche signal, that is, the charge pulse-height distribution (as shown in Figure 5.4a) approximates a negative exponential rather than a saturated curve. The insufficient gain of the MCP and the exponential pulse height distribution are the reason for the inability of conventional video rate ICCDs to act as true "photon-counting" detectors, despite the fact that their overall gain/noise ratio permits a clear detection of single photoelectrons.

Obtaining a saturated distribution (output signal magnitude vs. frequency of occurrence) for the multiplied output signals from the individual microchannels requires the use of a few MCPs in tandem to achieve gains on the order of 5×10^6 compared with a few thousand (Figure 5.4b) for a single MCP.

To reduce the effect of ion feedback in tandem configurations, the MCPs are often arranged with the channels slightly slanted on opposite slopes, that is, chevron shaped. The use of such in-tandem

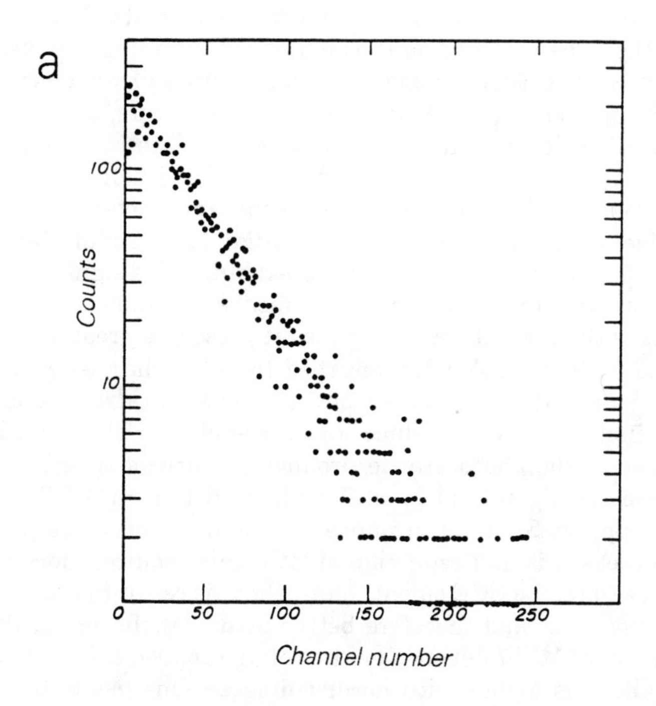

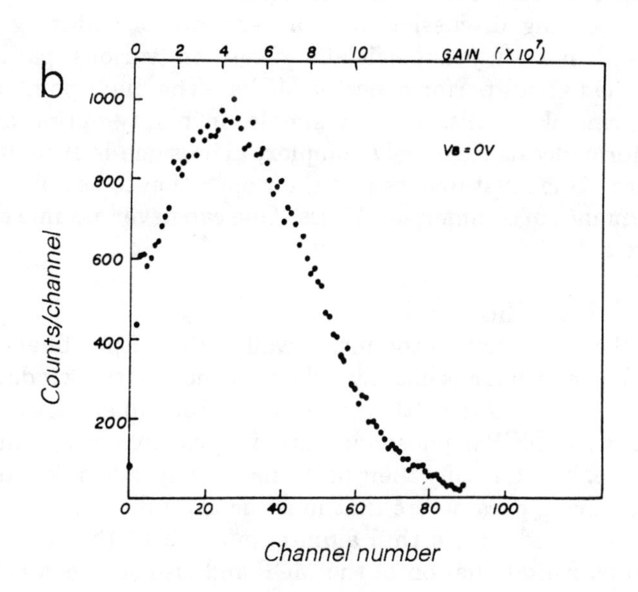

Figure 5.4 Pulse amplitude distributions. (a) Straight channel MCP. (b) Double MCP (Chevron configuration) at 1 kV.

configurations is rather counter productive for regular ICCDs since it reduces the dynamic range of the device, although it is very useful for operation with resistive anode or resistive anode array devices [5,6].

With high-performance slow-scan ICCD detectors, usually neither MCP nor phosphor saturation are a problem. The MCP gain can be controlled by varying the voltage across it. Operation in the 600- to 900-volt range produces nearly linear response characteristics, with the gain value proportional to an exponential function of the voltage (Figure 5.5). Higher gain can be obtained but with a significant added noise mostly due to ion feedback. Ion feedback is induced by electron collisions with residual gas molecules (at pressures greater than 10^{-6} Torr) and with gas molecules desorbed from the channel walls under electron bombardment. These positive ions will drift back to the channel input and will produce ion after-pulses. They will also be accelerated to the photocathode producing spurious signals.

Where the light level is sufficient, reduction in MCP gain will actually improve SNR performance (see SNR performance in section 5.5). The reason is that reduction of MCP gain requires more photons to produce the same CCD signal. More photons result in lower relative photon shot noise and therefore better SNR. Furthermore, the SNR performance of MCPs depends on the voltage applied across them, that is, typically it is higher with lower voltages. One has to be careful, however, to stay within the operational range of the MCP and to avoid overexposure of the photocathode to light.

The preceding discussion may be somewhat confusing to the reader as it has only superficially dwelt on various parameters affecting the overall performance of MCPs. The main point is that, while in principle MCPs are very simple, in reality optimization of their performance is extremely complex. The same is true of image intensifiers. One must understand their operation to be able to judge the performance of commercial ICCDs. One can never assume that "all ICCDs are alike."

5.3.4 Phosphor

The phosphor is coated on the inside wall of the output fiber-coupler window. The phosphor is usually either brushed on the window or, in certain cases, allowed to settle (gravitation) from an aqueous suspension. As with MCPs, phosphors are in principle very simple to manufacture, but in reality their performance depends on a multitude of manufacturing parameters that must be carefully controlled. The phosphor is coated with a thin aluminum overcoat that reduces ion feedback and contamination of the MCP and also acts as a mirror to reflect some of the backward-directed photon image. The most common phosphor used for scientific grade intensifiers is P-20, although P-43 is also used. Military-type intensifiers for night vision

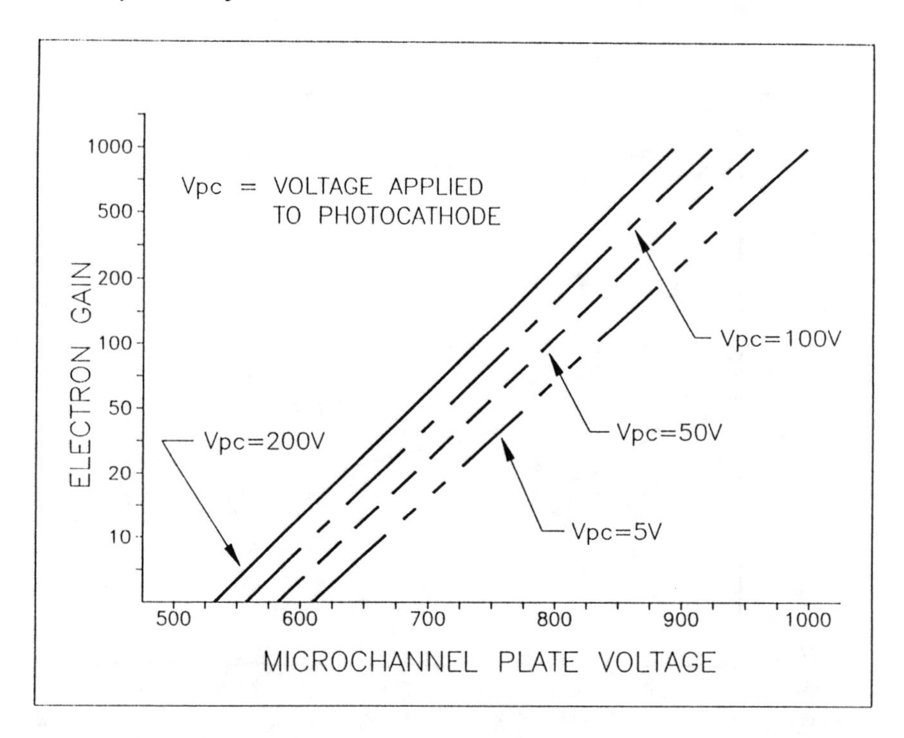

Figure 5.5 Typical MCP electron gain vs. applied voltage.

use slower phosphors corresponding to the response time of the eye.

Phosphor decay time is a complex parameter that cannot be described by a first-order kinetics model. It has a convoluted term due to a multiplicity of coexisting decay mechanisms. Usually a very fast decay time is followed by a substantially longer decay "tail." To further complicate matters, methods for measuring decay times vary by application and the detector interfaced to the intensifier. Manufacturers of image intensifiers usually define the decay time as the time it takes the phosphor to decay from 90% to 10% intensity. For the ICCD or the IPDA, one has to take into consideration the overall *integrated* signal between consecutive readouts. Furthermore, decay times strongly depend on the duration (pulse width) of the excitation pulse. As shown in Figure 5.6, the shorter the excitation pulse, the shorter the phosphor decay time [7,8].

For operation at scan repetition rates above 4 to 5 Hz, P-20 is usually unsatisfactory, and as illustrated in Figure 5.7, rare-earth phosphors are preferred [9,10]. Incidentally, Figures 5.6 and 5.7 demonstrate the great discrepancy between decay values reported by the intensifier's manufacturers and by ICCD users. Clearly, rare-earth phosphors are about 40 times faster than P-20 and are better suited

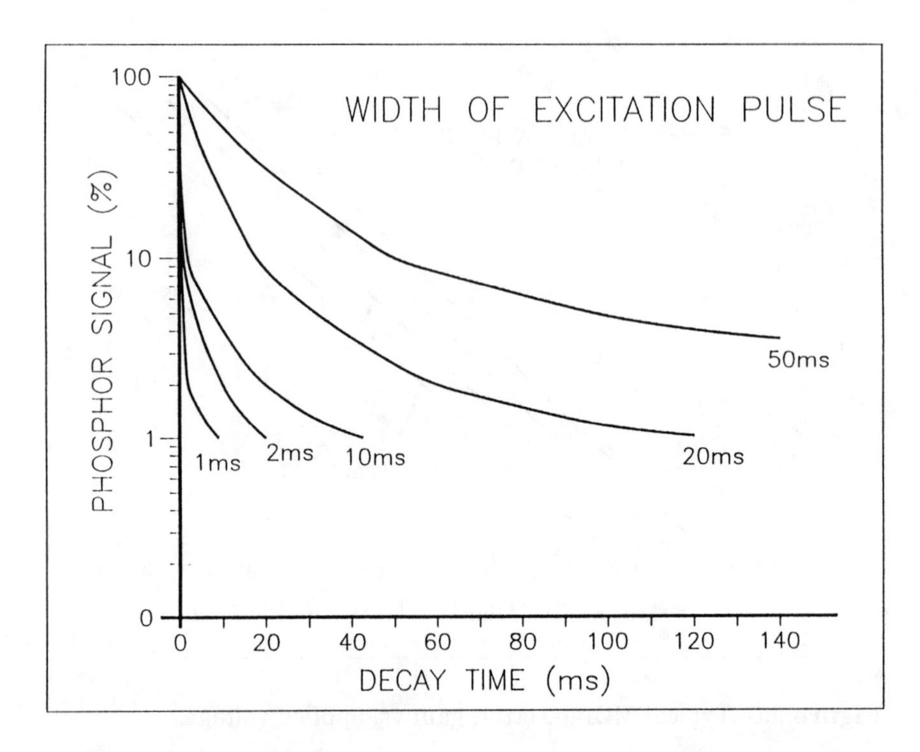

Figure 5.6 P-20 phosphor decay time as a function of the excitation pulse width.

for operation at repetition rates of up to 50 to 100 per second. Rare-earth phosphors are less efficient than P-20 and their fluorescence spectrum is more blue shifted, a disadvantage for operation with CCDs, which are more red sensitive. Nonetheless, overall efficiency is sufficient for optimal operation of ICCDs.

While there are also some reports [11] that decay time depends on the intensity of the excitation pulse, we have not observed a significant dependence. P-20 phosphors from different manufacturers have shown decay time variations greater than 10-fold. Obviously, the generic term P-20 is rather inaccurate and represents a class of phosphors with only a general resemblance in regard to physical characteristics. Another important parameter of phosphors is their linearity of response (transfer characteristics). Most phosphors are, unfortunately, nonlinear in response, especially at very low light levels. While phosphors can exhibit a low-light-level threshold that corresponds to a slope of less than 1 on a log/log input/output transfer curve, P-20 phosphor shows only slight nonlinearity at very low light levels with

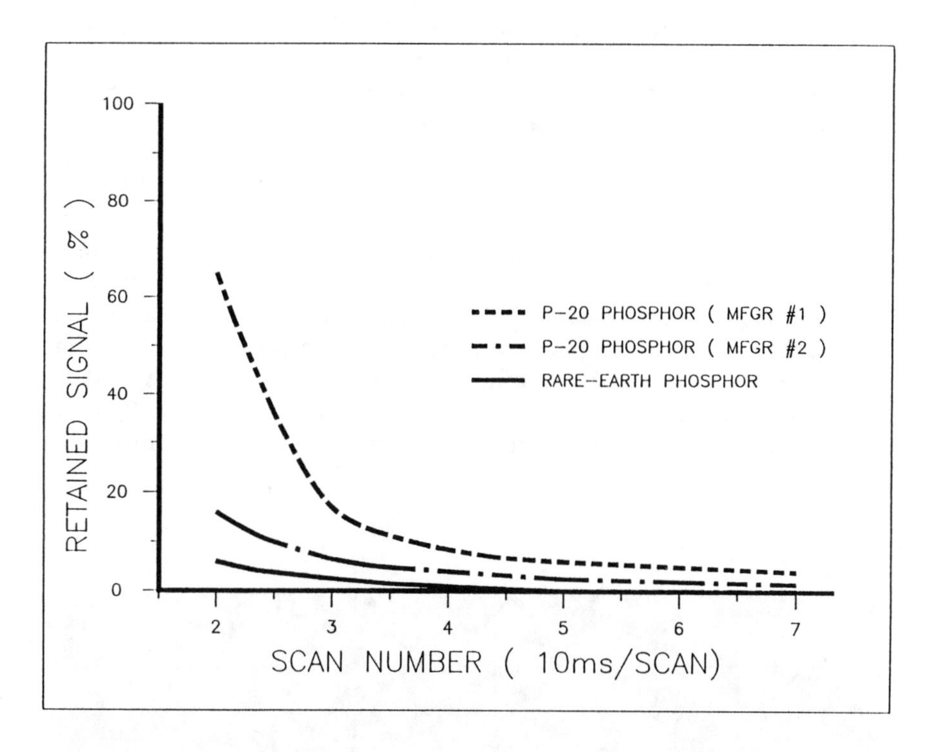

Figure 5.7 Phosphor decay time as a function of the number of consecutive CCD scans.

equivalent input signals of 1 to 4 photoelectrons, and rare-earth phosphors have demonstrated excellent linearity to a single photoelectron level. As previously mentioned, phosphor saturation is not a problem with high-performance ICCDs where a 16-bit maximum signal typically represents an input signal of 650 to 6500 photoelectrons/pixel.

5.3.5 Optical-Fiber Couplers

The final optical image emerging from the phosphor is conveyed to the CCD via two separate optical-fiber couplers. These are glass plates consisting of millions of individual glass fibers (4–6 μm in diameter) that are arranged in "perfect" input-to-output geometric registration. Each fiber transmits a small portion of the image. The fiber-coupler window of the CCD is set very close to the CCD but not actually touching it. Optical coupling oil or cement is used to ensure index-of-refraction matching between the CCD and the fiber-coupler windows.

There are a few mechanisms for loss of resolution through the fiber windows:

- Defective fibers, damaged or partially transmitting.
- Regions of low transmission due to the imperfect stacking processes of fiber bundles. These are often referred to as "chicken wire"; see Figure 5.8. These patterns are also typical of the MCP, which also consists of stacked hollow-fiber bundles.
- When optical-fiber reducers are used, for example, to reduce the phosphor image of a 25-mm phosphor (large-format intensifier) to the size of a standard CCD (typically 15–16 mm diagonally), spectral resolution can be further reduced by pin-cushion distortion that bends the slit image. Pin-cushion distortion is caused by manufacturing processes and by fiber-to-fiber cross talk at the output end of the coupler where cladding becomes very thin. The latter problem can be minimized if the fibers are EMA coated (extramural absorber coating). As shown in Figure 5.9, new developments are now under way to improve the uniformity and quality of fiber couplers.

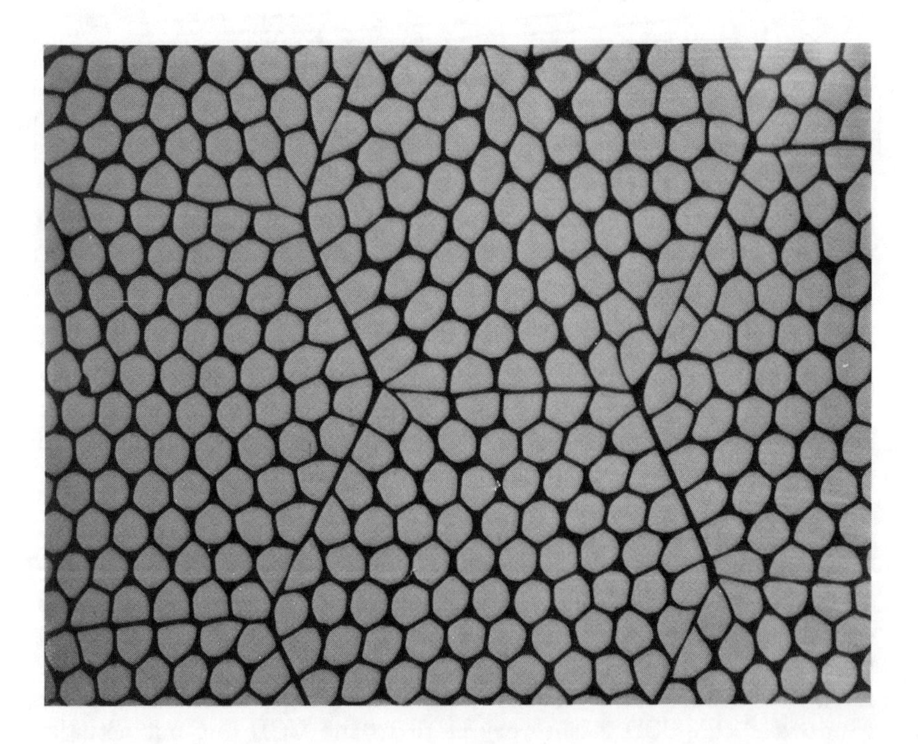

Figure 5.8 Microscope photograph of a typical fiber-coupler window, with an apparent "chicken-wire" pattern.

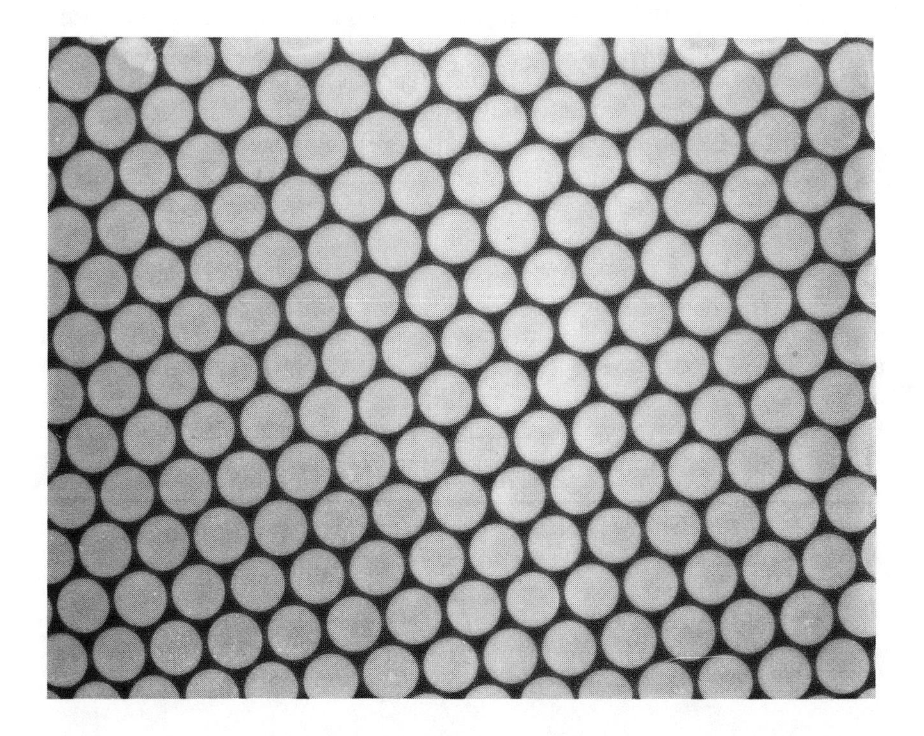

Figure 5.9 Microscope photograph of a newly developed fiber-coupler window, free of "chicken-wire" imperfections.

5.3.6 Optical-Fiber Minifiers

As mentioned in section 5.2, an optical-fiber reducer, commonly called a minifier, is needed to couple the image from a large format intensifier to a small-format CCD (Figure 5.3). Image intensifiers of 18-, 25-, 40-mm and even 75-mm diameter are now commercially available. There are two major advantages to image "minification." First, the signal flux per unit area is proportionally reduced, thus minimizing saturation of the MCP and therefore further increasing the dynamic range of the ICCD. Second, because the resolution of the CCD is far better than that of the intensifier (CCDs have typically 40–100 pixels/mm), minification will improve the overall resolution of the ICCD roughly proportionally to the demagnification ratio for a given CCD. However, transmission losses increase as the minification ratio increases, at a rate roughly proportional to the square of the demagnification ratio. The practical limit is around 2:1 [8]. In fact, losses as high as 95% have been reported for optical-fiber coupling of small-format CCDs to 40-mm image intensifiers. The losses in transmission are mostly due to thinning of the fiber cladding and the optical entendue of the fiber, which depends on the fiber's diameter and index of refraction. Overall transmission is also limited by the Lambertian-distributed nature of the light emitted from phosphors. Compared with collimated light, only a fraction of phosphor-emitted light is actually collected by the fibers.

5.3.7 Lens Coupling — an Alternative Coupling Configuration

Coupling between the image intensifier and the CCD array can also be accomplished with lenses to relay the phosphor image to the CCD array. The advantages of lens coupling are as follows:

- Flexibility — A single detector can be used either with a lens, as an ICCD, or without it, as a regular CCD.
- Only a single fiber coupler is used, thus reducing image imperfections due to "chicken-wire" defects.
- The CCD array can be maintained at cryogenic-level temperatures, since no optical fibers are necessary.
- Image size focused on the CCD can be varied.
- Signal flux incident on the CCD can be varied simply by selection of lens *f*/number (aperture size) values.

As a rule of thumb, when a system is designed with maximum flexibility for both low and medium light levels, its overall performance will be compromised. Similarly, the lens-coupled ICCD suffers from the following deficiencies:

- There is an unfavorable trade-off between lens size and throughput. For even 2 to 3% throughput, a relatively large size lens assembly is needed. With relay lenses consisting of two low *f*/number camera lenses, that is, back-to-back configuration, a bulky assembly makes interfacing to spectrometers difficult and cumbersome.
- Even with such fast relay lenses, for example consisting of *f*/1.2 camera lenses, only about 10% of the phosphor image is collected, compared with more than 60% transmission obtainable with optical-fiber couplers. This loss of transmission is sufficient to prevent single photoelectron detection, Figure 5.10. This estimation of detection probability is discussed further below.
- To compensate for the low transmission provided by relay lenses, image intensifiers must be operated at their fully allowable gain, that is, at high MCP voltages. Operation under such conditions considerably shortens the lifetime of the image intensifier. These voltage requirements are even more severe when low efficiency rare-earth phosphors are used.

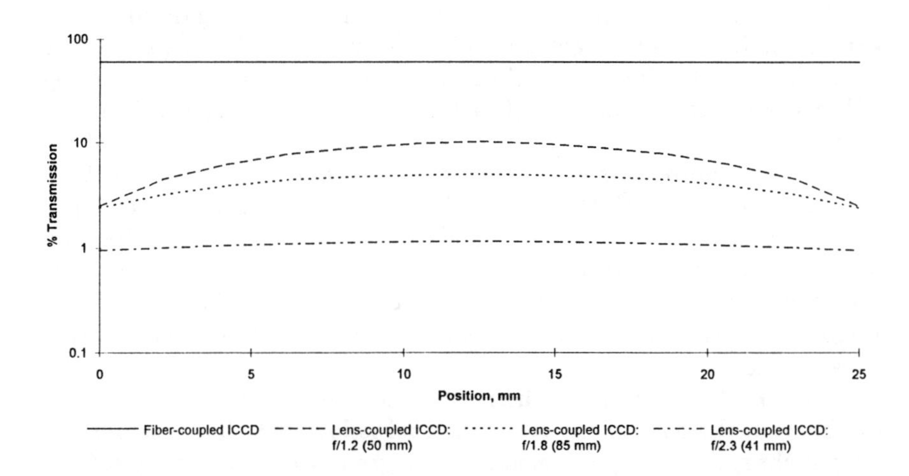

Figure 5.10 Vignetting (spatial variations) for various ICCD configurations. CCD array has dimensions of 1152 pixels × 298 pixels.

- Whereas with optical-fiber couplers uniformity across the image remains constant, relay lenses suffer from image vignetting, that is, transmission is reduced from center to edge of the image. Vignetting is especially severe with high-transmission relay lenses (Figure 5.10), so the user confronts the obvious dilemma of a compromise between vignetting and transmission. As a side note, when high demagnification ratio optical-fiber couplers (>2:1) are used, image vignetting can also become a problem. Additionally, resolution usually gets worse from center to edge of the image, making spectral calibration and analysis more difficult.
- The use of lenses usually creates a higher level of stray light, which reduces the intraspectral dynamic range of the ICCD (see section 5.4.5.3).
- The "advantage" of *f*/number selection to control signal flux is at least partially illusive because it is true only for high signal levels in which case usually a CCD detector rather than an ICCD is preferable in regard to SNR performance.

5.3.8 Effect of Reduced Aperture on Lens-Coupled ICCDs

It has been claimed that it is possible to increase the dynamic range of lens-coupled ICCDs by operating the MCP at maximum gain and reducing the light output between the intensifier and the CCD by partially closing the lens diaphragm, by virtue of the higher current density in the MCP. In fact, this strategy reduces the dynamic range. The reason is that as the MCP voltage is increased, the gain increases exponentially (it is an electron multiplier system) whereas the static channel current (sometimes called the strip current) increases linearly (Ohm's law). If, for instance, the MCP voltage is increased from 800 to 850 volts, the channel current will increase 6.3% but the output charge per photoelectron will double. Each time a photoelectron enters a microchannel, the output charge resulting from the multiplication process reduces the accelerating potential in that microchannel. The charge emitted from the microchannel in the direction of the phosphor discharges the microchannel capacitance. The potential is restored by the strip current, which recharges the microchannel capacitance.

During the time in which the microchannel capacitance is discharged, the gain of the microchannel is lowered. The greater the output charge for one photoelectron, the greater the gain reduction for the following photoelectrons, until the channel is recharged. Furthermore, because the recharging model is a simple distributed RC circuit, the greater the loss of gain, the longer it will take to recover it. Published data [12] show significant nonlinearity of photoresponse (about 12% at 2 photons per projected pixel). In contrast, fiber optically coupled detectors typically exhibit linearity better than 1% over the same range.

5.3.9 Estimation of the Detection Characteristics for Single Photoelectron Detection

In attempting single photoelectron detection, the user picks a threshold (S_{ph}) (normalized to the rms readout noise). Any signals above the threshold are considered photoelectron events. It has been shown above that the MCP gain is described by an exponential distribution, while the readout noise in the absence of a photoelectron signal is Gaussian. Thus the probability of a signal, s, is described by

$$p(s) = \frac{k}{G} e^{s/G} \tag{5.1}$$

where G is the average gain and k is a constant depending on system coupling efficiencies, the point in the system, and the units chosen for the comparison. The probability of detection is simply

$$P(\det) = \int_{s}^{\infty} p(s)\ ds \tag{5.2}$$

where the effect of noise on probability of detection has been ignored. The probability of false alarm is the probability of the noise being above threshold. Thus the probability of false alarm for one pixel is

$$P_{fa} = \int_{S_{th}}^{\infty} \frac{1}{\sqrt{2\pi N_R}} e^{-(x/2N_R)^2 dx} = \frac{1}{2} erfc(\frac{S_{th}/N_R}{\sqrt{2}}) \tag{5.3}$$

where N_R is the CCD readout noise (see section 5.5.2.2). These formulas were used to generate Figure 5.11 based on a 576 × 384 pixel (221,184 pixels) ICCD with the probability of false alarm calculated on a per-frame basis. The large number of pixels, even in a small CCD, requires that the probability of false alarm per pixel be extremely small, which in turn forces the threshold high, requiring a high gain between the photocathode and the CCD. To get 90% probability of detection requires the photoelectron signal to be 50 times the N_R. It was shown above that gains of this magnitude are only practical in fiber-coupled CCDs.

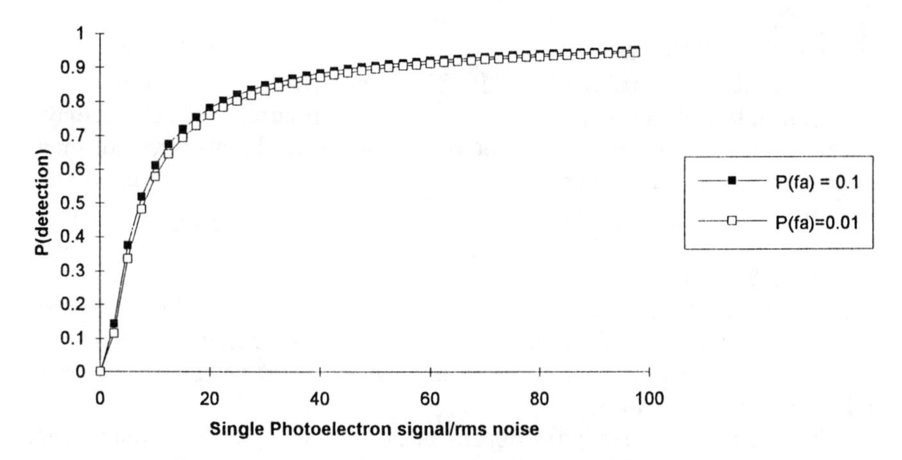

Figure 5.11 Detection probability characteristics of a fiber-coupled ICCD with a 576 × 384 pixel CCD array.

5.4 Detector Performance

5.4.1 Spatial Resolution

In principle the resolution of an ICCD depends on the combined resolution of the intensifier and the CCD array. In reality it is limited by the resolution of the image intensifier. The resolution of image intensifiers is determined by the following imaging processes.

5.4.1.1 Input Proximity Focusing

This describes the efficiency and accuracy by which the photoelectron image emitted from the photocathode is focused onto the MCP input. The proximity-focusing process is, in principle, very simple and assumes a linear acceleration of the photoelectrons along the electrical field. A certain degree of distortion is created by non-ideal focusing mechanisms that set a limit on both the resolution and the geometric registration of the image. The resolution, N, can be simply expressed as

$$N \propto \frac{E^2}{d} \tag{5.4}$$

where E is the electrical field and d is the gap between the photocathode and the MCP. The practical values for these parameters are typically 200 volts across the gap and d = 0.15 to 0.2 mm. At least one manufacturer is now experimenting with techniques that will allow increased voltages that improve resolution.

5.4.1.2 MCP Resolution

Basically, the resolution of an MCP is determined by the diameter of the individual channels, usually 12 μm. When constructed properly, there is very little cross talk between channels. However, the dead space between individual channels can produce adverse aliasing effects (Figure 5.12).

5.4.1.3 Output Proximity Focusing

The mechanism for focusing the output MCP image onto the phosphor is identical to that used in the input section. The main difference here is that the voltage across the gap is 5000 to 5500 volts and the gap width is 1.0 to 1.2 mm.

The limit on both parameters is set by the vacuum inside the intensifier, which determines the minimum gap that will not induce destructive internal discharges.

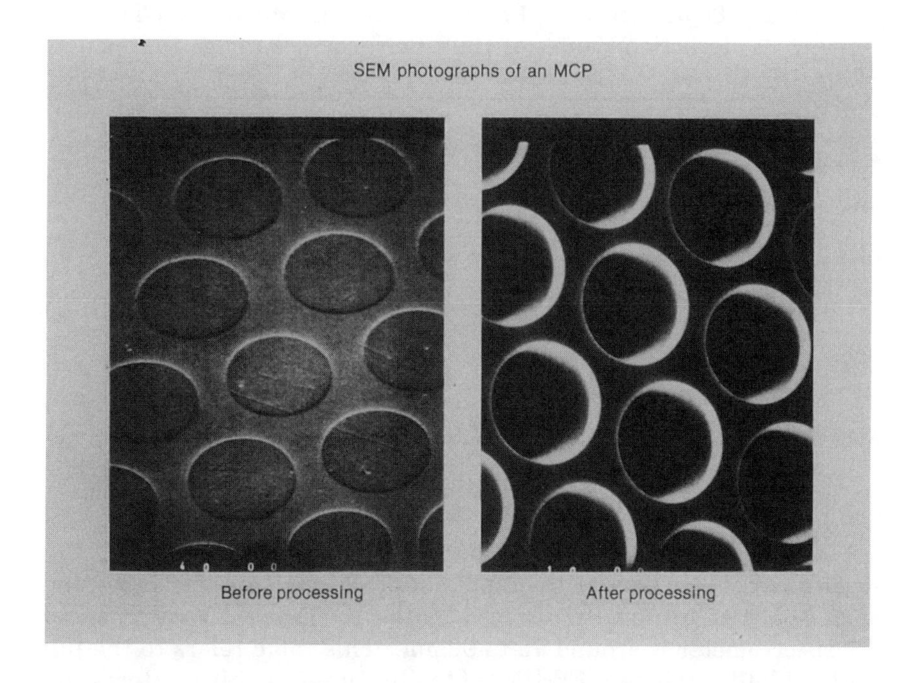

Figure 5.12 SEM photograph of an MCP.

5.4.1.4 Resolution and Modulation Transfer Function

To summarize, the overall resolution of image intensifiers depends on the electrical fields across the intensifiers, the phosphor quality, and the optical-fiber coupler quality. High-performance image intensifiers can produce spatial limiting resolutions as good as 30 to 35 line pairs per millimeter (lp/mm). Spatial resolutions always should be presented as curves of lp/mm versus modulation-transfer function (MTF) rather than only a "limiting resolution" value. Although a complete description of MTF is beyond the scope of this chapter, MTF is a measure of the degradation in resolution caused to an input image, that is, the lower the MTF, the worse the distortion. When evaluating spatial resolution it is always beneficial to consider the associated MTF values. In essence, MTF can be correlated to the degree of separation between two adjacent spectral lines. A complete separation between the two will be the equivalent to 100% MTF, whereas a partial overlap between them correlates to lower MTF values. Typical values of MTF versus spatial frequency for a "good" intensifier are shown in Table 5.1.

Table 5.1 Typical Spatial Frequency versus Modulation-Transfer Function Values of High-Performance Image Intensifiers

Spatial Frequency (lp/mm)	MTF (%)
2.5	98
7.5	80
16	50
25	20
35	8

5.4.2 Practical Resolution for Spectroscopists

Regardless of how narrow the input slit of the spectrometer is set, the narrowest line ultimately resolved by the ICCD at the focal plane of the spectrometer is around 75 to 90 μm. This value refers to the full width at half maximum (FWHM) of the line intensity. It is interesting to compare this value with that of a CCD element size, typically 15 to 35 μm. Let us assume that the CCD format is 576 $\times$ 384 pixels, with the 576-pixel axis coinciding with the wavelength axis. If a pixel width is 23 μm, then the FWHM of a spectral line occupies about 3.5 pixels. If we consider two lines to be resolved if their degree of overlap is 50%, then the real maximum number of resolution elements is 576/3.5 = 165. A 576-pixel CCD will cover 576 $\times$ 0.023 mm = 13.25 mm of the focal plane. Accordingly, the simultaneous spectral coverage of the spectrometer, *SSC*, and spectral resolution of the ICCD, *SR*, are

$$SSC = 13.25 \text{ mm} \times RLD \text{ (nm/mm or cm}^{-1}\text{/mm)}$$

where *RLD* is the reciprocal linear dispersion of the spectrometer.

$$SR = 13.25 \times \frac{RLD}{165} \tag{5.5}$$

For example, if a 150-groove/mm grating produces an *RLD* of 20 nm/mm, then

SSC = 13.25 $\times$ 20 = 265 nm spectral coverage
SR = 265/165 = 1.6 nm spectral resolution

If a 25-mm intensifier were used with a "minification" ratio of 1.6, then

$$SSC = 1.6 \times 20 \times 13.25 = 424 \text{ nm}$$
$$SR = 424/165 \times 1.6 = 1.6 \text{ nm}$$

The spectral line width at the CCD plane is now only about 50 μm instead of 75 to 90 μm.

Consequently, the spectral coverage was increased by 60% without any significant loss of spectral resolution. Even higher minification ratios can be obtained if the ICCD is to be used only as a spectroscopic detector. In Figure 5.13, the MTF curve for an optical-fiber-coupled ICCD is compared with that of a lens-coupled ICCD. Once again the superiority of the former is clear.

5.4.3 Photoresponse

The photoresponse of an ICCD depends primarily on the photocathode selected and to a lesser degree on the transmission of the input window. A nickel undercoating is often used to increase the photocathode conductivity for fast gating. If used, it will further reduce the photocathode response by acting as an absorption filter.

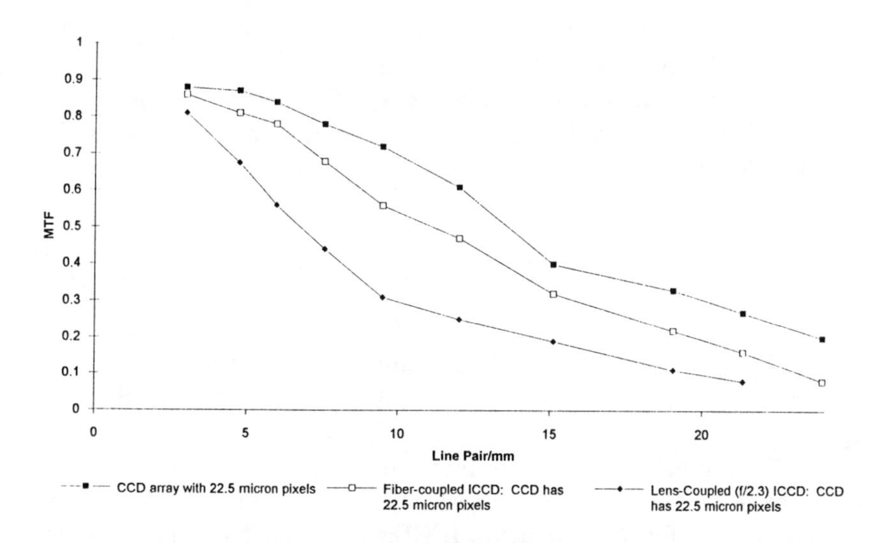

Figure 5.13 Modulation-transfer function versus resolution for a CCD, a lens-coupled ICCD, and a fiber-coupled ICCD.

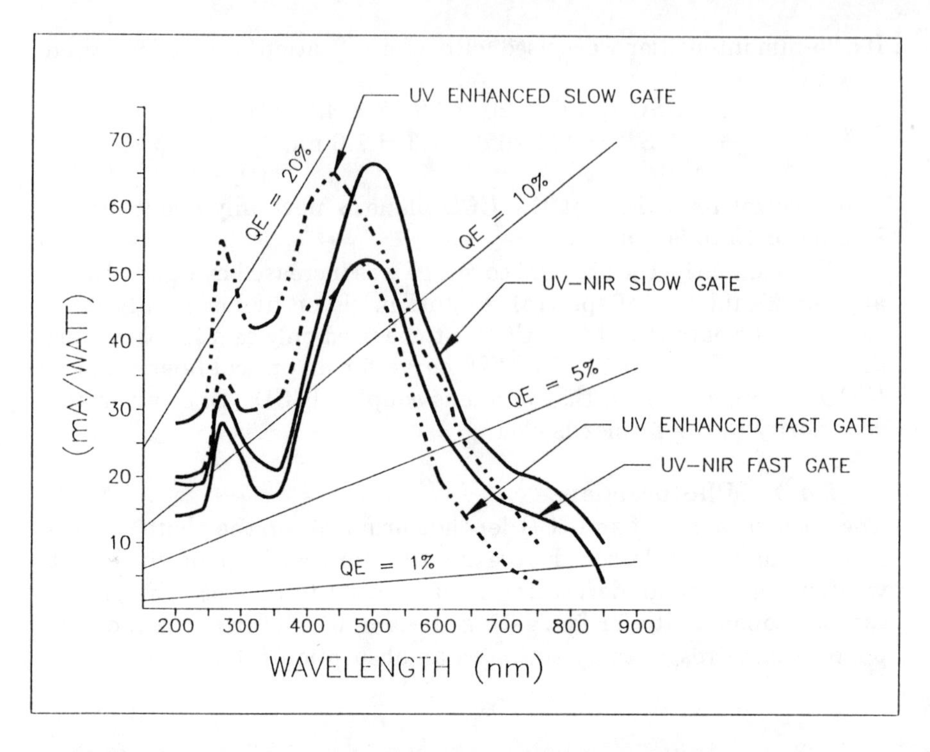

Figure 5.14 Typical response curves for photocathodes designed for the UV-visible and UV-NIR spectral range.

Figure 5.14 shows typical response curves for "slow-gate" (low nickel deposition) and "fast-gate" image intensifiers. (Fast-gate and slow-gate are discussed in section 5.4.6.) To convert responsivity R (mA/watt), to the more familiar QE (quantum efficiency), we use the following equation:

$$QE = \frac{R \times 123.9}{\lambda(nm)} \tag{5.6}$$

MgF_2 windows with bi-alkali photocathodes provide excellent photoresponse in the 120- to 500-nm region, although they are operable to 800 nm (Figure 5.15). ICCD detectors based on such intensifiers require appropriate interfacing to vacuum spectrometers, which ensures operation in vacuum without arcing in the HV section.

As shown for Figure 5.16, ICCDs can also be applied in the vacuum-UV to x-ray spectral regions, 0.1 to 290 nm [13,14], and for detection of electrons, (e.g., ESCA spectrometers) and ions (e.g., mass

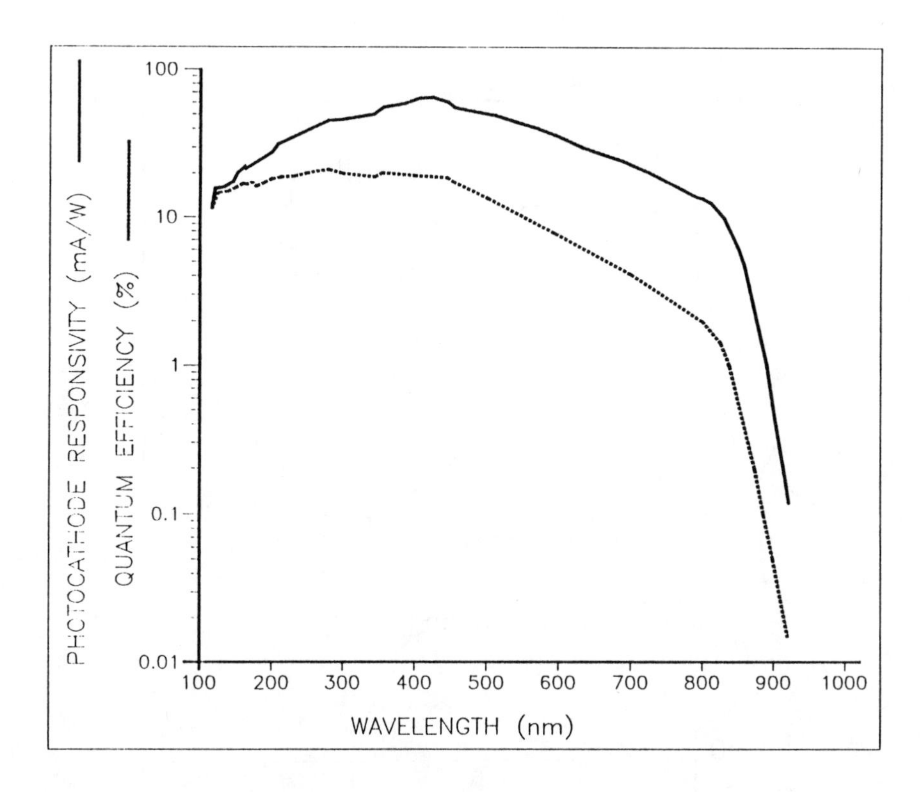

Figure 5.15 A typical response curve for a bi-alkali photocathode deposited on an MgF_2 window.

spectroscopy) [15,16]. Intensifiers used for these applications are windowless and the transducer is coated directly on the MCP. Typical transducers are Au, CuI, and CsI. The intensifier is typically sealed hermetically to a vacuum flange. In operation, the MCP is kept inside the vacuum spectrometer, while the fiber-coupler (with the phosphor deposited on the inside) outer end protrudes to the atmosphere so that it can be coupled to the CCD array. Fiber-to-fiber coupling of the intensifier and the CCD requires the two fiber couplers to be kept parallel to each other. This is essential because the optical coupling between the two cannot be permanent, since the intensifier must be occasionally baked (to 350°C) in place to desorb contaminants deposited over time on the MCP. We have successfully designed and constructed various ICCDs for operation with these high-energy intensifiers, with CCD formats of up to 1152 × 298 pixels.

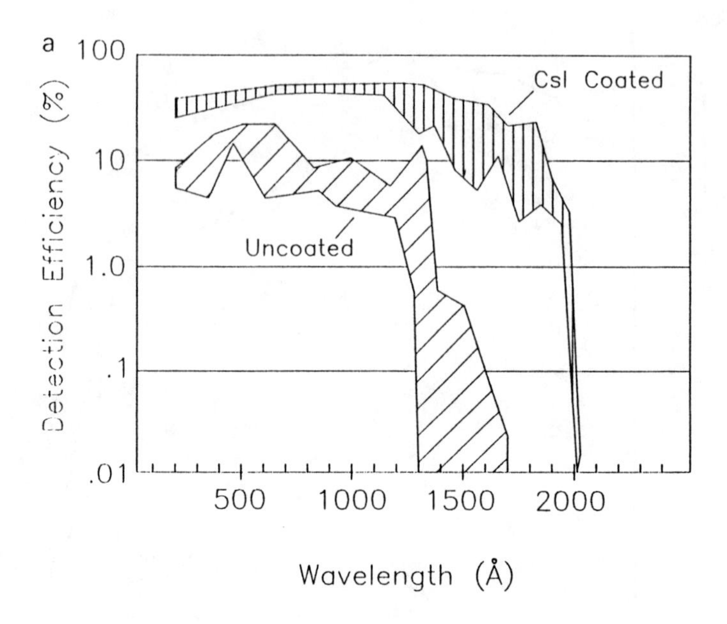

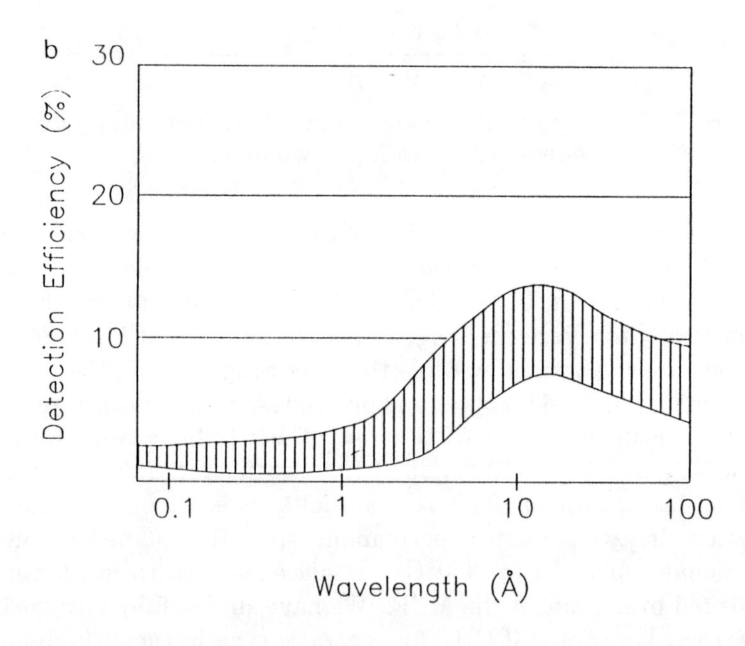

Figure 5.16 Typical response curves for windowless image intensifiers for (a) vacuum UV (VUV), (b) x-ray, and (c) electron detection.

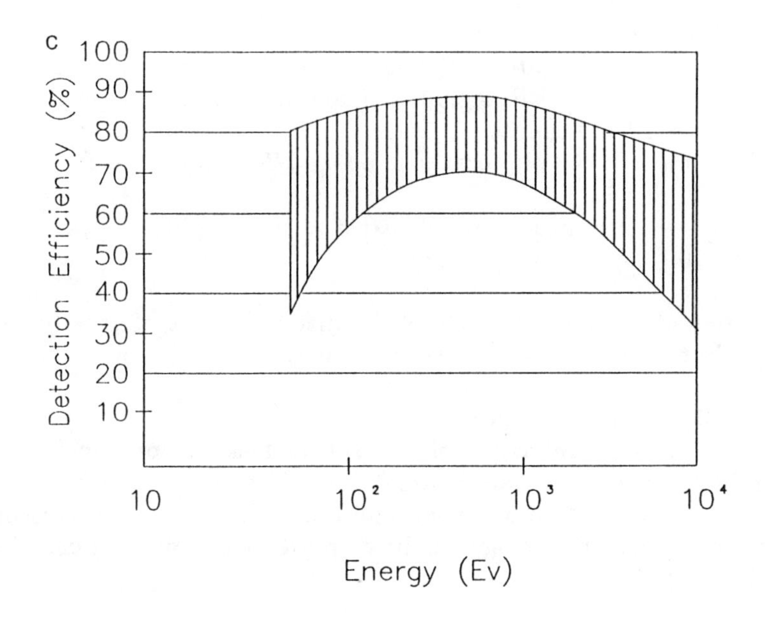

Figure 5.16 *(Continued)*

5.4.4 Photocathode Thermal Emission

Photocathodes have characteristic dark-charge background due to spontaneous thermal emission of photoelectrons. The intensifier industry refers to it as equivalent background illumination or EBI. As expected from their semiconductor energy transfer characteristics, NIR-enhanced photocathodes have significantly higher EBI levels compared with UV-enhanced ones. EBI levels can be reduced by photocathode cooling. A 20°C reduction in temperature reduces EBI by a factor of 10 to 20 for UV-enhanced photocathodes and of 50 to 100 for NIR-enhanced ones (approximately a factor of 0.8/°C).

Reduction in EBI level means that only the number of false events is reduced but not their magnitude, which depends only on the overall gain of the ICCD. EBI is a problem only when the ICCD is used in the CW mode. In the gate mode, EBI levels are negligible because the total time the photocathode is turned on is very short.

Example Assume the application to be laser-induced fluorescence, with the following experimental conditions:

- Super-pixel size = 200 pixels (super pixel = number of CCD pixels grouped or binned together and dedicated to a single wavelength; the term subarray is used elsewhere in this volume).
- EBI = 1 count/pixel/s
- CCD exposure time = 50 s
- Laser repetition rate = 200 pulses/s with 50 ns pulse width

  CW mode: S_{EBI} (EBI signal) = 1 count/pixel/s $\times$ 200 pixels $\times$ 50 s = 10,000 counts

  Gate mode: The photocathode opened 200 times/s, each time for 50 ns.

  S_{EBI} = 1 count/pixel/s $\times$ 200 pixels $\times$ 50 s $\times$ 200 pulses/s $\times$ 50 $\times$ 10^{-9} s/pulse = 0.1 count

Whereas in the CW mode EBI consumed most of the dynamic range, in the gate mode it is clearly negligible.

5.4.5 Dynamic Range

Dynamic range is, unfortunately, one of the least comprehensible and often ill-defined performance parameters. It is rarely described in the literature in terms that are meaningful to multichannel spectroscopy. In this chapter, terminology will be employed that has been described previously [17].

5.4.5.1 Single-Pixel Dynamic Range

This is defined as the ratio of the maximum signal detected by a single pixel to the minimum noise associated with its readout. The units are usually in A/D converter counts for the signal and root mean square (rms) counts for the noise. However, counts must be converted to meaningful units, usually electrons or photoelectrons. Single-pixel dynamic range is limited both by the well capacity of the pixel, typically around 500,000 electrons, and by the current flow through individual microchannels, usually not a limiting factor in the CW operating mode.

5.4.5.2 Super-Pixel Dynamic Range

This is defined as the ratio of the maximum signal detected by a super pixel to the noise associated with its readout. Super pixel is defined as a group of pixels or subarray that are binned together and read out as a single "giant" pixel by the CCD's preamplifier. It is possible to obtain a two- or three-fold improvement in dynamic range because the preamplifier node capacitance is usually higher than that of a single pixel. However, if all pixels in a preselected super pixel are accumulated in memory rather than in hardware, then dynamic range can be, in principle, equal to a single-pixel dynamic range multiplied by the square root of the size of the super-pixel.

5.4.5.3 Intraspectral Dynamic Range
Defined as the ratio of maximum and minimum signals that can be read out for one exposure of the array. Here these signals are read out by different pixels located on different regions of the array. Limitations are usually set by stray light interferences originating from the spectrometer or the lenses and from the detector's own windows as well as from wave-guide interferences between the SiO_2 overcoat and the silicon surfaces of the CCD.

5.4.5.4 Variable Integration Time (VIT) Dynamic Range
If different regions of the spectrum with varying intensities are acquired at different exposure times, a great increase in the overall dynamic range can be accomplished [17,18]. Ideally, the CCD array should allow random-access readout so that each spectral feature could be exposed to different lengths of time, that is, inversely proportional to its intensity. Unfortunately, the only random-access arrays now available are either linear PDAs with readout noise values that are about 300 times higher than those of CCD arrays or charge injection devices (CID) with both significantly higher noise than CCDs by a factor of 10 to 30 and limited commercial availability. Therefore, VIT processes can only be efficiently achieved by taking successive scans, each with different exposure time, to improve localized dynamic range.

ICCDs are operated at overall gains of 1 to 100 counts per photoelectron emitted from the photocathode, where 1 count is typically equal to 8 to 10 electrons in the CCD. With a 500,000-electron CCD pixel well size, it is possible to obtain a single pixel dynamic range of 65,000 (16 bits) if the gain is reduced to 1 photoelectron = 1 count. With pulse signals, the dynamic range is limited to approximately 16,000 due to microchannel saturation. If the gain is set at maximum, 1 photoelectron > 100 counts, the range is reduced to 650 or less. Super-pixel dynamic range can be twice or three times as high, depending on the well capacity of the horizontal shift register and the node capacitance of the on-chip preamplifier.

As previously explained, accumulation of individual pixels within a preset group ("software binning") can also proportionally increase the dynamic range. For example, if each pixel can detect 500,000 electrons, then 100 pixels, read out separately but accumulated in memory as a single pixel, will detect 5×10^7 electrons or 5×10^6 counts (if 1 count $\approx$ 10 CCD electrons). Since with ICCDs readout noise is rarely the limiting factor (see section 5.5), this mode of operation results in a greatly improved SNR performance.

The intraspectral dynamic range of the ICCD is hard to quantify because it also depends on other experimental parameters, for example, stray light in the optics. Typically it is less than 10,000.

5.4.6 Gating Performance

Gating, that is, turning on the intensifier for a preset period of time, is accomplished when a negative pulse of approximately −200 volts magnitude is applied to the photocathode, temporarily making it more negative than the MCP output. Only during the duration of the gate pulse is the ICCD active. The ability of the ICCD to reject light signals during its "off" periods is referred to as the ON/OFF ratio. ON/OFF ratio = "gate-on" sensitivity/"gate-off" sensitivity. It is typically 5×10^6 to 2×10^7 at best. Even though this ratio appears very high, it sets a practical limit on the intensity of the continuous background signal on which the transient signal of interest is superimposed. For example, assume a laser pulse inducing a fluorescence signal superimposed on a high background flame. Also assume that the CCD exposure (readout time) is 100 ms. If the background signal = 10^{11} photons/s = 10^{10} photons/exposure, and if the fluorescence signal intensity is only 100 photons then

$$\frac{\text{Background intensity}}{\text{Fluorescence intensity}} = \frac{10^{10}}{100} = 10^8 > 2 \times 10^7 \qquad (5.7)$$

Because the background/signal intensity ratio exceeds the ON/OFF ratio of the image intensifier, the background signal will interfere with the fluorescence signal. ON/OFF ratios are limited mostly by light transmitted through the semitransparent photocathode, inducing weak electron emissions in the MCP. This is the main reason why ON/OFF ratios are significantly lower in the UV compared with the visible and NIR spectral regions.

When the gate pulse is applied between the photocathode and the MCP input, the outer edge of the photocathode is turned on first. The pulse continues to propagate inwardly, turning on the center of the photocathode last. Conversely, the outer edge is turned off before the center is. This temporal behavior is referred to as the iris effect. The time from the application of the gate pulse to the time the intensifier reaches full gain operation and uniform resolution, that is, full open, is the irising time, which is approximately 2 to 3 ns for high-performance image intensifiers.

Unfortunately, different manufacturers have different definitions for gate time, some of which are quite meaningless for spectroscopic applications. For this discussion, gate time is defined as the time it takes for the intensifier to pass from a half-open to a half-closed status, that is, FWHM. High-performance conventional intensifiers should have gate times of 5 to 6 ns with a minimal iris effect.

Figure 5.17a shows the gate-time record for an image intensifier with a significant iris effect; Figure 5.17b shows the equivalent record for a "fast-gate" intensifier.

In each case only a diametrical 2.5-mm-wide strip of the photocathode was monitored. However, because the photocathode is basically radially symmetrical, this is a good presentation of the gate characteristics of the whole intensifier. The photocathode was exposed to sequential 200-ps pulses from an InGaAs laser, each pulse delayed by 2 ns with respect to the preceding pulse.

Clearly the iris effect is negligible for the "fast-gate" intensifier, whereas it is significant for the "slow-gate" intensifier. Iris effect can greatly deteriorate the accuracy of measurements when the measured transient phenomenon is on the same order or slower than the gate time, for example when measuring relatively slow fluorescence decay times are measured. It is far less important if the duration of the measured phenomenon is much shorter, for example picosecond absorption, as long as the intensifier is fully open after the irising time period and if the signal is monitored only after the irising time period. This can be easily accomplished using an appropriate synchronization timing scheme.

As previously discussed, fast gating requires a correct balance of photocathode constituents and nickel undercoating to improve photocathode conductivity. Additionally, it is necessary to minimize the capacitance between the photocathode and the MCP input, and to ensure a correct interfacing between the gate pulser and the intensifier.

Intensifiers with new designs, for example heavy mesh cathode underlays [19,20] in conjunction with gate pulsers capable of delivering subnanosecond gate pulses of up to 1 kV magnitude, can achieve subnanosecond gate times. Unfortunately, this technology is not yet fully available commercially, at least not for scientists with "nonmilitary" budget constraints.

5.5 Signal-to-Noise Ratio (SNR) Considerations

The usefulness of any detector is determined by its overall SNR performance. As the term signal-to-noise ratio suggests, it is the ratio between the response of the detector to the measured signal and the noise associated with this measurement. The higher the signal response, S, and the lower the noise, N, the better the performance. Ideally, SNR should be limited predominantly by the photon shot noise

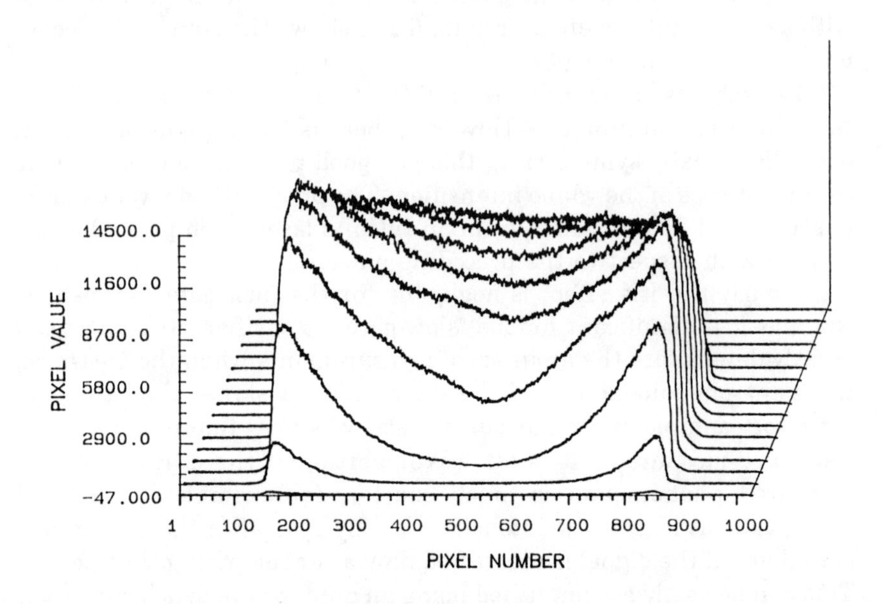

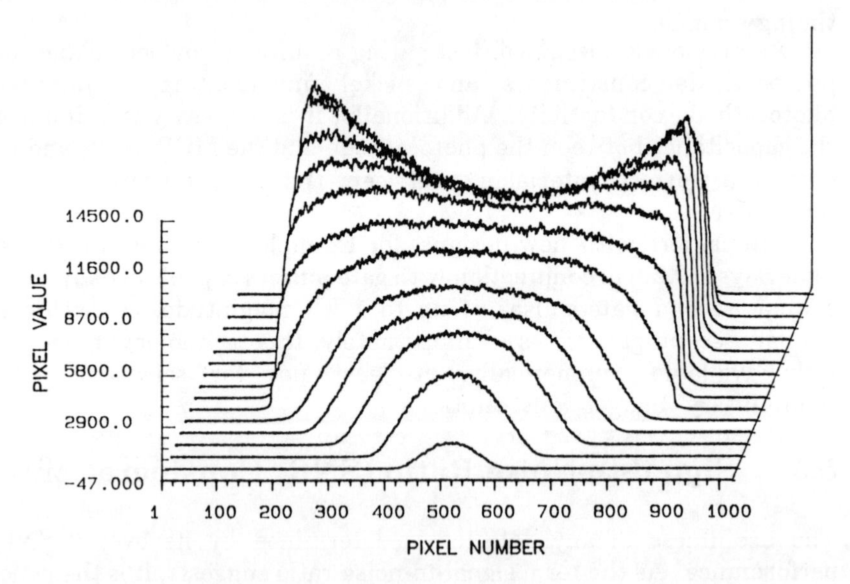

Figure 5.17a Gate-time record, 2-ns intervals. Upper: slow gate, 0 to 34 ns (opening). Lower: slow gate, 34 to 0 ns (closing).

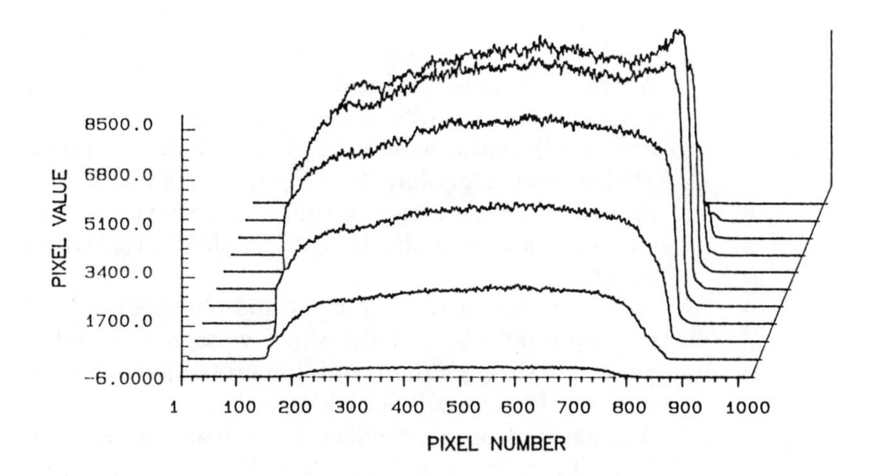

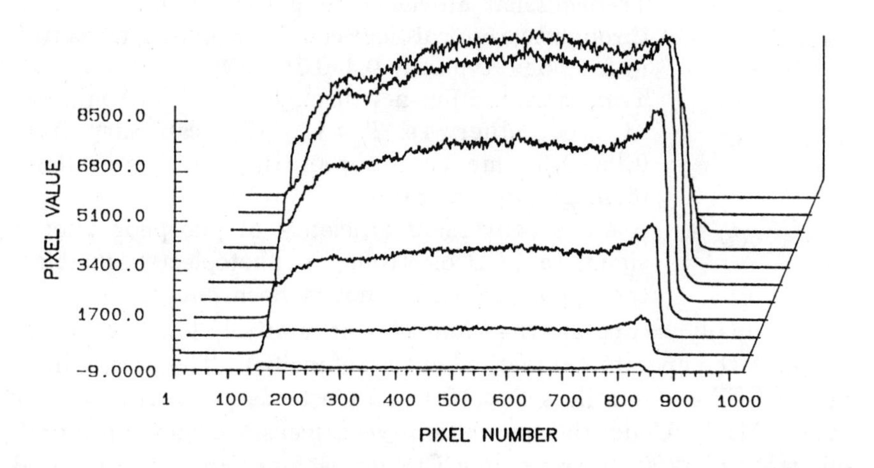

Figure 5.17b Gate-time record, 2-ns intervals, continued. Upper: fast gate, 0 to 20 ns (opening). Lower: fast gate, 20 to 0 ns (closing).

associated with the measured signal. The following section is rather lengthy and will be divided logically into three subsections: Signal Response, Noise, and SNR.

5.5.1 Signal Response, *S*

The signal response, *S*, can be generally defined by the following equation:

$$S = S_{ph} \times QE_{pc} \times t \times G \times \Delta V \times E_p \times T_f \times T_m \times QE_{CCD} \quad (5.8)$$

where

- S_{ph} = Average photon flux incident on the CCD or ICCD photocathode in units of photons/s/pixel
- QE_{pc} = Average QE value across the photocathode in units of photoelectrons/photon, (see Figures 5.14 and 5.15)
- t = Signal integration time in units of seconds
- G = Current gain (amplification) of the MCP, typically 1 to 3000
- ΔV = Difference between the output and phosphor voltage (typically 5-6 kV) and threshold voltage required to penetrate through the aluminum protection phosphor overcoat (typically 3.5-3.7 kV).
- E_p = Average phosphor efficiency for converting electrons to photons, in units of photons/eV (E_p = ~0.015 for P-20 and ~0.004 for rare-earth phosphors)
- T_f = Transmission efficiency of phosphor light emitted through two optical-fiber coupler windows, measured in fractions (typically 0.4–0.5)
- T_m = Transmission efficiency through optical-fiber minifier (if used, otherwise T_m=1) (T_m can vary from 0.05–0.6 mostly depending on reduction [demagnification] ratio)
- QE_{CCD} = Average conversion efficiency of phosphor photon signal to CCD electrons, in photoelectrons/photon (phosphor emission is not monochromatic)

Normally we operate the intensifier at relatively low *G* values to conserve the operational lifetime of the intensifier. The preamplifier of the CCD is typically set at 10 to 15 electrons per A/D converter count (ADU). Under these conditions we can easily achieve an overall efficiency of 1200 electrons (on CCD) for each photoelectron emitted from the photocathode when a P-20 phosphor is used. For rare-earth phosphors, *G* is raised by increasing ΔV by approximately 50 to 60 volts. If we assume that: a P-20 phosphor is used, no minifier is used, QE_{pc} = 0.1, QE_{CCD} = 0.2, and the rest of the parameters are as given above, then an S_{ph} signal of 10 photon/pixel will generate a digital response signal, S > 80 counts/pixel.

As will be shown later, the noise is basically independent of the signal acquisition time at least when operated in the gate mode. Therefore, the SNR performance is practically independent of time. For example, an S_{ph} of 10 photon/s/pixel with t = 1 s will produce a SNR similar to that of an S_{ph} = 1 photon/s/pixel with t = 10 s. It also becomes immediately apparent that the overall gain of the ICCD can be excessive, actually reducing both SNR and dynamic range. Fortunately, the gain can be varied to accommodate experimental requirements.

5.5.2 Noise, *N*

This section will define the various noise sources that contribute to the overall noise characteristics of the ICCD. No attempt has been made here to discuss these parameters in great depth. It is important, however, that the reader understands the origin of these noise terms, their overall effect, and when each of them becomes the dominant noise source.

5.5.2.1 N_{ADC} — A/D Converter Noise

This noise is associated with the quantization (digitization) process of the A/D converter. With properly designed ICCD interface electronics and commercially available high-resolution ADCs, N_{ADC} = 0.5 to 0.7 counts (ADC) rms (root mean square), a rather insignificant noise term. The theoretical value of $12^{-1/2} \approx 0.29$ ADU is not approached by commercially available ADCs.

5.5.2.2 N_R - CCD Readout Noise

The CCD readout noise is a complex parameter and depends on many variables. Basically, N_R depends strongly on the design of the on-chip preamplifier and the signal-shaping processes used. N_R also depends on the scan rate and to a lesser degree on the charge shift mechanism and direction. If the ICCD design is correct, the high-voltage section does not increase N_R.

To a greater degree, the readout noise depends on the thermal (Johnson) noise, a noise resulting from random molecular motions within the CCD and the preamplifier, especially at higher scan rates. At lower scan rates, flicker noise usually predominates. A good ICCD should exhibit N_R = 4 to 12 electrons rms.

A potential contributor is *KTC* noise, which is proportional to the on-chip preamplifier node capacitance. *KTC* noise is proportional to $(KTC)^{1/2}$ where K is the Boltzmann constant, C the capacitance, and T the absolute temperature. As described in Chapter 2, the effect of the

KTC noise can be removed using double correlated sampling, and all scientific CCD systems use this technique.

N_R is reduced as the temperature is lowered. Typically, for small super pixels further reduction in noise is minimal below −60°C to −70°C. It can be more pronounced for large super pixels, as shown in Table 5.2 for a 576 × 384 pixel CCD. Variations of noise as a function of super-pixel size depends also on the CCD's internal electronic design. For example, the 1024 × 256 CCD array made by EEV shows almost no increase in noise with super-pixel size.

5.5.2.3 N_D — CCD Dark Shot Noise

This noise term results from thermally generated electrons in the CCD. This dark charge signal decreases nonlinearly with decreases in temperature. Typically, dark charge decreases by a factor of two for a decrease of temperature of 6°C in the −20°C to −50°C range.

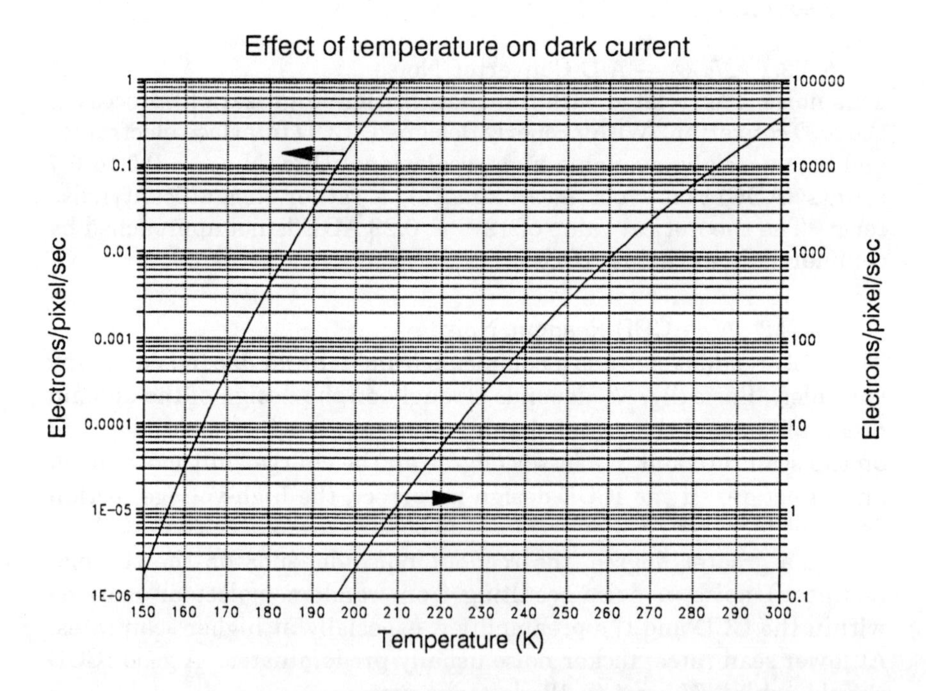

Figure 5.18 Typical dark charge versus temperature characteristics of a CCD (measured for a Tektronix 512 × 512 CCD).

Reduction accelerates as lower temperatures are reached (see Figure 5.18). Dark-charge signals 10- to 300-fold lower (depending on the design) can be achieved with CCDs based on the relatively new multipinned phase technology (MPP, discussed in Chapter 2). Unfortunately, well capacity of individual pixels in MPP-type CCDs can be significantly lower, thereby reducing the dynamic range of the CCD, but this is usually not a problem for spectroscopic applications where relatively large super pixels are used.

Consequently $N_D = (S_D)^{1/2}$, where S_D is the dark charge in electrons. As will be shown later, N_D can become significant and even dominant for large super pixels (spectroscopy) and with relatively long exposure times, especially when operated in the gate mode. MPP-type CCDs are then the device of choice.

5.5.2.4 N_{EBI} — Photocathode Dark Noise

This noise is a result of spontaneous thermal emission of photoelectrons from the photocathode according to equation (5.9).

$$N_{EBI} = (2 \times S_{EBI} \times t)^{1/2} \tag{5.9}$$

where S_{EBI} is the dark-charge signal in photoelectrons per second per pixel. The factor of 2 is due to the exponential distribution of secondary electrons. N_{EBI} is often the dominant noise term in CW work.

5.5.2.5 N_{ph} — Photon Shot and MCP Noise

The measured photon signal has an associated noise that is, at best, root-mean-square proportional to its magnitude. The MCP multiplication process further increases the noise by a factor of $2^{1/2}$, due to the nonideal distribution of secondary electrons produced by the MCP. As mentioned previously, this distribution is exponential rather than saturated. The result is described by equation (5.10).

$$N_{ph} = (2 \times S_{ph} \times QE_{pc} \times t)^{1/2} \tag{5.10}$$

5.5.2.6 CCD Fixed Pattern "Noise"

This term represents spatial rather than temporal noise. It is caused by CCD pixel-to-pixel offset and gain variations. These variations differ from one CCD to another. Fortunately it is easy to remove the fixed pattern response by simply subtracting a background and normalizing. Usually localized pixel-to-pixel variations are on the order of $\pm 5\%$ and as low as $\pm 1\%$ for super pixels.

5.5.2.7 Photocathode Fixed Pattern

Once again, this term usually represents spatial rather than temporal variations. Response variations across the intensifier are usually a result of nonuniformities in the photocathode consistency and thickness and spatial variations during the process of electron scrubbing of the MCP; that is, the more efficient the scrubbing, the lower the resultant MCP gain. High-performance intensifiers should have response variations of less than 10%. In addition to these variations, which are characteristically of low spatial frequency, there are other variations which show far higher spatial-frequency distribution, that is, very "sharp spike" in appearance. Most typical of these are dark (black) and white (hot) blemishes.

Dark blemishes are regions of very low response or no response. Normally these are the result of a one-time internal arcing, incomplete photocathode evaporation, or dead MCP channels.

White blemishes are regions with higher than normal responses. Unfortunately, white blemishes are often "hot spots," that is, regions with responses that vary with time. Hot spots are usually a result of partially broken channels or microscopic debris in the channels. In either case, intense localized electrical fields are created that can induce false signals. Whereas small dark blemishes somewhat deteriorate the quality of the image, hot spots are far more troublesome because it is difficult to compensate for their temporally varying behavior.

Let us now go back to the original problem of response variation. Response variations across the photocathode are a function of both location (spatial) and wavelength. A full digital-processing compensation for these variations is therefore difficult to achieve. Nevertheless, a meaningful compensation can be accomplished by combining "flat-field" and "relative radiometric" corrections. Flat-field correction requires a "perfectly" uniform illumination of the ICCD in order to store a normalizing response image. Because response variations are wavelength dependent, it is best to limit the spectral ranges covered simultaneously. If these ranges are small enough, a series of "flat-field spectra" can be obtained, each with a corresponding spectral notch filter. Later on, spectra obtained at different spectral ranges could be corrected by corresponding "flat-field spectra." Relative radiometric calibration provides wavelength-dependent correction for the specific photocathode used. It also compensates for other wavelength-dependent response variations, that is, grating efficiency, optics, and so on. This correction requires a radiometric light source whose spectra are used to create normalizing arrays for different spectral ranges. To achieve maximum correction for response variations, both flat-field and relative radiometric correction are often used in tandem.

5.5.3 Examples of Noise Performance for an ICCD

The overall noise, N_T, can be described simply as

$$N_T = (N_R^2 + N_D^2 + N_{EBI}^2 + N_{ph}^2)^{1/2}$$

In a series of examples, the effect of noise terms for both CW and gate measurements will be demonstrated.

$\boldsymbol{N_R}$ The readout noise of state-of-the-art CCDs is between 4 and 10 electrons rms for a single pixel. As shown before, it can be higher for large super pixels, up to 40 electrons or more, depending on the temperature and the mode of charge readout, that is, whether binning is performed along the columns or the rows of the CCD. Normally we set the preamplifier gain of the CCD so that 1 count (ADU) = 10 electrons. The readout noise of a well-designed ICCD should be around 1 count rms.

$\boldsymbol{N_D}$ The CCD cannot be cooled below −45°C without risk of damage to the optical cement. At −40°C, dark charge (S_D) is around 10 to 20 electrons/s-pixel. New CCD designs (MPP) can achieve a dark charge of approximately 0.07 electrons/s-pixel at −40°C, but the pixel well size is typically 100,000 to 125,000 electrons instead of 500,000.

Let us now examine the effect of N_D for imaging applications, where typically the super-pixel size = 1 pixel, and for spectroscopy, where the super pixel is typically 200 pixels or more. Assume a 20-s integration time.

Imaging example.
Super-pixel size = 1 pixel.

S_D = 1 pixel × 20e$^-$/s/pixel × 20 s = 400e$^-$ = 40 counts (1 count = 10e$^-$)
N_D = $(400)^{1/2}$ = 20e$^-$ = 2 counts rms

Therefore, for imaging, N_D is still rather small.

Spectroscopy example.
Super-pixel size = 200 pixels.

S_D = 200 pixels × 20e$^-$/s/pixel × 20s = 80000e$^-$ = 8000 counts
N_D = $(80000)^{1/2}$ = 283e$^-$ = 28.3 counts rms

It is obvious that both dark-charge signal and dark-charge noise become very significant. Clearly, for such applications, ICCDs, based on lower dark-charge (MPP-type) CCDs, will be superior.

Next let us observe the effect of N_{EBI} in CW and gate measurements.

CW Measurements. The effect of N_{EBI} is most significant for CW work, because the photocathode is always turned on. N_{EBI} will have a relatively smaller effect for imaging applications, but as shown below, it is most significant for spectroscopic work.

Assume S_{EBI} = 1 count/s/pixel, and recall that typically 1 photoelectron = 80 counts.

S_{EBI} = 1 count/s/pixel × 200 pixels × 20 s = 4000 counts = 50 photoelectrons (pe)
$N_{EBI} = (2 \times 50)^{1/2}$ = 10 pe = 800 counts rms

This translates to approximately 4000 counts of peak-to-peak noise. Undoubtedly this is a very high noise term. To demonstrate how dominant it is, let us assume that we try to measure a rather weak signal of 0.1 count/s-pixel (20 seconds integration time) at 80 counts/photoelectron.

S_{ph} = 0.1 × 200 pixels × 20 s = 400 counts, or 5 pe
$N_{ph} = (2 \times 5)^{1/2}$ = 3.16 pe, or 253 counts rms

In this case we can ignore N_R and N_D since they are relatively small. Therefore,

$$N_T \approx (N_{EBI}^2 + N_{ph}^2)^{1/2} = [(800)^2 + (253)^2]^{1/2} = 839$$

so that

$$N_T \approx N_{EBI}$$

and therefore

$$\frac{S}{N} = \frac{400}{839} \approx 0.5$$

Clearly, N_{EBI} totally dominates the overall noise of the ICCD and significantly deteriorates its SNR performance. It is possible to cool the photocathode to $-15°C$ with a special coolant circulation fixture. It is possible to further reduce the photocathode temperature to $-40°C$ using a closed-loop gas cooler. At $-15°C$ S_{EBI} is reduced by a factor of 10 to 20 for UV-enhanced photocathodes and 40 to 100 for NIR-enhanced photocathodes, approximately a factor of 0.8/°C. If we assume an average reduction factor of 40, then

$$S_{EBI} = \frac{4000}{40} = 100\,\text{counts} = 1.25 \text{ pe}$$

$$N_{EBI} = (2 \times 1.25)^{1/2} = 1.58 \text{ pe} = 127 \text{ counts rms}$$

Now the result is $N_T = (127^2 + 253^2)^{1/2}$ counts rms, obviously dominated by N_{ph} instead of by N_{EBI}. So, cooling the photocathode does help in reducing N_T and could result in an efficient operation in the CW mode.

Gate Measurements. This case was described earlier (see section 5.4.4) where it was shown that N_{EBI} is negligible. Therefore, for gate applications, usually $N_T \approx N_{ph}$; that is, the ICCD behaves as an almost ideal detector whose performance is dominated by the measured-signal noise rather than by its own noise.

5.6 Signal-to-Noise Ratio Case Studies

Now that both "signal acquisition" and "noise" have been analyzed, it is time to add all the different "ingredients" together and determine their overall effect on the resultant SNR performance of the ICCD. To do so we analyze a few general cases that cover most spectrometric applications. Furthermore, for each case, the relative performance of an ICCD will be compared with that of a liquid- nitrogen-cooled CCD (LN/CCD) detector. These comparisons will make it easier for the reader to determine which detector technology is more appropriate for his or her own application. But first we must clarify the differences between SNR performance and detectability.

SNR performance directly relates to the precision of the signal measurement; that is, the higher the SNR the greater the precision. Depending on the definition of precision, a SNR performance of SNR = 100 could translate to a precision of 1% in the measurement of a signal.

Detectability is defined here as the ability to detect a signal regardless of the precision of measurement. These differences will become clearer as we proceed.

5.6.1 Case 1 — Single Event: CW Detection of a Transient Phenomenon

Conditions.

S_{ph} = 100 photons/super pixel (200 pixels size)
Scan time: 100 ms
QE_{pc}: 0.1 @ 500 nm
QE_{CCD}: 0.5 @ 500 nm (back-illuminated CCD)
CCD gain: 10 electrons = 1 count
ICCD gain: 1 pe = 80 counts

LN/CCD.

S_{ph} = 100 photons × 0.5 = 50 e$^-$ = 5 counts
N_R = 10 e$^-$, and $N_D \approx 0$
$N_{ph} = 50^{1/2}$ = 7 e$^-$, so $N_T \approx$ 12 e$^-$ = 1.2 counts

The noise adds in quadrature *before* digitization so that the SNR will be

$$\frac{S}{N} = \frac{50/10}{(50+10^2)^{1/2}/10} \approx 4.1$$

ICCD.

S_{ph} = 100 photons × 0.1 = 10 pe = 800 counts
N_R = 1 count rms and N_{EBI} is negligible (20 counts or 0.25 pe)
$N_{ph} = (2 \times 10)^{1/2}$ = 4.5 pe = 360 counts rms

In this case, $N_T \approx N_{ph}$ and therefore

$$\frac{S}{N} = \frac{100 \times 0.1 \times 80}{(100 \times 0.1 \times 2)^{1/2} \times 80} = \frac{800}{360} \approx 2.2$$

Case 1 is typical of transient spectroscopy where only a single event can be detected at a time. The ICCD can be operated either in the CW or in the gate mode. Note that in this case it was assumed that the CCD has a far superior *QE* compared with the ICCD. This is not necessarily always true. The ICCD can actually have higher *QE* values, especially in the 200-to-450 nm range.

5.6.2 Case 2 — Multiple Events: CW Detection of Multiple Transient Phenomena

This case does not require analysis in detail, because the effect of N_{EBI} which dominates such experiments has been discussed previously (sections 5.4.4, 5.5.2.4). Basically, CW operation under such conditions is not recommended for the ICCD unless the photocathode is sufficiently cold.

5.6.3 Case 3 — Single Event: Gated Detection of a Transient Phenomenon

This case is generally very similar to Case 1, because N_{EBI} will not be significantly lower when the overall scan time is so short. Still, although the calculated SNR performance will not be improved, the actual detectability is far superior compared with CW because the probability of detecting a "false" thermoelectron emitted from the photocathode or a "cosmic" event is negligible. Therefore, all detected events are real. This is not true for the CW case.

5.6.4 Case 4 — Gated Integration of Multiple Transient Events

Very often it is possible to integrate a large number of consecutive transient events on the CCD chip or in memory. This is true for most pulse laser spectroscopic applications, for example fluorescence, Raman, and picosecond absorption. This mode of operation requires a stable sample and laser. For picosecond and femtosecond spectroscopic work, it is better to utilize detectors such as dual-beam devices capable of detecting simultaneously reference and sample spectra. Dual-beam operation compensates for pulse-to-pulse intensity variations.

Conditions.

Pulse laser repetition rate = 100 pulses/s
S_{ph} = 1 photon/pulse, a rather weak signal
Laser (and event) pulse width = 50 ns
Signal integration time: 50 s
Spectral region: UV
$QE_{pc} = QE_{CCD} = 0.2$
Super-pixel size: 200 pixels
S_D: 1 count/s/pixel for ICCD; negligible for LN/CCD
CCD gain: 10 electrons = 1 count, for both detectors
ICCD overall gain: 1 photoelectron = 16 counts (reduced gain to increase dynamic range)

LN/CCD.

S_{ph} = 1 photon/pulse/super-pixel × 100 pulses/s × 50 s × 0.2 = 1000 e$^-$ = 100 counts

$N_{ph} = S_{ph}^{1/2} \approx 32e^- = 3.2$ counts rms.

In this case, $N_T \approx N_{ph}$, so that

$$\frac{S}{N} = \frac{100}{3.2} \approx 32$$

ICCD.

S_{ph} = 1000 pe = 16,000 counts

$N_{ph} = (2 \times 1000)^{1/2} \approx 44.7$ pe $\approx$ 720 counts rms

S_D = 1 count/s/pixel × 200 pixels × 50 s = 10,000 counts = 100,000 e$^-$

$N_D = (100{,}000)^{1/2} \approx 317e^- \approx 32$ counts rms

$S_{EBI} \approx 0$ (section 5.4.4), and therefore $N_{EBI} \approx 0$, whereas N_R = 1 count rms

$N_T \approx N_{ph}$

Therefore

$$\frac{S}{N} = \frac{16{,}000}{715} = 22$$

It is quite obvious that the LN/CCD detector outperforms the ICCD under the conditions set above, despite the enormous gain advantage of the ICCD. Worse yet for the ICCD case, even though N_D is insignificant, S_D is so high that it consumes most of the available A/D dynamic range. Significant improvements in SNR can be achieved by using an MPP-type CCD as previously discussed.

5.6.5 Case 5 — Gated Integration of Multiple Transient Events Superimposed on an Intense Continuous Background

In many spectrometric applications, a very weak scattered signal is superimposed on a very intense background. Examples include:

- Thomson scattering requires detection of very weak laser signals that are superimposed on highly intense plasma emission.
- Fluorescence and Raman signals from intense flames and rocket plumes are typical for various combustion research studies.
- In-vivo fluorescence signals can be measured from tagged biological tissues and can reveal essential information concerning their nature, for example, carcinogenic versus benign. The problem is

that fluorescence signals may occur at the same spectral region as that of the incident laser light scattered from the samples.

- Real-time detection of oil spills from an airplane can be performed in daylight by monitoring the laser-induced fluorescence signals. Again, the background signals are orders of magnitude more intense than the fluorescence signals and must therefore be "removed."

The weak analyte signal must be discriminated from the intense background signal; failing to do so will result in very poor SNR performance. The ICCD can reduce, and in certain cases even eliminate, the effect of background signals through temporal discrimination.

If the analyte signal and the background signal "temporally coexist," that is, cannot be discriminated by sampling at different times, then gating will at least drastically reduce the sampling time of the background signal. This is achieved because scan time of the CCD is very long compared with the transient phenomenon lifetime. Gating the ICCD synchronously with the transient event ensures a very short exposure of the detector to the interfering background signal. For applications where the two signals do not "temporally coexist," the background signal can be "removed" even more efficiently.

An example is fluorescence decay studies where the decay time of the laser-induced fluorescence signals is far slower compared with the duration of the excitation pulse. By synchronizing the gating delay time to ensure that the ICCD is open only after the excitation signal has ceased, it is possible to detect only the desirable fluorescence signal. Consequently, a far superior SNR performance is achieved.

Conditions. All conditions are identical to those described in case 4, except that the signal is superimposed on a continuous background signal with a flux that is 0.05% (in terms of photons/s/pixel) of the analyte signal. Actually, the background signal, S_{BG}, is quite high: 10,000 photons/s/super-pixel; however, the integrated signal reaching the CCD is very small, because the photocathode is turned off at all times except during the transient pulse duration.

LN/CCD.

Total S_{BG} = 10^4 photon/s/pixel × 50 s × 0.2 = 10^5 e^- = 10^4 counts

S_{ph} = 10^3 electrons = 100 counts

N_{SB} = $(10^5)^{1/2}$ = 316 e^- = 31.6 counts rms

N_{ph} = $(10^3)^{1/2}$ = 32 e^- = 3.2 counts rms

Therefore, $N_T = (10^5 + 10^3)^{1/2} \approx 31.6$ counts rms, and

$$\frac{S}{N} = \frac{100}{31.6} = 3.16$$

ICCD.

$S_{BG} = 10^4$ photons/s/pixel $\times$ 50 s $\times$ 50 $\times$ 10^{-9} s/pulse $\times$ 100 pulses/s $\times$ 0.2 = 0.5 pe

S_{ph} = 1000 pe = 16,000 counts

$N_{BG} = (2 \times 0.5)^{1/2} = 1$ pe

$N_T = N_{ph} = (2 \times 1000)^{1/2}$ pe = 44.7 pe = 715 counts

Therefore,

$$\frac{S}{N} = \frac{16{,}000}{715} \approx 22$$

In the presence of large background the ICCD clearly can detect signals with a reasonably good SNR performance, whereas the LN/CCD failed miserably. In some situations, background intensity levels are actually orders of magnitude higher than given here, further emphasizing the relative advantage of the ICCD. The relative advantage of ICCD detectors is even higher when lower dark-charge CCDs are used and/or when they are utilized for imaging applications where single pixels are used, and therefore N_D is negligible.

5.6.6 Case 6 — "Software Binning": Accumulation of Multiple Spectra in Memory

For low-light-level signal detection with CCD detectors it is advantageous to electronically bin (group) the pixels oriented in the spectrometer's slit-height direction. Binning is accomplished by dumping the charge from all the pixels included in the "bin" (or super pixel) into the on-chip preamplifier capacitor before reading them. Consequently, the integrated charge is read out only once instead of n times (bin = n pixels). Where the SNR is limited by N_R, the signal becomes n times larger whereas the noise remains constant, and therefore the SNR improves by a factor of n. For example, if the bin = 100 pixels, the SNR will improve 100-fold. Additionally, it is also possible to bin horizontally, that is, in the wavelength direction, in which case the charge is first dumped and integrated on the shift register and then read out. In fact, simultaneous binning in both directions is also possible. As previously discussed (noise section), the actual readout noise, N_R, does not remain constant as the size of the super pixel increases, especially at moderate temperatures. It actually increases slightly with the size of the super pixel. Still, binning is the most important factor contributing to the excellent SNR performance of CCD detectors.

Although "electronic hardware binning" is advantageous for low-light-level detection, it is rather restrictive for medium-light-level signals and becomes highly disadvantageous when the CCD is used to detect high-light-level signals. As you recall, the well capacity of individual sensor pixels is about 250,000 to 500,000 electrons. The well capacity of a shift register pixel is usually 1 to 1.5 $\times$ 10^6 electrons and that of the preamplifier node capacitance is typically 6 $\times$ 10^5 to 1.5 $\times$ 10^6 electrons (it is lower for lower noise CCDs and is usually below 1 $\times$ 10^6 electrons). Consequently, for vertical binning (along the slit), the maximum allowable signal per sensor pixel is at most equal to 1 $\times$ $10^6/n$, where n is the number of binned pixels. It is similar for horizontal binning along the wavelength axis.

For example, if the bin = 200 pixels, then the maximum allowable signal per pixel is only 5000 electrons, or 1% of its well size. For signals that are larger than 100 electrons, $N_T \approx N_{ph}$. Because $N_{ph} = (S_{ph})^{1/2}$ and SNR = S_{ph}/N_{ph}, it is obvious that a larger S_{ph} results in better SNR performance. Therefore, if the signal magnitude is higher than 5000 electrons/pixel (for the previous example) binning becomes disadvantageous because it restricts the magnitude of S_{ph}. This disadvantage becomes worse for ICCD detectors because N_R is always negligible (no advantage to binning) and because the gain is so high (1 photoelectron $\geq$ 800 CCD electrons). Actually binning (assuming n = 200 pixels) restricts S_{ph} to approximately 6 photoelectrons per pixel.

At this point we turn from hardware binning to software binning. Software binning combines the charges from all the individual pixels in the bin in computer memory. Now as each pixel can store up to 5 $\times$ 10^5 electrons, the maximum stored signal (S_{ph}) for the entire bin is 1 $\times$ 10^5 $\times$ n (n = 200 in the above example) instead of only 1.5 $\times$ 10^6 electrons with "hard binning" (the preamplifier storage capacity). So, for the above example of n = 200 pixels and assuming QE = 0.2 for both detectors and the ICCD operated in the gate mode, the following SNR performances can be expected.

LN/CCD.

Hardware Binning.

Maximum $S_{ph} = 7.5 \times 10^6$ photons $= 1.5 \times 10^6$ e$^-$ $= 1.5 \times 10^5$ counts

If $N_T = N_{ph}$ (best case), then $N_T = (1.5 \times 10^6)^{1/2} \approx 122.5$ counts rms and

$$\frac{S}{N} = \frac{1.5 \times 10^5}{122.5} \approx 1225$$

Incidentally, unless dynamic range can be sacrificed (e.g., with lower preamplifier gain) this SNR can be achieved only with 18-bit ADC performance.

Software Binning.
Maximum S_{ph} = 200 × 5 × 10^5 e^-/pixel/0.2 = 5 × 10^8 photons = 1 × $10^8 e^-$ = 1 × 10^7 counts
Again, $N_T \approx N_{ph}$ and therefore $N_T = (1 \times 10^8)^{1/2} = 10{,}000e^-$ = 1000 counts; consequently,

$$\frac{S}{N} = \frac{1 \times 10^8}{(1 \times 10^8)^{1/2}} = 10{,}000$$

ICCD.

Hardware Binning.
Again, assuming 18-bit ADC performance (which is very difficult to achieve with ICCD detectors), the maximum S_{ph} is computed as follows:

$$S_{ph} = 1.5 \times 10^6 \text{ electrons} = 1.5 \times 10^5 \text{ counts.}$$

If 1 pe = 16 counts, then S_{ph} = 9375 pe.
If $N_T = N_{ph} = (2 \times 9375)^{1/2}$ = 137 pe = 2192 counts,

$$\frac{S}{N} = \frac{150{,}000}{2192} \approx 68$$

Even if 1 pe = 1 count (so that 16 times more photons can be collected), or $S_{ph} = 1.5 \times 10^5$ counts, $N_T = N_{ph} = (150{,}000 \times 2)^{1/2}$ ≈ 548 so that

$$\frac{S}{N} = \frac{150{,}000}{548} \approx 387 (only)$$

Software Binning. The maximum signal in CCD electrons is S_{ph} = 1 × 10^8 electrons = 1 × 10^7 counts = 6.25 × 10^5 pe, where 1 pe = 16 counts and 1 count = 10 e^- in the CCD. Then
$N_T \approx N_{ph} = 2 \times 6.25 \times 10^5$ pe = 1790

Therefore,

$$\frac{S}{N} = \frac{1 \times 10^7}{17{,}890} = 559$$

If 1 pe = 1 count, again to allow detection of a larger photon signal, then

$S_{ph} = 1 \times 10^7$ counts $= 1 \times 10^7$ pe, and

$$\frac{S}{N} = \frac{1 \times 10^7}{4472} = 2236$$

Whereas both detectors clearly show a significant improvement in their SNR performance when "software binning" is used, the LN/CCD performs better even when 1 pe = 1 count, a nonpractical setting. The reason for this difference is obviously the higher gain of the ICCD, which allows better detectability but reduces the upper end of the dynamic range and hence reduces the maximum SNR.

5.7 Photon Counting with Intensified CCD Detectors

If an ICCD detector has efficient intensifier-to-CCD coupling and a cooled, low-readout-noise CCD, then the single photoelectron signal will exceed the peak-to-peak CCD readout noise. Under these conditions photon counting operation is possible.

Photon counting advantages:

- Total elimination of CCD readout noise from the spectrum, resulting in substantial improvement of the visual impression of the spectrum, as well as improvement of the quantitative results.
- Elimination of the noise component arising from the broad pulse height distribution of the intensifier. The small percentage of photocathode events that generate large CCD charges are weighted equally with other counted events. This applies equally to photoelectrons, thermally induced electrons (EBI), and ion feedback. Not only does the signal have a narrower probability distribution, but the noise sources also have a narrower probability distribution; consequently, small signals are more easily discriminated from the noise.
- Recovery of most of the resolution lost in the intensifier. The point spread function of the intensifier is typically 75 to 85 μm/FWHM. Individual centroid finding of the detected photons

locates their centers to better than 10% of the FWHM. Note that this cannot be done after an image has been accumulated in the normal fashion, because after the signal from multiple photons has been allowed to overlap, the center of each photon can no longer be determined.

Photon counting disadvantages:

- A percentage of photoelectrons are below threshold and are lost completely. This problem is severe in noisy CCD systems (e.g., uncooled and/or video rate cameras) and on cameras with lossy intensifier-to-CCD coupling (e.g., lens-coupled ICCDs). For photon counting, a cooled CCD should be used, fiber-optically coupled to an image intensifier. In this type of camera, up to 93% of photoelectrons can be detected.
- The light flux is limited to levels below 1 photoelectron per pixel CCD readout. Note that this is the limitation on the brightest portion of a spectrum (unless some portions of a spectrum are allowed to become nonlinear).
- The circuitry and/or software required to implement photon counting is more expensive than that used for conventional analog or digital readout of an ICCD.

5.7.1 Principles of Operation

Three steps are usually used in photon counting.

Step 1. CCD Readout and Signal Thresholding.
The spectrum is repeatedly read out from the CCD and compared with a threshold set just above the peak-to-peak CCD readout noise. Pixels that are above the threshold represent detected photoelectrons. This process produces a binary image (each pixel value = 0 or 1). Thresholding can be done by analog or digital circuits or by software. For optimum performance, the threshold level should be set adaptively, based on the distribution of CCD readout noise and a predefined "false alarm" rate. An adaptive threshold will track small deviations in the DC offset of the system (e.g., dark-current fluctuations, analog offset drifts) and any variations in the noise distribution. The noise distribution can be measured once, before readings start, by digitizing spectra with the photocathode turned off. Alternatively, it can be deduced from analysis of the histogram of pixels on each spectrum acquired.

Step 2. Reducing Each Event to One Count, and Centroid Finding.
After thresholding, individual photoelectrons will appear in the spectrum as small contiguous regions of pixels with value = 1. Photoelectrons that were highly amplified and that resulted in a large signal will result in larger regions. Photoelectrons whose signal was

just barely above threshold may result in regions with just one pixel. Each of these regions needs to be reduced to just one pixel. This is done by software that locates all the adjoining pixels above threshold and finds their center. Then the group of pixels is replaced by just one at that center location. This can be done most accurately by referring back to the original (prethresholding) digital spectrum. The center of mass is located in it after the photoelectron is initially located in the thresholded spectrum. This is called centroid finding. Because the center of mass of the signal peak caused by a photoelectron is close to the location of the photoelectron, centroid finding recovers much of the image resolution lost in the intensifier. With centroid finding, it is even possible to achieve resolution below the level of a single CCD pixel. Experimentally, resolution of about 1/10 pixel has been achieved.

Step 3. Spectrum accumulation from multiple readouts.
Once each photoelectron is represented by just one count in a single pixel, the individual spectra are accumulated to build a spectrum with a wider dynamic range. If resolution greater than the level of individual CCD pixels is desired, then photon counts must be accumulated in two-dimensional arrays in memory that have correspondingly higher resolution than the CCD. Note that for spectroscopy, if the CCD is read out two-dimensionally (so that more individual photoelectrons can be discriminated per readout), only the wavelength coordinate of each photon needs to be kept, and photon-counting spectra can be accumulated in one-dimensional arrays. If multistripe spectroscopy is done (e.g., dual beam), then the coordinate perpendicular to the wavelength axis can be used to determine to which of several one-dimensional arrays to assign a given photon count. If two-dimensional readout of the CCD is used (e.g., imaging spectroscopy) many photons can be detected at a given wavelength on each CCD readout.

5.7.2 Single versus Multiple MCP Intensifiers

Before cooled CCD cameras were available with fiber optic coupling to image intensifiers, the discrimination of individual photoelectrons from camera-readout noise was more difficult. One solution to this problem was to use double microchannel plate (DMCP) image intensifier tubes. DMCPs have typically 100 times more gain, so they can overcome and readout noise of even uncooled video rate cameras. In addition, they can be operated so that individual photoelectrons will temporarily and locally saturate the second MCP, limiting the intensity of the signal at the CCD and producing a narrower pulse-height distribution. Whereas the pulse-height distribution from a single MCP image intensifier is exponential, the distribution from a camera with a double MCP intensifier can be bimodal, with the lower peak resulting from CCD read noise and the upper peak due to photoelectrons. If the two peaks are separable, then photoelectrons can be discriminated more easily from the noise.

The double MCP approach has some residual problems with ion feedback, even though it is reduced by the use of the "chevron" configuration. This is addressed by some intensifier manufacturers by offering triple MCP intensifiers. These intensifiers can be saturated by individual photoelectrons (thus achieving a narrower pulse-height distribution) with less voltage across each of the stages. In addition, the angle between successive microchannel plates, which forms the "chevron" of a double microchannel plate, now forms a "z-stack." This additional angle also helps suppress ion feedback.

The additional gain and narrower pulse height distribution of multiple MCP intensifiers were an advantage for discrimination of individual photoelectrons from CCD readout noise before cooled slow scan CCD cameras were fiber optically coupled to image intensifiers. However, the advantage of multiple MCP intensifiers has been diminished because a cooled, fiber-coupled single MCP ICCD can detect 93% of the photoelectrons. Consequently, heroic efforts to recover the last 7% are hard to justify.

Single MCP intensifiers are available (on average) with higher QE than those with multiple MCPs because single MCP intensifiers are used with many types of systems, including those where only time discrimination is needed. Thus, when there is a manufacturing yield spread (typically 2:1) in QE of intensifiers, the lower QE units can be used for these applications and the higher QE units can be saved for applications requiring the highest QE (e.g., photon counting). With multiple MCP intensifiers, there is not market for lower QE units; thus the overall average is what camera manufacturers must use. This difference in QE more than makes up for the slightly higher percentage of detectable photoelectrons that can be achieved.

When using an ICCD for CW work (which is usually where photon counting is applied), EBI is the dominant noise source. Cooling the photocathode dramatically reduces the EBI level of an intensifier. In principle, this advantage should apply to multiple as well as single MCP intensifiers. In practice, in our laboratory we find that cooling multiple MCP intensifiers can make them unstable. We are not certain of the reason for this problem, but it may have to do with mechanical changes in the MCP stack when the temperature is reduced. As a result of this problem, we have not been able to develop cameras using these intensifiers with cooled photocathodes. This puts multiple MCP intensifiers at a further disadvantage compared with single MCP intensifiers, when used for photon counting. For all these reasons, double and triple MCP-intensified CCD cameras will in general yield lower performance than single MCP versions.

The final limitation on multiple MCP intensifiers is their cost. Because single MCP intensifiers are made in high volume for military night-vision applications, but multiple MCP intensifiers are not, double MCP intensifiers are currently priced about 2.5 times higher than single MCP intensifiers. Triple MCP intensifiers are only marginally commercially available, and are 8 to 10 times the price.

5.7.3 Photon Counting Statistical Issues

5.7.3.1 SNR Improvement

The total noise of an ICCD camera without photon counting is the sum (in quadrature) of the CCD readout and dark noises, plus the photon and pulse-height distribution statistics of the intensifier. The thresholding inherent in the photon-counting method effectively eliminates the CCD readout and dark noise (if the photocathode is cooled). Thus, the photon shot noise and pulse-height distribution noise remain as the dominant noise sources. Photon counting can eliminate the latter but not the former. This reduces the noise by a factor of $2^{1/2}$.

5.7.3.2 Minimum Detectable Signal

A major advantage of photon counting occurs at the lower limit of detection, where the number of events is very low: EBI < 1 event/pixel readout. Although the number of EBI events is low, the amplitude of an individual event can still be very high. Thus if a threshold is set above the EBI, it must be quite high. For a 250,000-pixel ICCD and a typical false alarm probability, the threshold is about 18 equivalent photoelectrons. With use of photon counting, all EBI events are reduced to the same amplitude, and the threshold can be set as low as 4.

On a typical ICCD with photocathode cooled to −20°C and with CCD pixel size 22 × 22 μm, the EBI rate is typically 0.005 event/pixel/s. For exposures < 200 s, the EBI will average below 1 event/pixel/s and the above analysis is appropriate. Above this exposure time, the statistics will make a transition to the analysis in the limit of large numbers of events per pixel (presented in the prior section) and the advantage will be less. Note that if binning is used (particularly for spectroscopy where binning ratios of 100 or more are common), this transition exposure time will become proportionately shorter. If the photocathode is not cooled, the EBI rate will be 10 to 50 times higher and the transition time will be proportionately shorter. If EBI can be further reduced, the transition exposure time will become proportionately longer. EBI reduction is currently limited by the lower operating temperature limit of commercial intensifiers (−20°C) and by the characteristics of the photocathodes.

If EBI is effectively eliminated (by gating), then the statistics of the EBI will no longer be the limit of detection. In that case, the probability of detection will be simply given by the chance of a photoelectron being detected at all. Photon counting does not affect this. Photon counting will still improve the SNR of signals that are detected, as discussed above, but it will not improve the probability of detection itself.

5.7.4 Linearity

Photon counting as described above assumes that either one or zero photoelectrons will be detected at a given pixel per CCD readout. As the average number per readout approaches one, a phenomenon similar to pulse pileup on single channel photon counters occurs, causing the response of the system to begin to be nonlinear. On the basis simply of single pixel statistics, nonlinearity reaches 1% at an average flux rate of approximately 0.2 photoelectron pixel per CCD readout time.

However, the intensity from a single photoelectron can be spread over typically 10 to 25 pixels (on a CCD with 25×25 μm pixels). To avoid spatial overlap of photoelectron responses, the expected number of photoelectrons per pixel per CCD readout should be $< 1/20$). This ratio can be improved some by better peak-separation algorithms that can allow at least some overlap, but as the complexity of these algorithms goes up, the frame rate that can be achieved is reduced. This partially offsets the benefit of the better algorithm because longer exposure times mean more photons per pixel per CCD readout.

The nonlinearity due to multiple photons being interpreted as one event can be modeled (either analytically or by Monte Carlo methods) or it can be experimentally measured. In either case, the result can be inverted to provide a linearity correction. This will have its limits too, but it can be used to slightly extend the photon flux range of useful operation.

5.7.5 Effective Elimination of Cosmic Ray Spikes

The detection rate of the cosmic rays varies considerably but is about 0.2/cm^2/s. Thus in a few tens of seconds of exposure time, these events will be quite common. The amplitude of these events can be extremely high, typically hundreds to even thousands of times the CCD readout noise. Thus if an image is being autoscaled on the basis of its minimum and maximum pixel values, the scaling can be totally inappropriate. This is true on intensified CCD cameras when light levels are very low. Photon counting will effectively eliminate this problem because cosmic ray events will be counted just as any others. Thus their large magnitudes will not give them any undue influence. The flux rate of cosmic ray events is at least two orders of magnitude lower than the rate of EBI events, even with a cooled photocathode. Thus, once cosmic ray events are counted the same as EBI and photoelectron events, they will become insignificant.

5.7.6 Advantages for the Detection of Higher Energy Photons

In the discussion provided above of pulse-height distribution, we have focused on the amplitude distribution due to nonuniform amplification of electrons from the photocathode in the microchannel plate. There is another contribution to pulse-height broadening when the number of photoelectrons per photon can increase beyond one. For multi-alkalai photocathodes and wavelengths below about 290 nm, it is possible to obtain more than one electron per photon. For x-ray intensifiers with phosphor converters or scintillating fiber-optic input windows, the number of visible photons per x-ray photon can be substantial. This can lead to multiple photoelectrons per x-ray photon. Because photon counting treats all events the same independent of their amplitude, this source of signal variation will also be suppressed. In these cases, without photon counting, the standard deviation of a signal due to the pulse-height distribution width can exceed that due to the original photon statistics. In these cases, the SNR improvement at large signal levels from photon counting can be more than 30% higher than calculated above. In cases where the number of photoelectrons per photon is large, it may even be possible to discriminate photon events from thermal (EBI) events.

5.7.7 Implementation Issues

5.7.7.1 Maximum Allowable Flux Rate and Required Exposure Time

One limit of photon counting is the maximum photon-flux rate that can be handled without deviations from linearity. As discussed above, this is about 0.05 photoelectron per pixel per CCD readout. Thus many readouts will be required to accumulate a significant image. For the brightest area in an image to accumulate 10 photoelectrons over the course of a series of exposures, there should be at least 500 exposures analyzed and accumulated. Even at video readout rates (25–30 frames/s) this is 15 to 20 s. For low-light-level imaging, exposure times are often long, but this example illustrates the minimum total exposure time that can be used with photon counting if linearity is to be preserved. Unless they can also be used as conventional integrating ICCD cameras, the minimum total exposure time of photon-counting systems reduces their flexibility. This is another advantage of cooled CCD cameras with cooled photocathode, single MCP intensifiers. The cooled CCD has a large dynamic range that is not filled as quickly by the excess gain of a double MCP intensifier. Photon-counting systems based on CCD cameras that operate at video rate only do not have the flexibility to perform the desired integration. Unless they provide external frame accumulation hardware, they will be limited in exposure time flexibility.

In spectroscopy, the minimum useful total exposure times can be shorter, because signal will be binned along the spectral line. If 100-pixel-high super pixels are formed after photons have been individually located and their centroids determined, a maximum of 0.05 photoelectron per pixel per readout will translate into signal accumulating at up to 5 photoelectrons per super pixel per CCD readout. To achieve 10 photoelectrons per super pixel would require only two readouts. If more pixels can be combined vertically into one super pixel, the rate can be even higher.

Note that even without any signal, the EBI will limit the maximum exposure time that can be used in photon counting. With a cooled photocathode and a CCD with 22 $\times$ 22 μm pixels, the EBI rate is about 0.005 event/pixel/s. If we want to limit the number of events per pixel per readout of the CCD to 0.02, this would limit the individual exposure time to four s. As the signal level is added, the frame rate must increase to keep the number of events per pixel per CCD readout below the limits discussed above for non-linearity.

5.7.7.2 Binning the CCD during Readout

In ICCD cameras where an individual photoelectron may span more than two pixels in each direction, binning the CCD array may improve performance of the system. This will happen in two ways. First, it will concentrate the signal from each photoelectron into fewer pixels, increasing the probability of exceeding the CCD readout noise. Second, it will reduce the number of data points to be processed per CCD readout, allowing faster frame rates, which will accommodate higher signals without nonlinearity. While this approach may reduce the ultimate resolution that can be achieved by centroid finding, this trade-off may be acceptable in some applications. In spectroscopy, it may be useful to bin the CCD slightly (e.g., 2:1 parallel to the spectral lines). This will provide an additional speed advantage without giving up resolution in the wavelength direction where it is desired. With new 25-mm format CCD arrays, where the shift register is parallel to the wavelength axis, binning of all pixels along the slit (column) and binning by two along the wavelength axis (rows) could provide scan rates of 100 to 200 spectra/s.

5.7.7.3 Light Collection Duty Cycle

It is clear from the discussion above that high frame rates will allow linearity in photon counting at higher light levels (these higher light levels are still extremely low — below 1 photoelectron/pixel/s even for video readout rates). For most scientific CCD systems there is a conflict with light collection duty cycle. This is because the intensifier must be turned off during CCD readout to prevent smearing. As the frame rate gets higher for a given pixel rate the fraction of the time that the intensifier must be turned off increases. On a scientific CCD camera with 1 MHz pixel rate and 576 × 384 pixels, each readout takes about 250 ms. Thus a frame rate of two per second (500 ms per frame total) reduces the light collection duty cycle to 50%. Since the whole goal of a photon-counting system is better performance in low light levels, this low-duty cycle is a problem.

There are at least three solutions to this problem. In the first, a frame transfer CCD is used to allow readout and exposure to occur simultaneously and, other than the very short charge shifting period, the intensifier can be on all the time. A similar solution can be developed using an interline CCD. In this case, because half of the area of the array is covered, the signal per photoelectron is lower and a few more photoelectrons will fail to be discriminated above the CCD readout noise.

For gated work based on low-repetition-rate pulsed laser, it may be possible to read out the whole CCD between laser pulses. Thus no exposure time is lost. For the case of a 576 × 384 CCD readout at 1 MHz (given above), 2 × 2 binning could be used to bring the readout time below 100 ms (the time between pulses on many lasers). The advantages and disadvantages of binning have been discussed above.

5.7.8 Photon-Counting Summary and Discussion

If an ICCD detector is already being used in a laboratory for its gating or low minimum detectable signal, then photon counting is a useful technique to add. It is useful for CW or gated work where the light levels are very low. Photon counting can improve the resolution of an ICCD system by almost an order of magnitude, and subpixel resolution has been achieved.

The SNR improvement due to photon counting is < 40%, but the minimum detectable signal (MDS) level can be lowered substantially. This MDS improvement is largest in moderate exposure time work where EBI is low but not zero. For gated work, where EBI is zero, MDS is not improved. For very long exposures where EBI becomes substantial, the MDS improvement due to photon counting is again reduced. For work with moderate exposure times (typically up to 200 s in imaging applications), photon counting makes a low-MDS detector even better.

Photon counting is not appropriate for all experiments. For low-light-level CW work in which reduced photon shot noise is more important than minimum detectable signal, the higher QE of back-illuminated CCD cameras gives them an advantage.

Nonlinearity limits the average number of photoelectrons that should be detected by a photon-counting system per pixel per CCD readout to about 0.05 for 1% nonlinearity. Thus substantial numbers of frames must be processed and accumulated to acquire an image of even minimum total exposure time per useful frame.

The range of total exposure time for which photon counting substantially reduces the MDS is limited. The minimum total exposure must include enough frames to get an image with a useful signal level (many frames are required even for a signal as small as 10 photoelectrons per pixel!). On the high end, the accumulation of EBI limits the MDS improvement to 30%. For slow scan systems (frame rate < 1/s), these two limits are similar and there is no exposure time for which MDS is greatly improved while a minimum SNR level is also achieved. Video-rate camera systems, if built with extremely low noise to keep the percentage of photoelectrons detected high, hold more promise. They must also be coupled with cooled photocathode intensifiers or EBI will be prohibitive. Such a system has not yet been offered commercially.

In imaging photon counting, the large minimum total exposure time can be prohibitive for many applications. For flexibility in a camera system, nonphoton-counting operation should also be possible.

By using binning techniques, photon counting can achieve useful results for spectroscopic detection, though the problems are similar to those discussed above for imaging. Photon counting may also have broader applicability to spectroscopy then imaging, because it can achieve useful signal levels in much lower total exposure times. Many spectroscopic applications can benefit from the resolution improvement offered by photon counting, independent of the MDS level improvement.

Better photon-counting systems can be anticipated in the future. Recent advances in CCD amplifier design have substantially reduced the readout noise of CCD arrays at high readout rates. This promises to make the design of high-performance photon-counting systems with faster frame rates possible for the first time. It is also possible that intensifiers that can be cooled to lower temperatures will become available. If this happens, EBI is expected to decline, and photon-counting systems will be able to improve MDS levels at longer total exposure times. If both these advances occur, the applicability of photon counting systems can be significantly widened. Competing against this advance will be the improving performance of cooled CCD arrays. As their readout noise continues to decline, their MDS levels drop, and the detection advantage of intensified systems is diminished.

5.8 Summary of the Relative Advantages of Each Detector

CCD.

1. Wider simultaneous wavelength coverage at a significantly lower cost
2. Longer lifetime and excellent resilience to accidental overexposure to light
3. Excellent visible-to-NIR (to 1 μm) spectral response, and recently, good UV response
4. Excellent spectral/spatial resolution
5. Better SNR performance for medium- to high-light-level signals in the CW mode
6. Extremely low readout noise and dark-charge noise

ICCD.

1. More stable response in the UV, because there is no need for coating with unstable chemical transducers (New developments of UV antireflection coatings of thinned back-illuminated CCD arrays are reducing this advantage)
2. Usable in the vacuum-UV to x-ray range
3. Usable for electron detection (e.g., electron spectroscopy for chemical analysis) and ion detection (e.g., mass spectrometry)
4. Single photoelectron detection of transient phenomena
5. Detection of weak transient signals that are superimposed on intense background signals
6. Provides temporal information, for example, decay times of luminescence spectra
7. Minimal interferences from cosmic events
8. Continuous variable gain adjustment
9. Excellent detection of single photoelectron per pixel signals when longer integration times are feasible

5.9 ICCD Applications

The following section presents two examples, out of many, that illustrate the superiority of ICCDs compared with other detectors for specific situations.

5.9.1 Spectroscopy Studies of Subpicosecond Dissociation Dynamics

A UV-enhanced ICCD (Princeton Instruments, Inc., model ICCD-576B/G) detector was used in the gate mode to monitor the Raman spectra obtained with a Nd:YAG pulsed laser. Figure 5.19 is the resonance Raman spectrum of the CD3 [1100]–[0100] hot band

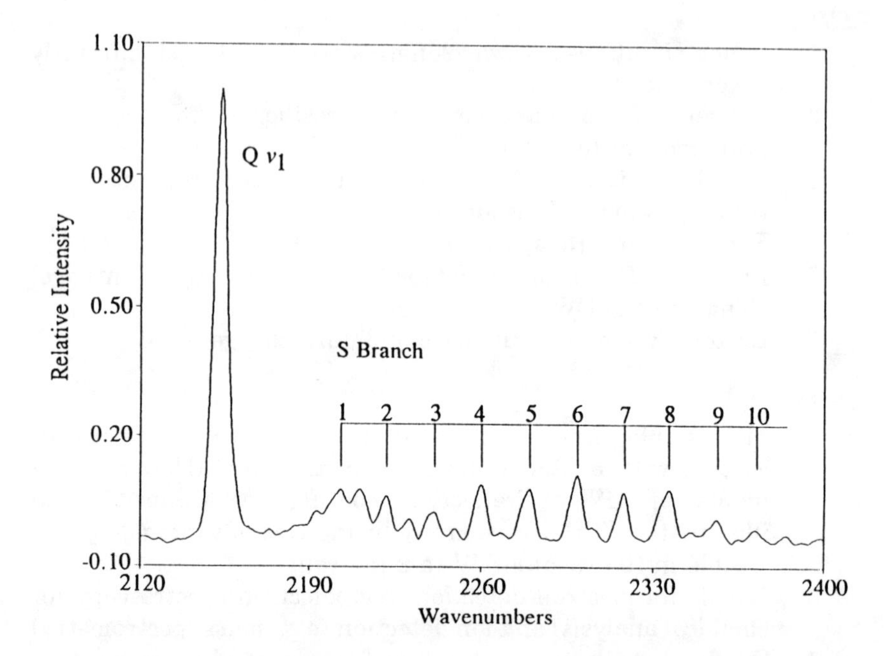

Figure 5.19 Resonance Raman spectrum of the CD3[1100]-[0100] hot band vibration obtained with 211.342-nm excitation.

vibration obtained with 211.342 nm excitation. The spectral features shown are the Q, R, and S branches of the [1100]–[0100] vibration. This spectrum was obtained as part of a study of the predissociation of the methyl radical Rydberg 3s state [21,22]. In this case the probe laser is at 211.342 nm, which is resonant with the R branch of the [0100]–[0100] vibronic transition in CD3. The goal of this work is to gain insight into the dynamics occurring on the Rydberg 3S state. This was accomplished experimentally by measuring the intensities of the rotational Raman lines as a function of excitation wavelength. The change in intensity of rotational features as a function of wavelength is modeled to extract the linewidth of the resonant level and therefore the excited-state lifetime.

A rough comparison between the performance of a PMT with a scanning monochromator and a ICCD with a spectrometer follows.

The PMT monochromator system was used with a 300-μm slit and provided only 20 wavelength data points. It required 16-fold less time to obtain the data with the ICCD, that is, 15 minutes versus 4 hours, even though the ICCD obtained over 500 data points. Furthermore, whereas the PMT/monochromator system acquires data from each wavelength at different times, the ICCD acquires all data points simultaneously and therefore the data are far less susceptible to temporal fluctuations.

The main advantages of the ICCD for this application are:

1. Excellent sensitivity due to the multiplex advantage
2. Increase in SNR by "gating out" of the background signals
3. Higher precision due to simultaneity of the data acquisition

5.9.2 Imaging Spectroscopy: Two-dimensional Fluorescence Lifetime Imaging

A "home-made" ICCD was used in the gate mode to time resolve fluorescence images excited by laser pulses. Following the timing scheme of Figure 5.20, the laser excitation of the sample and the

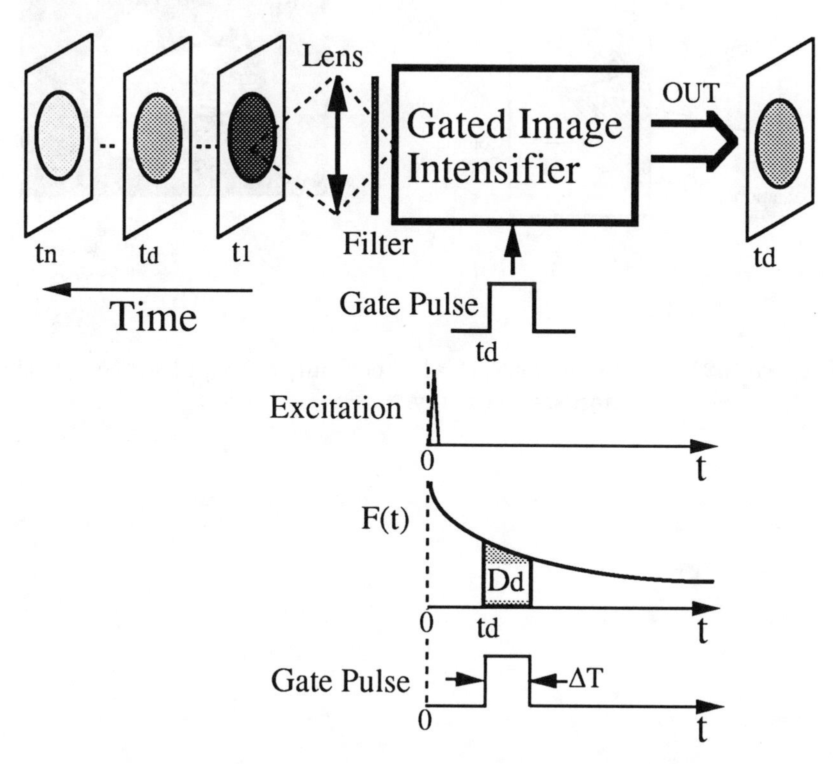

Figure 5.20 Timing scheme used for the study of time-resolved fluorescence images excited by laser pulses.

gating of the ICCD were fully synchronized via the use of an optical-fiber delay line.

Figure 5.21 shows the model for the spatial deposition of the samples, in this case quinine sulfate (a), coumalic derivative (b), and the overlap of both (a+b).

Figure 5.22 shows the time-resolved fluorescence image pattern obtained. Clearly, this system represents a new tool for performing imaging fluorescence lifetime studies in the nanosecond time range.

Note that this "homemade" detection system was far less optimized than the one used in the former application; the ICCD used lens coupling rather than optical-fiber coupling.

This application is ideal for ICCD detectors: it takes full advantage of key features of the ICCD including high sensitivity, gatability, and imaging.

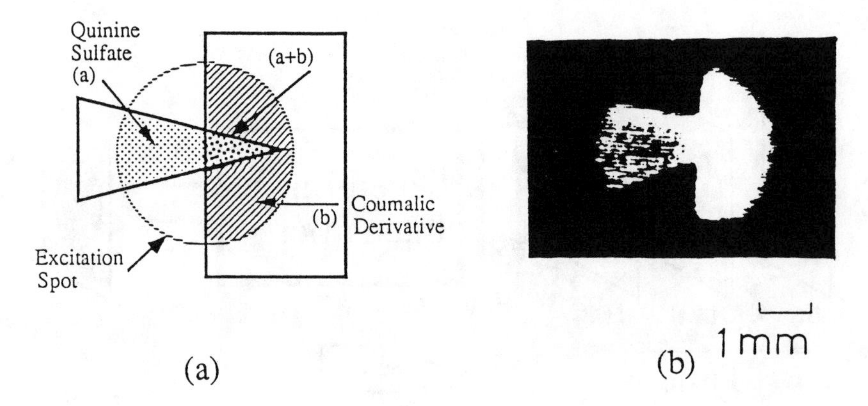

Figure 5.21 Model for spatial deposition of samples used in the fluorescence study.

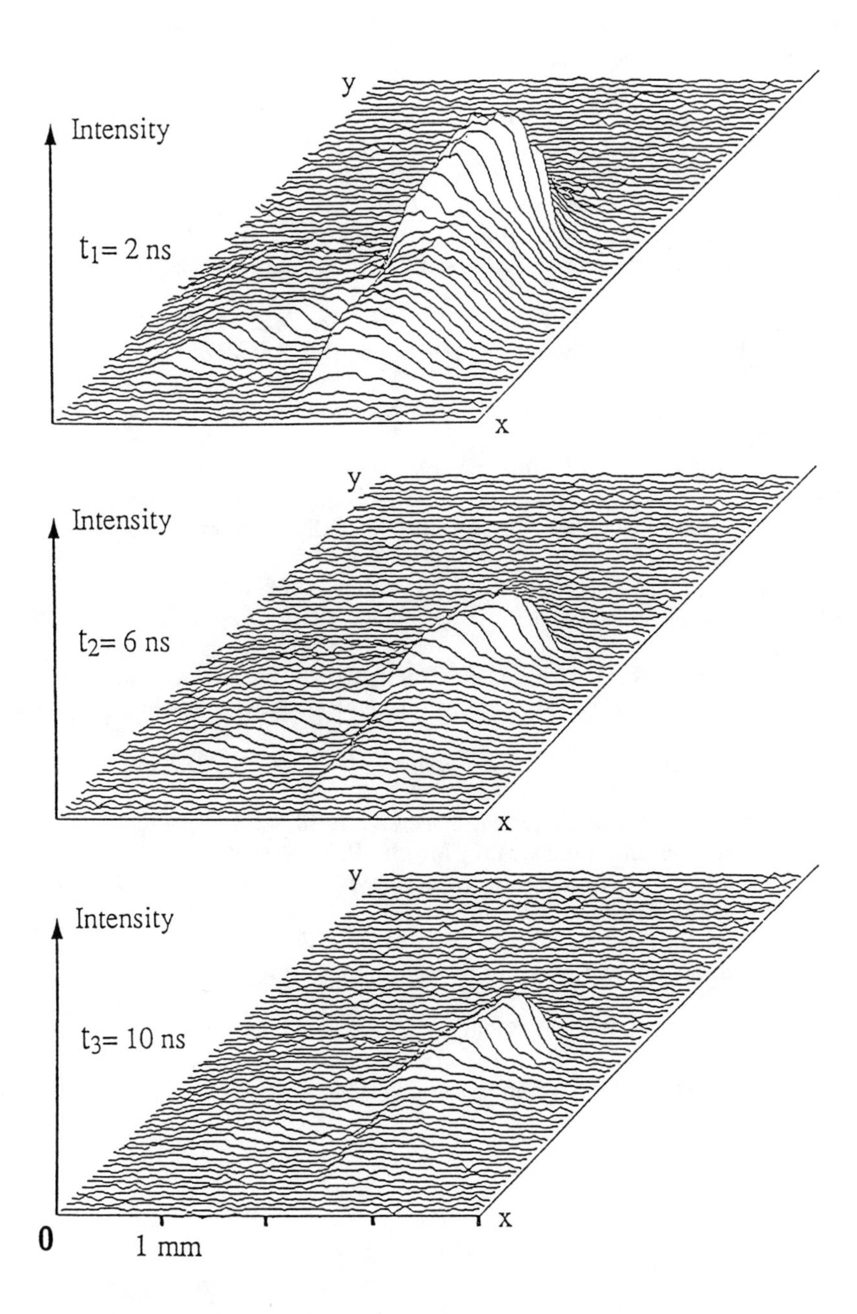

Figure 5.22 Time-resolved fluorescence image pattern.

References

1. Janesick, J.; Elliott, T.; Collins, S.; Marsh, H.; Blouke, M.; Freeman, J. *Proc. SPIE* **1984,** *501,* 2.
2. Murray, C.A.; Dierker, S.B. *J. Opt. Soc. Am.* **1986,** *A3,* 2151.
3. Pemberton, J.E.; Sobocinski, R.L. *J. Am. Chem. Soc.* **1989,** *111,* 432.
4. Pelletier, M.J. *Appl. Spectrosc.* **1990,** *44,* 1699.
5. Sams III, B.J. *Rev. Sci. Instrum.* **1991,** *62,* 595.
6. Siegmund, O.J.W.; Maline, R.F. In *Multichannel Image Detectors,* Talmi, Y., Ed. American Chemical Society, Washington, D.C., 1982.
7. Fraser, G.W.; Pearson, J.F.; Lees, J.E.; Feller, W.B. *Proc. SPIE* **1988,** *982,* 98.
8. Hiller, B.; Paul, P.H.; Hanson, R.K. *Rev. Sci. Instr.* **1990,** *61,* 1808.
9. Snelling, D.R.; Smallwood, G.J.; Sawchuk, R.A. *Appl. Phys. Lett.* **1989,** *28,* 3226.
10. Snelling, D.R.; Smallwood, G.J.; Parameswaran, T. *Appl. Opt.* **1989,** *28,* 3233.
11. Torr, M.R. *Appl. Opt.* **1985,** *24,* 793.
12. Dunham, M.G.; Donaghue, D.W.; Schemp, W.V.; Beardsley, B. "Performance Factors for Intensified CCD Systems," Photometrics Advanced Technologies, Tucson, Arizona.
13. Marconi, M.C.; Rocca, J.J. *Appl. Phys. Lett.* **1989,** *54,* 2180.
14. Marconi, M.C.; Rocca, J.J.; Krause, G.J. *J. Phys. E. Sci. Instrum.* **1989,** *22,* 849.
15. Boettger, H.G.; Griffin, C.E.; Norris, D.O. In *Multichannel Image Detectors,* Talmi, Y., Ed. American Chemical Society, Washington, D.C., 1978.
16. Hill, J.A.; et al. *Intl. J. Mass Spectrosc. Ion Proc.* **1989,** *92,* 211.
17. Talmi, Y. *Appl. Spectrosc.* **1982,** *36,* 1.
18. Lepla, K.C.; Horlick, G. *Appl. Spectrosc.* **1990,** *44,* 1259.
19. Thomas, S.; Shimkunas, A.R.; Muger, P.E. In *Proceedings of the 19th International Congress on High Speed Photography and Photonics,* Cambridge, England, September 1990.
20. Thomas, S.; Trevino, J. *Proceedings of the 19th International Congress on High Speed Photography and Photonics,* Cambridge, England, September, 1990.
21. Westre, S.G.; Kelly, P.B.; Zhang, Y.P.; Ziegler, L.D. *J. Chem. Phys.* **1991,** *94,* 270.
22. Westre, S.G., private communication, 1992.

6
CTD Detectors for Planar Separations and Electrophoresis

Mark Baker and M. Bonner Denton

Department of Chemistry, University of Arizona
Tucson, Arizona

6.1 Introduction

A simple scan of the chemical literature reveals the revolutionary impact that scientifically operated charge-transfer device (CTD) detectors are having on analytical chemistry. These detectors have proven themselves to be a nearly ideal choice for many applications pertaining to analytical spectroscopic detection [1–3]. They have shown a wide applicability for atomic emission, fluorescence, and chemiluminescence analysis [4–7].

Applications of analytical imaging that have benefitted from the advent of CTD technology include chromatographic plates and planar electrophoretic gels. The features that have made CTDs so successful for analytical spectroscopies are directly applicable to the spatial imaging analysis of gels and plates. Other imaging techniques that were employed in the past have fallen far short of the ideal imaging system for these two separation classes; in the absence of the technologies described herein, qualitative analysis is cumbersome and quantitative analysis is impractical.

This chapter illustrates the advantages of using scientifically operated solid-state array detectors for imaging electrophoretic gels and chromatographic plates.

6.2 Thin-Layer Chromatography, the Development of a Quantitative Technique

A primary responsibility of many analytical labs is the qualitative and quantitative analysis of a variety of species, often in complex matrices. This requires separation of the various sample components by some chromatographic technique followed by detection of the separated compounds. Most commonly the means of separation and quantification is gas chromatography (GC) or high-performance liquid chromatography (HPLC). Both techniques separate organic molecules through differential migration of the sample molecules as they traverse the chromatographic column. Detection of the molecules is achieved as the separated components pass through the detector in a sequential manner. The high sample throughput of GC and the high selectivity afforded by HPLC have made these modes of separation very useful analytical tools.

A variation of liquid chromatography known as thin-layer chromatography (TLC) is also utilized for separations. While it is probably best to consider modern TLC as a complimentary technique to HPLC rather than a competitive one [8], TLC does offer some significant advantages:

1. TLC, unlike HPLC, allows for the parallel processing of samples, often resulting in more than an order of magnitude of increased sample throughput.
2. With TLC, the separation process can be continuously monitored as it occurs on the plate surface, allowing for optimization of the separation process, whereas in HPLC, separated components are observed only after they exit the column.
3. Since plates are relatively inexpensive and used only once, TLC is ideally suited for problem separations that leave material adsorbed to the stationary phase [8].
4. The separation process leaves the separated components dispersed on the chromatographic plate, allowing for detection by a variety of imaging techniques such as fluorescence and absorbance spectroscopy as well as a number of others [9,10].

When large numbers of samples need to be processed in a timely fashion, modern TLC can be an attractive alternative to the use of HPLC. However, TLC is essentially used only for simple routine qualitative screening and is considered by many chromatographers to be a semiquantitative technique at best.

Over the past two decades a number of technical advancements have been made in TLC, making it more competitive with or complimentary to HPLC as well as improving quantitative performance [9]. One of the most important advances is the introduction of fine homogeneous particle layers, resulting in what is now termed high-performance thin-layer chromatography (HPTLC). These particle layers result in fast and efficient separations of complex mixtures of organic molecules. The homogeneous nature of the stationary phase has not only improved separation efficiency but has significantly reduced the background, thereby increasing the potential of HPTLC as a sensitive and quantitative technique.

Figure 6.1 illustrates the effect of plate structure on the detector baseline [8] and shows that layer homogeneity reduces background fluctuation allowing for improved sensitivity and precision. The comparison shown in Figure 6.1 is made between an HPTLC plate

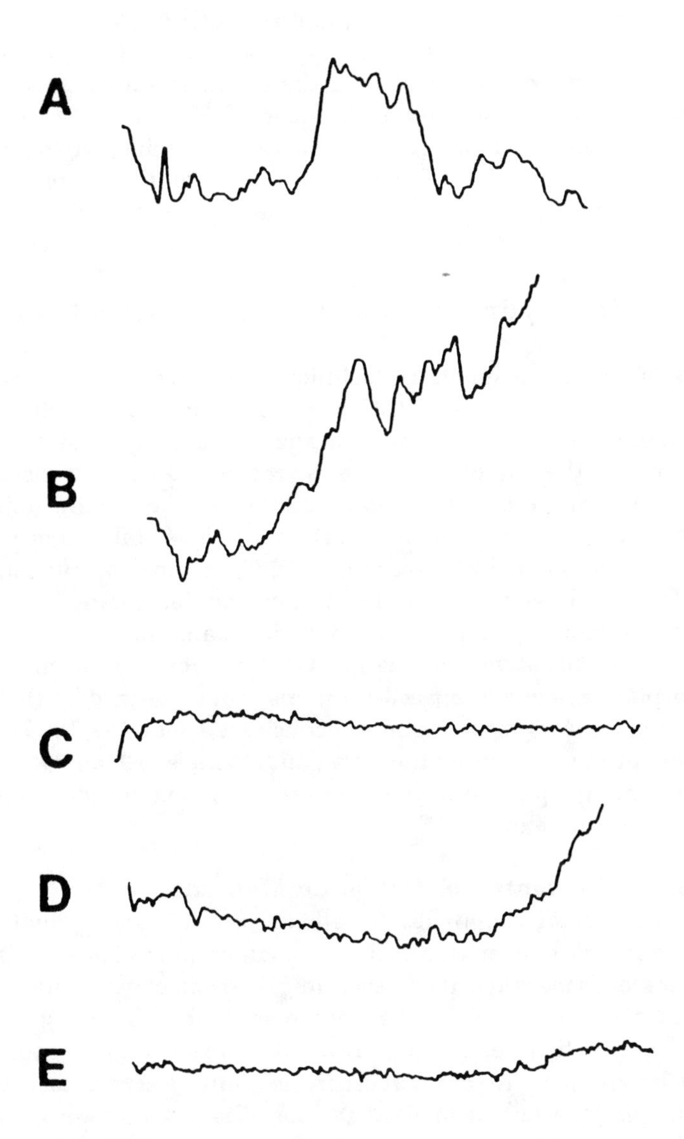

Figure 6.1 Influence of plate structure and contaminants on the baseline. See text for detail. (Reprinted with permission from reference 8.)

cleaned by solvent development (6.E) and a conventional plate before cleaning (6.A), after solvent elution (6.B), scanned in dual wavelength (λ_1 = 250 nm, λ_2 = 275 nm) mode (6.C), and after multiple solvent developments (6.D).

The ability to deliver small quantities of sample precisely and accurately to the chromatographic plate is crucial for quantitative analysis. This need has resulted in the development of some rather sophisticated sample delivery techniques [9,11]. These techniques have often employed automated spotters capable of delivering nanoliter-size quantities of sample solution to the plate without disruption of the silica surface. With the development of these sample delivery systems, practical quantitative analysis by HPTLC is a reality.

6.3 Thin-Layer Chromatographic Detection

The availability of an analytical technique that possesses the powerful separation capabilities of HPLC while establishing a high sample throughput is attractive. When an analysis is performed by either HPLC or GC, the samples to be analyzed as well as the necessary quantitative and qualitative standards are sequentially injected, separated, and detected. This cumbersome process takes place at the expense of time, and in the case of HPLC, of solvent consumption. When HPTLC is employed as the separation technique, numerous samples are processed in parallel. Individual samples of unknowns, as well as the quantitative and qualitative standards, can be spotted on a single plate at equally spaced intervals along one end of the plate and then chromatographed simultaneously. As many as 100 samples could be processed simultaneously on a single 10 cm × 15 cm chromatographic plate offering two orders of magnitude increased sample throughput over HPLC.

6.3.1 Conventional Detection Methods

While a number of techniques have been used to extract qualitative and quantitative information from chromatographic plates, by far the most common is mechanical slit, scanning densitometry. Commercially available slit-scanning densitometers operate by observing a small portion of the light emanating from the chromatographic surface defined by an aperture. The aperture is scanned across each predefined sample lane in a sequential fashion, thereby providing data on each of the separated sample components. For high precision work, the aperture size remains constant during the scanning process and is defined by the largest spot diameter. This has the effect of reducing the sensitivity of sample components represented by smaller spot diameters. Precision is affected by the ability to reproducibly align the aperture with the predefined sample lanes. The complete scanning process can take several hours to complete depending upon the number of sample lanes to be scanned and the various instrumental parameters. This is the rate-limiting step in the planar chromatographic process. These devices have been constructed for operation in the ultraviolet (UV) and visible spectral regions and allow for operation in either single- or dual-beam absorbance and fluorescence modes of detection.

Ideally, a detection system should take full advantage of the parallel processing nature of HPTLC. Detection in HPTLC utilizing mechanical slit-scanning devices not only reduces sample throughput but also reduces the advantages afforded through automation. Simultaneous recording of the entire plate image is ideal.

Film has a two-dimensional format that is well matched to the dimensions of the HPTLC plate, thereby allowing for simultaneous image integration of a large portion, if not all, of the chromatographic surface. The resulting high-resolution image of the plate encodes qualitative information in the spot or band position, and quantitative information in the darkness of the film. The recording of the image on film allows for the long-term storage of the chromatographic results, which can be useful for verification and compliance with governmental standards.

Unfortunately, film has several disadvantages associated with its use for quantitative imaging. Film has a nonlinear response to exposure, making the extraction of quantitative information difficult. It also has a somewhat limited intensity response range [12]. This constrains the range of sample concentrations that can be quantitatively captured on film typically to less than two orders of magnitude. The use of film requires a time-consuming and labor-intensive development process. These factors coupled with the fact that the photographic image itself must be scanned to extract quantitative numbers makes film an impractical medium with which to work for quantitative analysis of chromatographic plates.

6.3.2 Chromatographic Imaging with CTD Detectors

CTD detectors offer a unique solution to the analysis of modern HPTLC plates. CTDs have a two-dimensional format well matched to the HPTLC plate. The matched format allows simultaneous image acquisition of the entire plate surface, thereby taking full advantage of the parallel processing nature of HPTLC. These devices are solid-state imagers whose readout may be digitized and the resulting image may be stored in computer memory. This feature, to be discussed in more detail later, has implications with respect to data storage, interplate qualitative analysis, image processing, field-illumination correction and contrast enhancement for improved qualitative analysis.

Video-rate-operated CTDs are showing considerable promise for a wide variety of applications in analytical chemistry including thin-layer chromatographic imaging [13–15]. In addition, CTDs can be operated in scientific modes, with slow-scan speeds, special readout electronics, and cryogenic cooling. As described in Chapters 2 and 3, the dark-count rate and read noise are greatly reduced, making them excellent detectors for low-photon-flux chemical imaging analysis [3]. Typically, the scientific CTD is manufactured without masking the device, taking full advantage of the available photoactive area, unlike devices for standard video applications. The 16 bits of digital resolution commonly afforded by scientific CTD imaging systems offer greater than four orders of intraimage dynamic range, far exceeding the 256 levels of resolution offered by most video-based systems. The high spatial resolution, large linear dynamic range, high sensitivity from the soft x-ray to the near-IR, and the ability to integrate signal for long periods of time has made the scientifically operated CTD a nearly ideal detector for a wide variety of chemically related imaging applications. The remainder of this chapter will concentrate on the use of scientifically operated CTDs for the imaging of chromatographic plates and electrophoretic gels.

6.3.3 Chromatographic Fluorescence Imaging

A chromatographic imaging system for HPTLC must be capable of making sensitive and precise fluorescence and absorbance measurements of the separated components. We have investigated in our laboratories the ability of both the scientifically operated charge-injection device (CID) and charge-coupled device (CCD) array detectors to make sensitive and quantifiable fluorescence measurements on HPTLC plates after chromatographic separation.

The analysis of a group of substances known as aflatoxins is of considerable interest to commercial food and livestock feed producers. Aflatoxins are highly toxic, extremely carcinogenic substances excreted by the fungus *aspergillus flavus*. These compounds have been found in peanuts, chili peppers, and livestock feeds. Aflatoxins are highly fluorescent materials that when illuminated by long-wave UV light, emit in the blue to green portion of the visible spectrum. Typically, these substances have been analyzed by HPLC or planar chromatographic techniques with fluorescence detection.

Preliminary investigations into the sensitivity of detection of aflatoxins utilizing both classes of scientifically operated CTDs are presented in Table 6.1. Both detectors demonstrated similar sensitivities for all aflatoxins studied. As shown in Figure 6.2, excellent linearity was achieved over greater than two orders of magnitude for aflatoxin G2.

Figure 6.3 shows the fluorescence images of a TLC plate after the separation of several aflatoxins; a single-dimensional separation was carried out on a 100-μm silica gel HPTLC plate. A quantitative measurement from this plate is demonstrated in Figure 6.4 by a single-lane plot of intensity versus pixel number of 500 pg each of aflatoxins G2, G1, B2 and B1. The data plotted in Figure 6.4 were obtained by

Table 6.1 Preliminary Detection Limits for Aflatoxins: CID and CCD

Aflatoxin	CCD (pg)	CID (pg)
G2	10	30
B2	140	100
B1	150	180

summing in computer memory 26 rows perpendicular to the direction of sample migration and plotting the summation along the direction of sample migration. Summing a lane in this manner means that the entire width of the spot is included, which results in a more accurate plot of the separated components while simultaneously reducing plot scatter and increasing sensitivity. Thus, the area under the curve in the plot becomes representative of concentration regardless of spot shape or sample distribution.

The excellent sensitivities observed for all the aflatoxins investigated and the high linearity of the aflatoxin G2 plot indicate the use of scientifically operated CTDs for quantitative HPTLC imaging to be very promising. Further reductions in background plate fluorescence and improvements in plate homogeneity must be made to exploit the full potential of the scientifically operated CTD and to justify the current high cost of these imaging systems over video-based systems. We are currently investigating prechromatographic plate-cleaning techniques, low or nonfluorescing plate backing materials, novel illumination techniques, and the effect that variations in layer thickness have on the observed background intensity. We found that the commercially manufactured plates used in this investigation have fluorescent impurities in or on the support material. These represent the ultimate limiting factor in sensitivity, not the noise associated with the background fluorescence or source intensity. These plate inhomogeneities appear as fluorescing spots in the recorded image and can be seen in the high-contrast image at the bottom of Figure 6.3. We do not know the nature of these fluorescent impurities at this time; however, they cannot be removed by brushing the plate surface or by immersing the plate in spectroscopic-grade methanol. The impurities appear to be adsorbed to the support material or entrapped between the silica particles. We have also found that the severity of these background fluorescing impurities varies significantly among plate manufacturers.

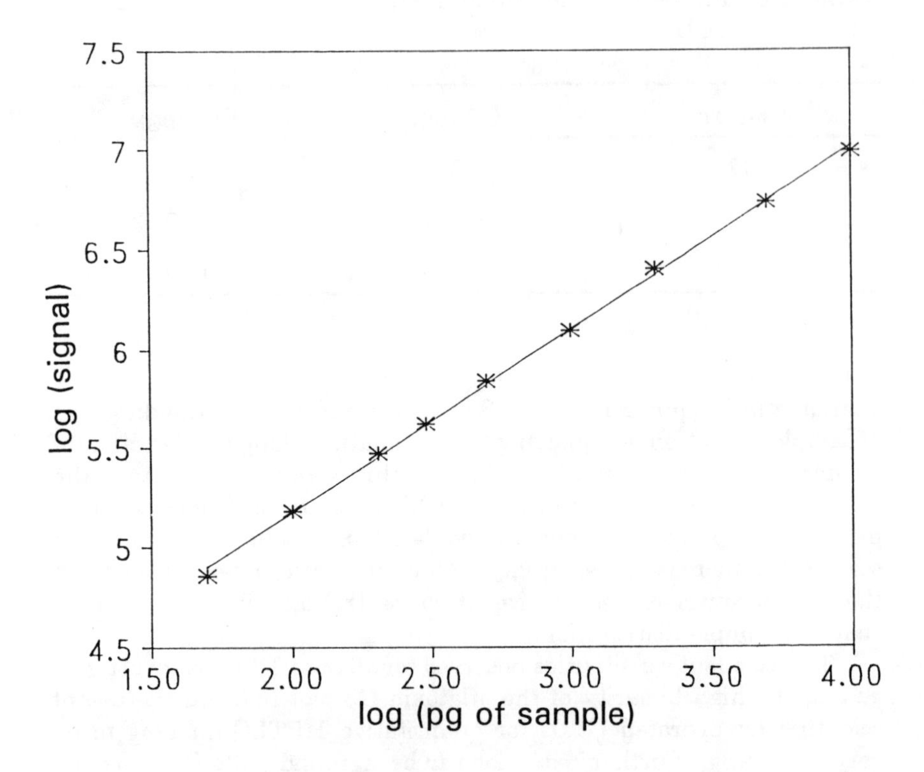

Figure 6.2 Working curve obtained for aflatoxin G2 using the CCD for chromatographic fluorescence detection. The integration period was 0.3 s.

6.3.4 Qualitative Image Enhancement

The direct digitization of the chromatographic image into computer memory affords advantages of image processing that can be used to aid in the qualitative analysis of chromatographic plates. Figure 6.3 (top) was obtained by bracketing (setting the maximum and minimum gray levels of the video output display to encompass a specified digital range) the entire digital range of the image about the 256 gray levels of the display. When display of the image is done in this manner, much of the gray level range is dedicated to the background fluorescence, leaving only a small portion of the gray levels left for the often minute fluctuations in intensity due to the low concentrations of aflatoxins. This results in a rather low contrast image which is difficult to qualitatively analyze. Figures 6.3 center and bottom were obtained by rescaling the gray scale level of the monitor to more accurately bracket the digitized information of interest.

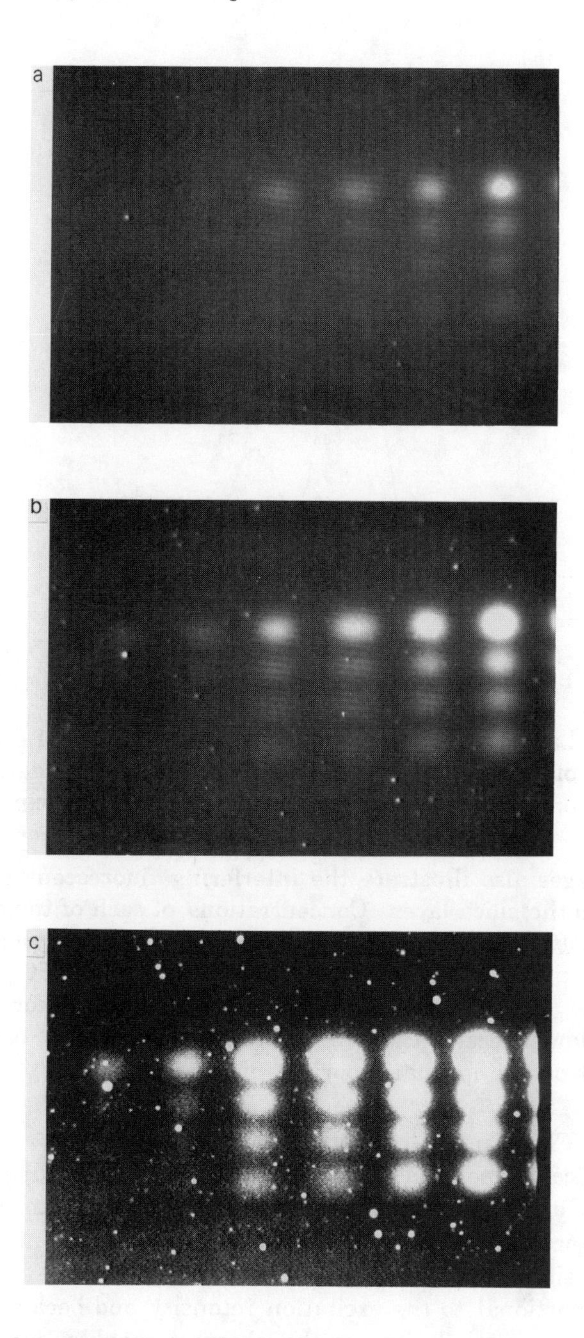

Figure 6.3 Bracketed images of an aflatoxin HPTLC separation. Top: bracketed by 256 gray levels of the display. Center: bracketed about 400-4000 digital levels. Bottom: bracketed about 770-1000 digital levels.

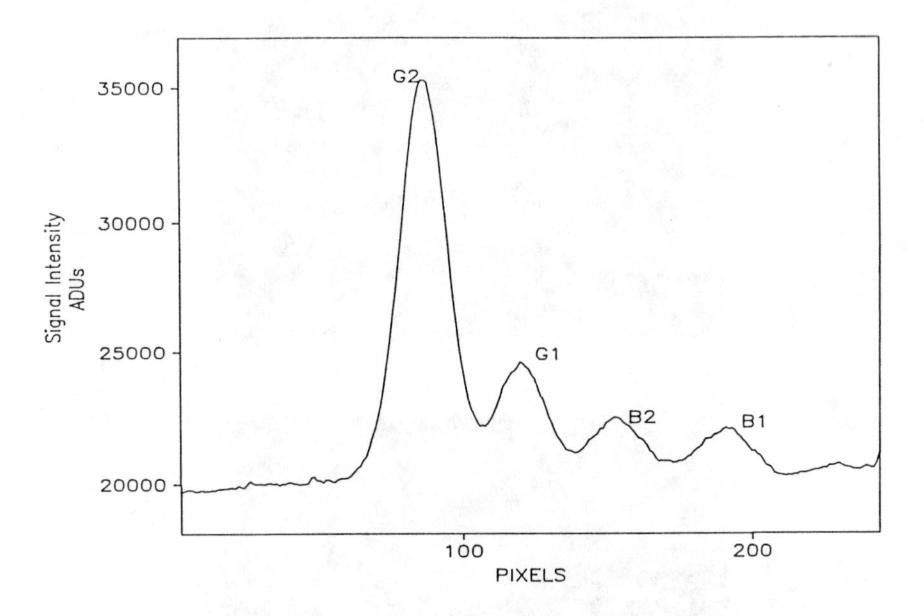

Figure 6.4 Lane plot of intensity versus pixel number of 500 pg each of aflatoxins G2, G1, B2, and B1. The detector was operated at -100°C with a 1.5 s integration time.

The images also illustrate the interfering fluorescent impurities embedded in the silica layer. Concentrations of each of the aflatoxins investigated (G2, G1, B2, B1, top to bottom) are from right to left: lane 1: 50 pg; lane 2: 100 pg; lane 3: 500 pg; lane 4: 750 pg; lane 5: 1 ng; lane 6: 2 ng. This procedure allows the researcher to distinguish the low concentrations of aflatoxins from the background fluorescence on the left-hand portion of the images.

6.3.5 Flat-Field Correction

The scientifically operated CTD detector easily allows for the implementation of a sophisticated field correction technique, known as flat-fielding, to account for spatial fluctuations in source intensity and detector sensitivity. Because the observed fluorescence intensity is directly proportional to the excitation intensity and because samples are distributed spatially across the chromatographic surface, even illumination of the plate is required to extract the most precise and accurate quantitative information. For high precision work, which does not require the use of expensive optics or elaborate illumination designs, the technique of flat-fielding can be employed.

This technique was applied to the imaging of aflatoxins in a manner described by Sutherland and co-workers for the fluorescent imaging of electrophoretic gels [16]. An orange filter (Hoya Optics sharp cut filter O-56) was placed in the position normally occupied by the HPTLC plate. When exposed to UV light, this filter fluoresces at orange wavelengths, providing a spatial map of UV illumination intensity, Figure 6.5a. The information obtained by imaging this fluorescent material was used to correct each pixel in the analytical image, allowing for the extraction of accurate quantitative information. In Figure 6.5b, the data from replicate 1-ng samples of aflatoxin G2 are plotted, uncorrected and corrected, demonstrating the effectiveness of the flat-fielding procedure.

The mathematical relationship used to obtain this correction appears below. Typically, images are corrected for fixed pattern electronic offsets by the subtraction of a bias exposure which is an image generated by simply reading out the CTD with the shutter closed.

$$\text{Adjusted original image} = \text{Original image} - \text{Bias image}$$

$$\text{Adjusted flat-field image} = \text{Flat-field image} - \text{Bias image}$$

$$\text{Corrected image} = \frac{\text{Adjusted original image}}{\text{Adjusted flat-field image}} \times (\text{Adjusted flat-field image})_{\text{average}}$$

As seen in these equations, each individual value in the adjusted original plate image is ratioed against the corresponding value in the adjusted flat-field image, thereby correcting for the illumination intensity or detection sensitivity. The result is a corrected image of the plate, where the emission is independent of spatial location.

It is the ability of today's computers to process large amounts of digitally stored information, coupled with the scientifically operated CTD's capability to obtain and store in computer memory an image in a relatively short period of time, that makes practical flat-field correction of images possible. No longer is the instrument designer constrained to use elaborate illumination schemes and expensive optics that result in uniform illumination in order to obtain accurate quantitative information. One should note, however, that under circumstances leading to very uneven illumination or detection, the dynamic range offered by the CTD is reduced.

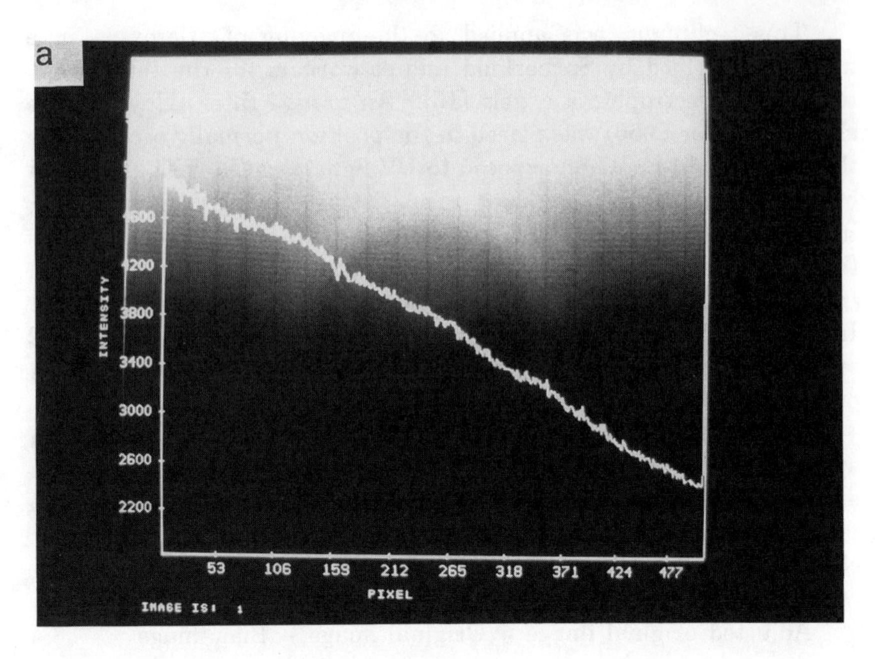

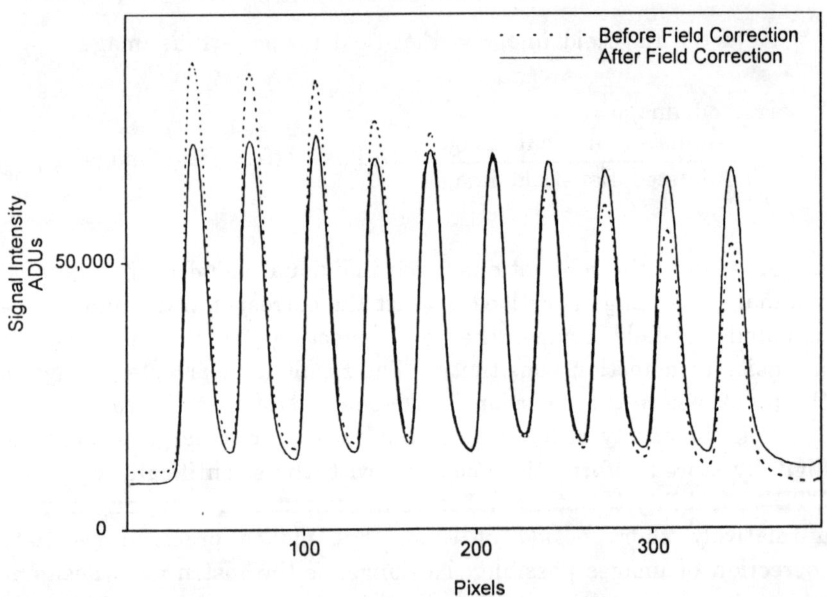

Figure 6.5 The illumination source was modified to give a skewed illumination across the HPTLC plate surface. a) Plot of illumination intensity across the surface. b) Line plots of uncorrected (dashed) and corrected (solid) data.

6.3.6 Absorbance and Spectroscopic HPTLC Detection

Scientifically operated CTD detectors are ideally suited for low-photon-flux analysis such as the sensitive, quantitative fluorescence imaging of chromatographic plates. Unfortunately, the majority of compounds separated by HPTLC are not fluorescent. An alternative may be UV absorbance detection. From a detection standpoint, UV absorbance measurements are very different from fluorescence measurements. For sensitive UV absorbance measurements one observes a small decrease in the intensity of the reflected radiation due to the absorption by the sample, resulting in dark spots on a bright background. Absorbance measurements are high-light-level measurements where the dominant source of noise is due to photon shot noise, which results in detection limits typically two to three orders of magnitude less sensitive than fluorescence detection. With scientifically operated CTD detectors, the low dark-current rate and low read noise become insignificant in comparison with the noise associated with the background. Assuming that photon shot noise is the dominant source of noise in an absorbance measurement, the following equation shows the dependence of the minimum detectable absorbance signal on full well capacity that theoretically can be measured:

$$A_{min} = -\log \left[1 - \left(\frac{8}{Q_{sat}}\right)^{1/2} \right]$$

where Q_{sat} is the full-well capacity or saturation level of the detector [17]. The minimum detectable absorbance signal observable with a scientifically operated CTD is obtained by allowing the background signal to just reach the point of saturation of the device. It is at this point that the best signal-to-noise ratio for an absorbing species will be obtained. As described by this equation, a CTD with a full-well capacity of 500,000 charge carriers would theoretically be capable of detecting an absorbance as small as 1.7 milliabsorbance units. Because the read noise or dark count is of lesser importance, the highest performance CTD is not required to make quality absorbance measurements.

Scientifically operated CTDs do offer two features not commonly found in video-based imaging systems: high precision and UV sensitivity. One of the limits to the attainment of high-precision absorbance measurements is the 8-bit resolution offered by video-based imaging systems, inadequate to take full advantage of the CTD's dynamic range. Bilhorn [18] describes well the digitization requirements of a CTD imaging system for HPTLC. Assuming that the dominant source of noise in an absorbance image is due to photon statistics only and is not affected by outside factors such as plate inhomogeneity, the noise level at the device saturation level is simply the square root of the full-well capacity. (This is actually an approximation, since noise measurements are bounded by the saturation level.) The minimum detectable signal is twice this noise level. For a device with a full well capacity of 500,000 charge carriers, the minimum detectable signal would be two times the noise level of 700 charge carriers. A minimum of 10 bits of digital resolution is required to make a measurement of this precision, two bits more than is commonly seen in video-based imaging systems.

Cosgrove and Bilhorn investigated the use of a cooled slow-scan CCD for spectroscopic detection and absorbance imaging of HPTLC plates [19]. The CCD used for this study was a Thompson TH 7882 that had been coated with Metachrome II, a UV enhancement phosphor coating, as described in Chapters 2 and 3. Separation of photographic couplers, which absorb in the UV, was achieved on a 5 × 10 cm Merck Silica Gel 60 F_{254} HPTLC plate. CCD images of the HPTLC plates with separated photographic coupler produced peak-shape and sensitivity results similar to those obtained with a mechanical slit scanning densitometer (TLC Scanner II, Camag; Muttenz, Switzerland) but with considerable savings in time. The scanning densitometer required 6 min to complete data acquisition on three chromatographic lanes over a 6-cm elution. In comparison, the CCD imaging was capable of acquiring data simultaneously on all three chromatographic lanes in one 40-s exposure, excluding processing.

System precision was investigated using a stable (> 7 days) red printing ink adsorbed to a Merck Silica Gel 60 HPTLC plate. Thirty-two replicate reflectance measurements were made on each of three consecutive days with the CCD system. Similarly, a second set of 32 measurements was also made on each of these days with the scanning densitometer in order to assess stage movement error. The scanning densitometric analysis was carried out at 568 nm. CCD imaging was accomplished using white-light illumination and broad-band detection over 4-s integration periods. Both modes of detection utilized partial band detection (4 mm) of a 251-ng sample of red printing ink deposited on the plate in a 5-mm band. Results of this precision study are summarized in Tables 6.2 and 6.3. These measurements indicate that greater than an order of magnitude improvement in intraday precision is achievable with the use of a scientifically operated CCD-based imaging densitometer in comparison with mechanical slit-scanning densitometry. It was determined that reproducible stage movement in

Table 6.2 Relative Standard Deviation (%) Values from the Precision Study Data.

	Day 1	Day 2	Day 3
Scanning densitometry, X-directional movement	0.60	0.41	0.46
Scanning densitometry, no X-directional movement	0.27	0.20	0.19
Imaging densitometry	0.027	0.17	0.018

Reprinted with permission from Reference 19.

densitometer was the greatest contributor (4:1) to the total instrument error.

Cosgrove and Bilhorn have indicated the potential use of CCD imaging for the acquisition of spectral information of separated components on HPTLC plates. Spectral acquisition of separated components can aid in the qualitative identification of the separated sample components and help discriminate between poorly resolved sample components. A flat-field spectrograph (UFS 200, Instruments SA; Edison, N.J.) was placed between the camera and plate to acquire spectroscopic information. Spectral acquisition and plotting capabilities are demonstrated using a 1.1-μg sample of anthracene applied to a Merck HPTLC Silica Gel 60 plate in an 8-mm band length, shown in Figure 6.6.

6.4 Planar Gel Electrophoresis

The other planar separation technique commonly employed in biochemistry labs is electrophoresis. Electrophoresis utilizes the action of an electric field to produce high-resolution separations of sample molecules in either a free solution or in an anticonvective medium such as polyacrylamide or agarose gels. Electrophoresis has found extensive application in bioanalysis and is routinely used for the separation and analysis of complex mixtures of proteins and nucleic acids. Along with the development of today's growing biotechnology industry comes the need to qualitatively and quantitatively analyze large numbers of these complex mixtures for the development or quality control of a product [20]. While electrophoresis has established itself as the method of choice for the high-resolution separation of

Table 6.3 Statistical Evaluation of the Precision Study.

Source	Scanning (% RSD)	Imaging (% RSD)
Pooled *RSD* of intra-day measurements[a]	0.50	0.21
Pooled *RSD* of inter-day measurements[a]	0.46	*b*
Total estimated *RSD* for 3-day study[a]	0.68	*b*
Pooled *RSD* of intra-day measurements with no *X*-directional movement	0.23	-
Estimated *RSD* of stage movement	0.44	-

[a] Data collected with *X*-directional stage movement between measurements.
[b] Configuration changes between days prevented the calculation of this statistical information.
Reprinted with permission from reference 19.

these complex biological mixtures, continued development of detection methods for the separated components is required to meet the future needs of this marketplace.

A number of these techniques result in the separation of single- or multiple-sample mixtures dispersed as bands or spots in a two-dimensional planar gel medium. Visualization of the separated components in the gel medium for both qualitative and quantitative analysis is accomplished by a variety of techniques, which include staining and autoradiography. Staining, whether it is dye staining, silver staining, or fluorescent derivatization, can allow for sensitive detection but does not directly yield quantitative information. The extraction of quantitative information from the stained gel must be accomplished with the use of a measurement technique such as mechanical scanning densitometry or film imaging followed by scanning. Autoradiographic imaging of gels, or a related process known as fluorography, also requires film and is used for the analysis of radiolabeled samples. Autoradiography involves overlaying the slab gel with an x-ray film. Emission of high-energy particles from the separated sample components will produce corresponding dark exposures on the film. This process tends to be extremely sensitive but very slow, on the order of hours to weeks, depending on the level

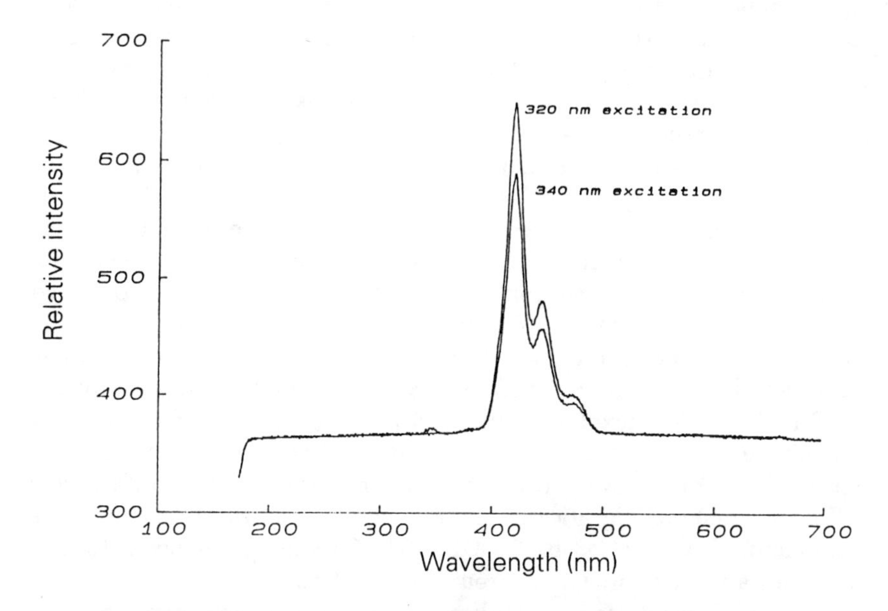

Figure 6.6 Fluorescence imaging of anthracene. (Reprinted with permission from reference 19.)

of radioactivity. The resulting developed film image must then be scanned to obtain quantitative information.

The same problems that apply to the analysis of chromatographic plates with film and mechanical slit-scanning instruments apply to gel scanning as well. While film meets many of the needs required of an electrophoretic imaging system, it falls far short of what would be considered ideal. Scientifically operated CTDs are much more sensitive to exposure level [21] with an equivalent film ISO rating on the order of 20,000 to 100,000. The increased sensitivity of the scientifically operated CTD in comparison with film means lower intensity UV sources can be utilized to illuminate the gel, thus minimizing problems associated with the degradation of photosensitive compounds. This also means that less sample material needs to be loaded into the gel initially to achieve a detectable amount of material of each of the individual sample components after separation. The decreased sample size may in some instances reduce the problems associated with sample overloading [22]. Concentrated samples of molecules can produce localized perturbations in the electric field, resulting in altered mobilities. These changes in mobility with respect to concentration can make qualitative analysis difficult if not impossible. Scientifically operated CTDs offer many of the advantages associated with film but without the associated disadvantages, making them an attractive replacement for film. Unlike film, the quantitative gel image is acquired without the need for a cumbersome development process. This feature allows for ease of adjustment of various imaging parameters such as focus, magnification, and exposure level as well as significantly shortens the overall analysis time.

Electrophoresis places additional demands on an imaging system to be utilized for quantitative and qualitative gel analysis. Two-dimensional gels have produced exceptionally high-resolution protein separations, often resulting in literally thousands of separated components over wide concentration ranges. Scientific CTDs have been produced in a variety of formats capable of producing quantitative images of large gel areas without appreciably degrading the high-resolution separation. Scientific CCDs have been routinely manufactured in 512 $\times$ 512 pixel and 1024 $\times$ 1024 pixel formats. While these formats do not remotely compare with the high-resolution imaging capabilities afforded by photographic film, more recently manufactured CCDs are making significant improvements. Ford Aerospace has manufactured a 4096 $\times$ 4096 CCD with 7.5-μm-square pixel dimensions and a total photoactive area of 9.4 cm^2 [21].

The high-resolution, two-dimensional gel image can produce complex protein patterns that can present the researcher with a deluge of information when trying to make intergel or intragel comparisons for qualitative analysis. As mentioned previously, the output of CTDs is digitized so that the representation of the gel image resides in computer memory. By allowing the computer to perform the tedious components of routine gel analysis, considerable savings in time can be realized.

The wide range of sample-component concentrations encountered in two-dimensional gels and standard concentrations utilized in one-dimensional gels requires an image sensor capable of a wide linear dynamic range. The scientifically operated CTD's excellent linearity over the entire full-well capacity coupled with the 16 bits of precision commonly found with these devices provides the analyst with greater than four orders of intraimage dynamic range. This allows, as with HPTLC imaging applications, maximum flexibility in quantitation and contrast enhancement. For those rare instances when this dynamic range is not sufficient, variable integration times can be utilized to extend this range.

An example of the application of integration-time bracketing to extend dynamic range is shown by Sutherland and co-workers [16] in Figure 6.7. The data for this plot were acquired with a modified video-CCD-based imaging system equipped with a mere 8 bits (256) of digital image resolution. The system was modified to use liquid nitrogen for device cooling, to reduce dark current and allow for extended integration times. Three exposures of 2, 6, and 20 s were required to record the fluorescence of all samples, and fluorescence intensities have been scaled accordingly.

The straight line with a slope of unity indicates that the integrated fluorescence is a linear function of the amount of DNA, from 2 to 120 ng per band, and illustrates the effectiveness of the technique. While exposure bracketing is also possible with photographic film, the nonlinearity and limited dynamic range of the media make its use difficult.

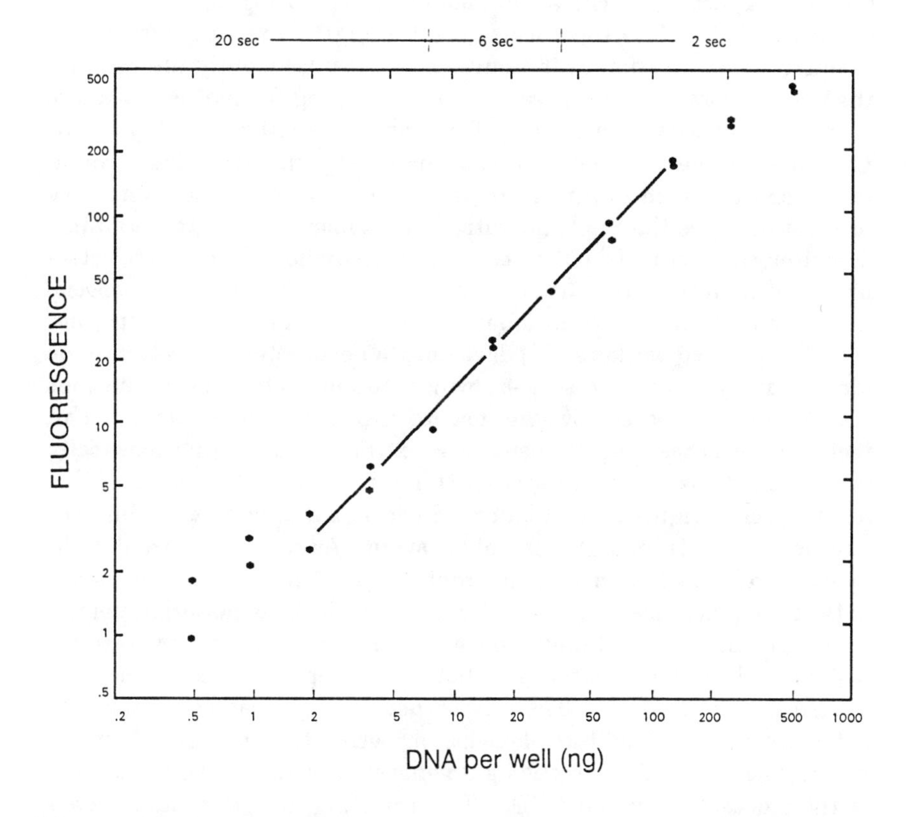

Figure 6.7 Integrated fluorescence in a lane as a function of the amount of DNA loaded into each well prior to electrophoresis. (Reprinted with permission from reference 16.)

6.4.1 Fluorescence Quenching Measurements in Agarose Gels

For the separation of very large DNA or RNA molecules, agarose gels are employed. As the DNA or RNA molecules migrate through the large pores of the agarose gel under the action of an electric field, separation of the molecules occurs via a sieving process [20]. Visualization of the sample components is typically achieved by staining with ethidium bromide (EB). EB reacts with the DNA or RNA strands by insertion between the bases. When the stained gel is illuminated with medium-wavelength UV light (approximately 310 nm), the EB-stained molecules fluoresce strongly at orange wavelengths. Quantitation of the separated components can be accomplished by mechanical scanning either of the gel directly or of its photographic image.

The actual staining process can occur either before or after electrophoretic separation. The advantage of incorporating the EB stain in the gel matrix before separation (prestaining) is that it allows one to monitor the separation as it occurs (in situ) subsequently allowing for the optimization of the separation time. Tagging the molecules before separation can have the effect of altering the native mobility of the molecule in the gel material and increasing the total background fluorescence, which in turn degrades sensitivity [23]. The alternative to prestaining is the implementation of a somewhat lengthy staining/destaining process. In either case, preelectrophoretic or postelectrophoretic staining, the hazardous nature of EB requires special handling and disposal techniques, since it is a mutagenic compound.

Chan and co-workers [23] developed a novel alternative technique for the analysis of agarose gels, which circumvented the limitations and hazards imposed by the use of EB for visualization. This technique was based on the use of a scientific grade, thermoelectrically cooled, CCD-based camera system (Photometrics Ltd., Tucson, Ariz.) for gel image acquisition. Figure 6.8 shows a diagram of the imaging system and electrophoretic chamber layout. As seen in Figure 6.8, the agarose gel, located in the electrophoresis chamber, is sandwiched between a quartz plate (top) and fluorescent plexiglas material (below). The gel chamber was illuminated with a 30-watt UV lamp operated at 254 nm. Locations in the gel that are occupied by DNA material absorb the 254-nm light. The unabsorbed UV light passes through the gel material, causing the plexiglas material to fluoresce at visible wavelengths. The fluorescent light generated by the plexiglas material is then detected by the CCD. The resulting image indicates dark underexposed areas at the locations of absorbing DNA material.

DNA fragments generated by the Hind III (purchased from BRL, Gaithersburg, MD) restriction digest of lambda DNA were used for system evaluation. Typical imaging times for this system were on the order of 5 s. Detection nonuniformities were corrected, as described

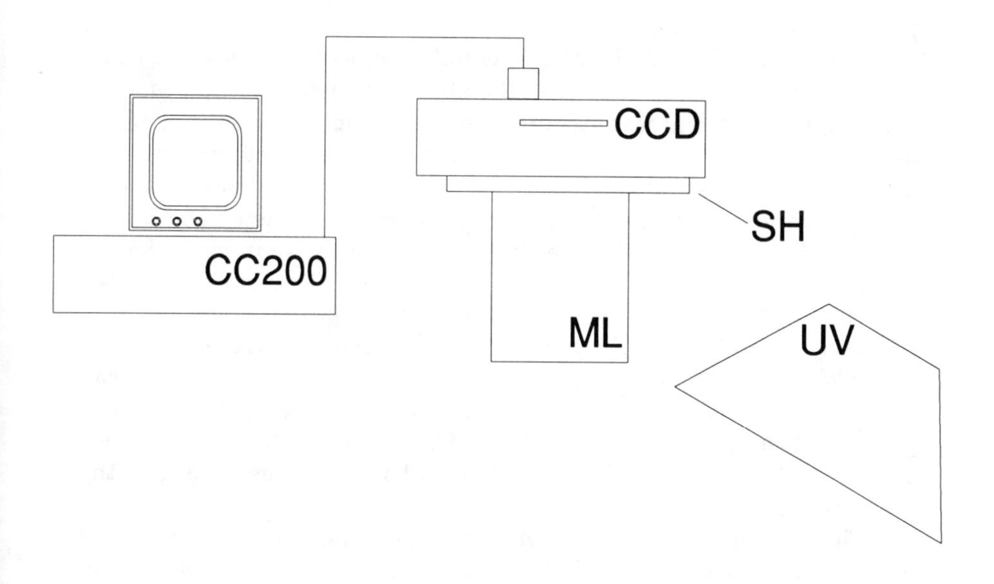

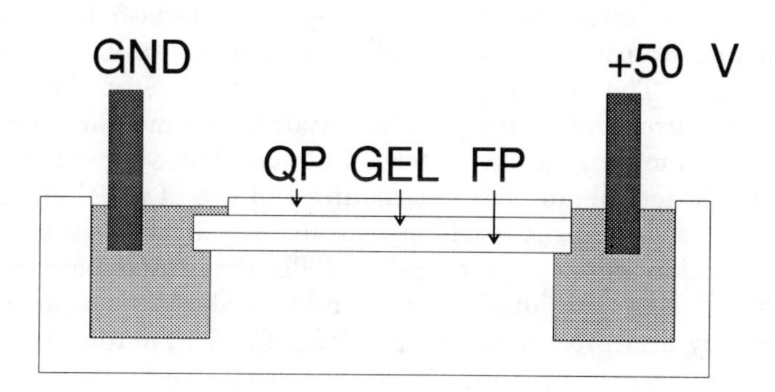

Figure 6.8 Experimental setup. SH=shutter; ML=macro lens; UV=UV lamp; GND=electrical ground; QP=quartz plate; FP=fluorescent plate; GEL=agarose gel; CC200=camera data system. (Reprinted with permission from reference 23.)

previously, using the technique of flat-field correction. The flat-field image was acquired with the loaded gel in the electrophoresis chamber prior to separation. The ability of this system to monitor the separation process as it occurs is demonstrated in Figure 6.9. The lane plots of emission intensity versus sample component location at various times during the separation process were smoothed with an 11-point Savitsky-Golay convolution filter. These plots indicate that for complete separation of all six DNA sample fragments, separation times in excess of 1 hr were required. The ability to quantitate the amount of separated material was accomplished and demonstrated in a plot of total band absorbance versus mass of DNA in that band, Figure 6.10. Sensitivity results indicated a detection limit of approximately 5 ng of double-stranded DNA per sample band. This level of detectability was found to be approximately five times less sensitive than with the use of EB but did not require the handling of a hazardous material or the need for a time-consuming staining/destaining process.

The data in Figure 6.11 demonstrate the changes in sensitivity and mobility compared with EB-prestained DNA samples that one can expect to achieve with this system. Peak directions for the EB-prestained samples are in the positive direction, indicating fluorescence emanating from the sample as opposed to absorption observed with the unstained DNA.

The work of Chan and co-workers [23] demonstrates the ability of the scientifically operated CTD detector to make reasonably sensitive, in situ, quantitative measurements of DNA material in agarose gels without the need for a fluorescent derivatization process. They have suggested the applicability of this technique for other related electrophoretic techniques such as the imaging of pulsed-field electrophoretic gels and preparative scale work that requires un-derivatized sample recovery. The ability of the scientifically operated CTD to generate digitally processable image information in a timely manner makes its use as a monitoring device for gel electrophoresis practical. The fact that this technique does not require the use of a UV-sensitive device and is a high-light-level measurement suggests that its use in conjunction with video-operated CTD array detectors may also be useful for routine analysis with only modest losses in quantitative accuracy and precision. It should be noted that the higher UV light levels needed to produce similar sensitivity to those achievable with the scientifically operated device may result in sample degradation.

6.4.2 Two-dimensional Gel Imaging

The sensitive nature of the scientifically operated CCD has prompted researchers to reevaluate visualization techniques that can exploit the advantages afforded by this means of detection. Initial investigations were conducted by Jackson, Urwin and Mackay [24]. They used the fluorophore 2-methoxy-2,4-diphenyl-3(2H) furanone (MDPF) for the visualization of proteins in two-dimensional polyacrylamide gels. The reaction of MDPF with proteins produces a product that on illumination by UV radiation fluoresces at visible wavelengths, allowing for sensitive and quantifiable detection. Background emissions from the

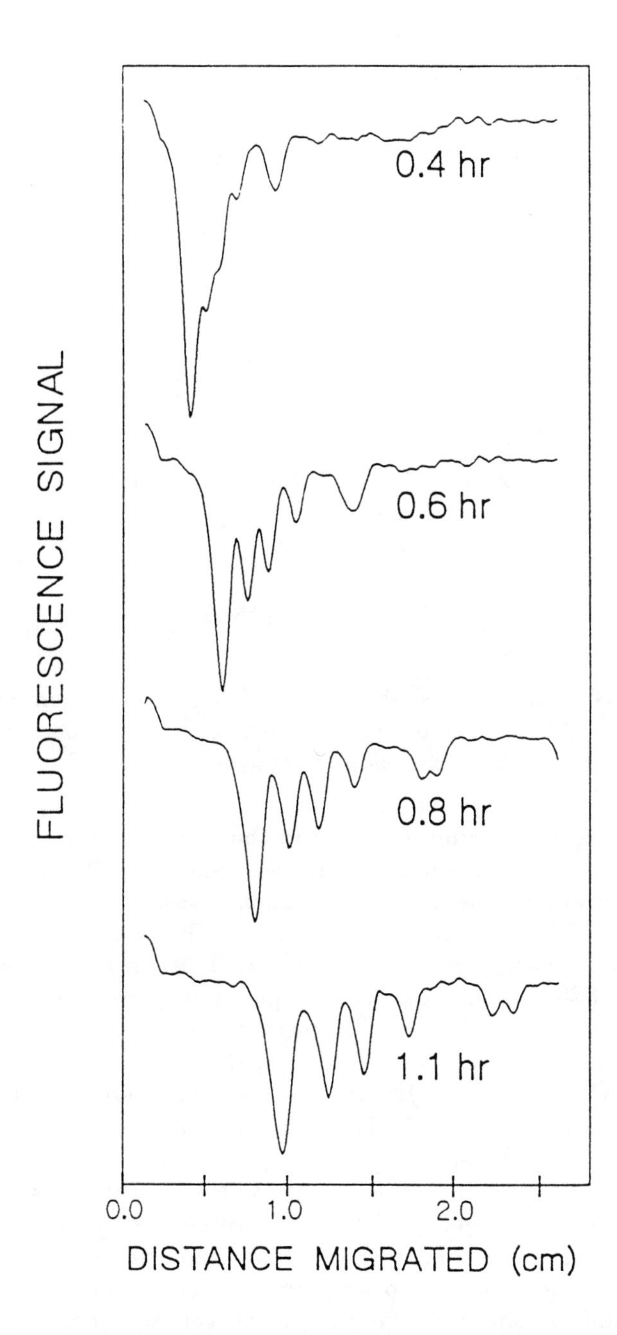

Figure 6.9 Fluorescence signals versus position in gel at various times during electrophoresis. (Reprinted with permission from reference 23.)

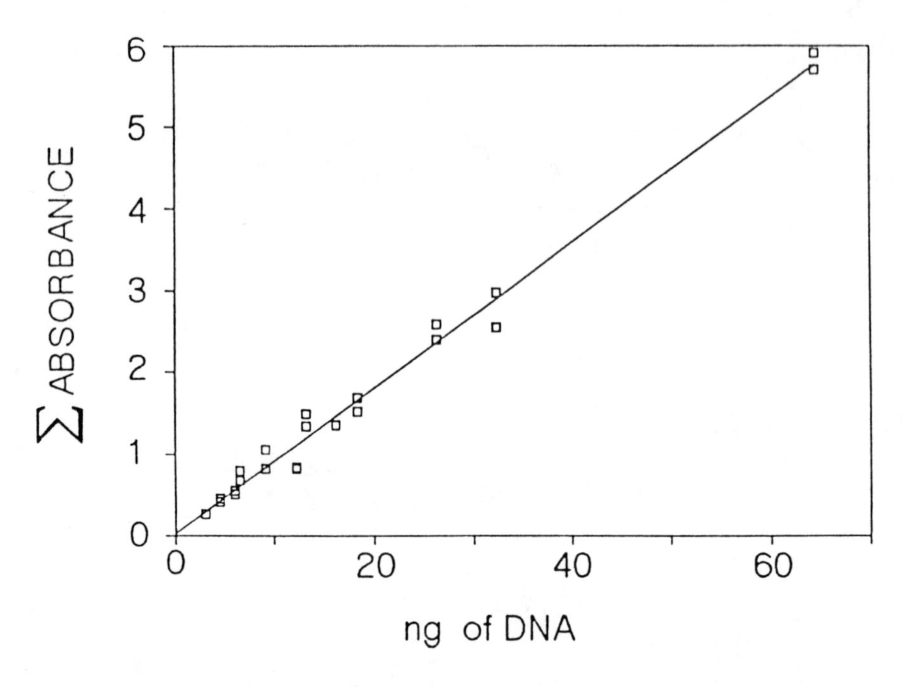

Figure 6.10 Integrated absorbance signal versus weight of DNA in each band and least-squares best-fit line. (Reprinted with permission of reference 23.)

tagging reagent are minimized due to the fact that unreacted MDPF and its hydrolysis products are nonfluorescent [25]. This is important for exploitation of the integrating capabilities of the scientifically operated CCD in order to achieve sensitive fluorescence detection.

The CCD used for the imaging of MDPF-stained two-dimensional gels was a 385 × 578 pixel array, liquid nitrogen cooled to -140°C. Illumination of the gels for CCD imaging was provided by a 150-watt tungsten-filament lamp with a 385-nm filter. Emission monitoring with the CCD employed a 480-nm filter and *f*/2.8 lens. CCD exposure times were on the order of a few seconds to 3 min. For purposes of comparison the gels were also photographed. The illumination source utilized for photographic acquisition was a UV light box with a 7000 μwatt/cm^2 surface intensity. The photographic exposure times of the two-dimensional gels, under those conditions with type 55 Polaroid film, a Wratten 8 filter (Kodak), and an *f*/4.5 aperture, was 40 s. Quantitation of the photographic negatives was performed on a Chromoscan 3 densitometer (Joyce-Loebl, Gateshead, UK).

Several images generated with the CCD-based imaging system are shown in Figure 6.12 [24]. The images show (a) IM9 cell proteins

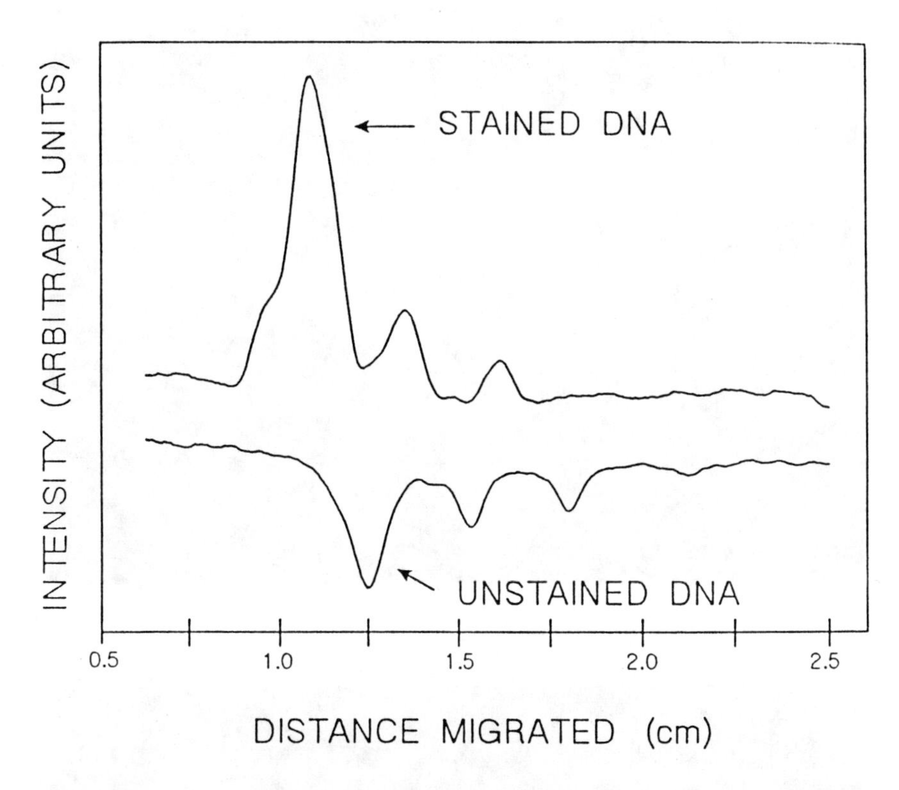

Figure 6.11 Intensity versus location of bands for stained and unstained DNA on same gel. (Reprinted with permission from reference 23.)

labeled with MDPF after two-dimensional electrophoresis, (b) IM9 cell proteins labeled with MDPF in the IEF gel after focusing and then electrophoresed in the second dimension, and (c) soluble proteins from *Neisseria gonorrhoeae* labeled with MDPF in the IEF gel after focusing and then electrophoresed in the second dimension. The sizes of the gels shown in (a), (b) and (c) were 141 mm $\times$ 94 mm, 110 mm $\times$ 78 mm, and 121 mm $\times$ 81 mm, respectively. The areas of the gels imaged by each pixel were 0.24 mm square in (a), 0.20 mm square in (b) and 0.21 mm square in (c). The quantity of protein analyzed for the gels shown in (a) and (b) was that prepared from 10^6 IM9 cells and 0.7×10^6 IM9 cells, respectively. For the gel shown in (c), 100 μg of protein was analyzed. Exposure times for CCD images (a) and (b) were 60 s, and the time for image (c) was 30 s. Images obtained with the CCD indicated similar and identifiable spot patterns in comparison with those obtained with film.

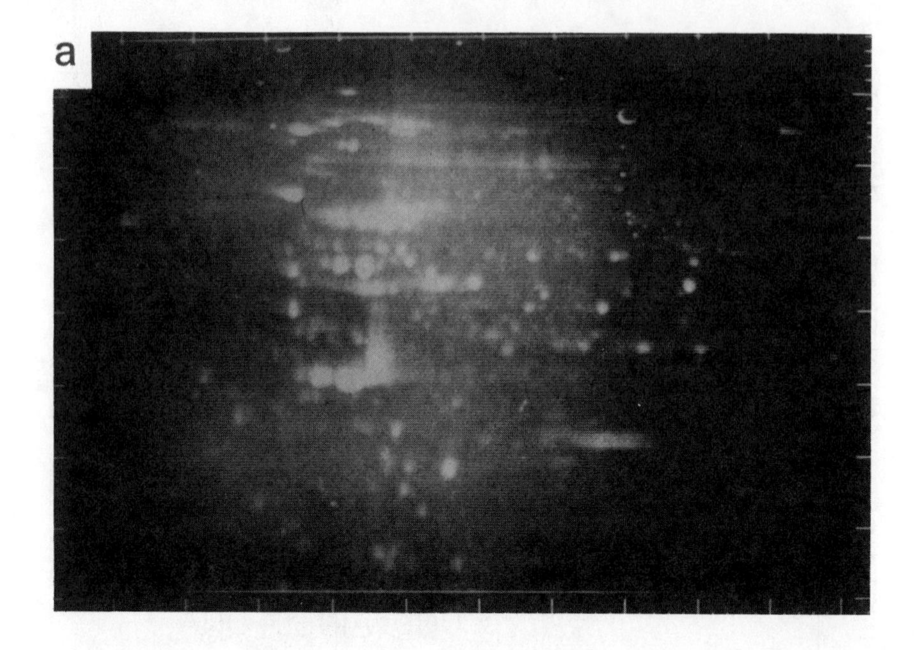

Figure 6.12 Photographs of the CCD images of fluorescent 2-D gels. Images a, b, and c are described in the text. (Reprinted with permission from reference 24.)

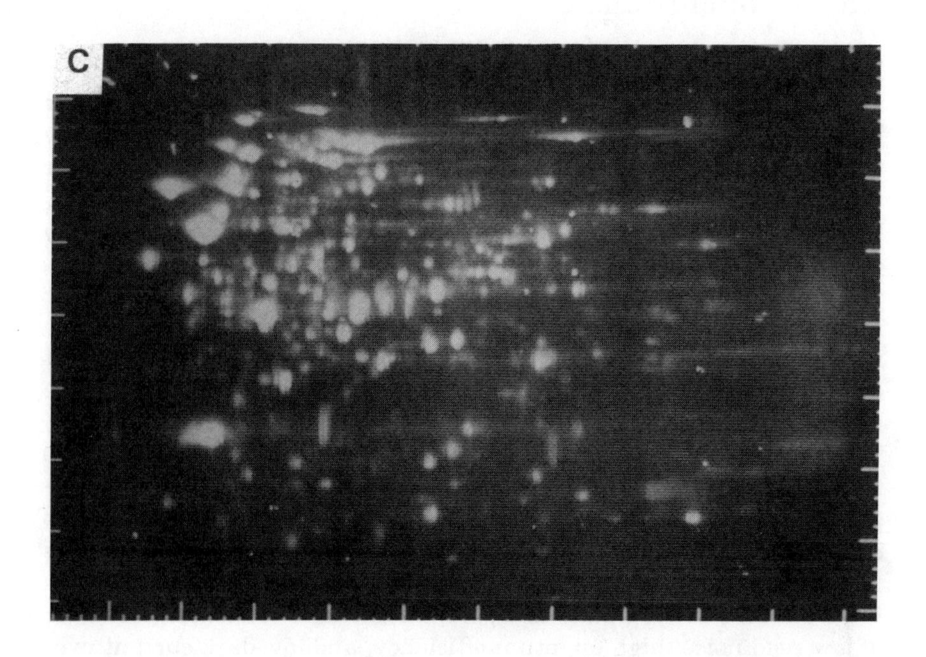

Figure 6.12 *(Continued)*

Figure 6.12c illustrates the ability of the CCD to obtain high-resolution images of complex spot patterns in two-dimensional gels. Images obtained with the CCD system demonstrated some loss of resolution due to the somewhat large gel area imaged by individual pixels, about 0.2×0.2 mm. Preliminary investigations into sensitivity indicate that both film and CCD detection have a similar lower limit of detection of 50 ng of protein in a single spot with a linearity extending to at least 300 ng.

This investigation illustrates the potential for sensitive fluorescence detection of proteins in two-dimensional gels using the fluorophore MDPF. The high quantum efficiency of the CCD allowed for sensitive data acquisition of protein images with a relatively low intensity excitation source and exposure times of 1 min or less.

The digitally stored nature of the CCD image and the often complex nature of the two-dimensional gel spot patterns indicate a need for computer-aided analysis routines. This is especially true if a large number of changes in the gel of interest are to be examined and compared with other two-dimensional gel patterns.

6.5 Conclusions

The scientifically operated CTD is a powerful tool for the detection of separated material in gels and on chromatographic plates, and further information is available from additional references [26-31]. It is easy to see the advantages afforded by the use of this technology over techniques utilizing mechanical slit scanning and film. Unfortunately, the high cost of scientific CTD imaging systems (approx. $25,000) may currently be prohibitive and make alternative technologies, such as video CTDs and scanning technologies, reasonable and practical for many imaging applications. With further time and development of this technology and the recent competition among manufacturers, the cost of the scientific CTD systems should be reduced to a level that is more affordable.

It is not just the maturation of CTD technology that must change for the use of a scientific CTD system to be justifiable for chromatographic or gel imaging. If one is to exploit the characteristics of these devices that have made them highly successful low-photon-flux imagers for astronomical and spectroscopic detection, careful attention needs to be directed toward plate and gel manufacture. The characteristics of low read noise, high quantum efficiency, and low dark current over extended integration times are most beneficial in an analysis where the researcher is trying to detect or observe a small amount of photoemissive material (chemiluminescence, fluorescence, etc.) under low background conditions. Thus, if scientific CTDs are to be practical for chromatographic plate analysis, further improvements in plate homogeneity are required. Further investigations into the reduction of background emissions in both gels and on chromatographic plates will improve the performance of this technology.

Scientific CTDs are commercially available in a variety of imaging systems and levels of sophistication. With only modest trade-offs in device imaging quality and system performance a researcher can possess a scientific CTD-based imaging system that fulfills many of his or her needs at competitive costs. These systems use different cooling techniques, thus operating at higher temperatures and dark currents, and can be supplied with lower grade, lower cost, but still adequate CTDs. An example of the excellent performance that can be achieved with these systems was discussed in this chapter through the work of Chan and co-workers [23]. These devices are excellent imagers for the analysis of gels and plates, especially in light of current limitations in background emissions. One drawback in all levels of sophistication among the scientifically operated systems lies in the lack of adequate software specifically tailored to the chromatographic or gel analysis and an integrated imaging system complete with optics and illumination sources.

Eventually, as system cost decreases and the technology grows to meet the needs of the biotechnology and analytical industry, these devices will significantly improve on the analysis of not only gels and plates but many other imaging technologies. In our lab we are continuing to investigate techniques for improving background emissions on plates and new ideas for both spatial and spectroscopic imaging of gels and chromatographic plates. We believe that we can circumvent limitations such as degraded resolution and cost by employing special readout techniques and system designs. As scientific CTDs are currently revolutionizing analytical spectroscopy, so also do we believe that a similar impact can be made on imaging applications in chemistry and biology.

References

1. Bilhorn, R.B.; Denton, M.B. *Appl. Spectrosc.* **1990**, *44*, 1538.
2. Pomeroy, R.S; Kolczynski, J.D.; Sweedler, J.V.; Denton, M.B. *Mikrochim. Acta [Wien]* **1989**, *3*, 347.
3. Epperson, P.M.; Sweedler, J.V.; Bilhorn, R.B.; Sims, G.R.; Denton, M.B., *Anal. Chem.* **1988**, *60*, 327A.
4. Epperson, P.M.; Jalkian, R.D.; Denton, M.B. *Anal. Chem.* **1989**, *61*, 282.
5. Jalkian, R.D.; Denton, M.B. *Proc. SPIE* **1989**, *1054*, 91.
6. Jalkian, R.D.; Ratzlaff, K.L.; Denton, M.B. *Proc. SPIE,* **1989**, *1055*, 123.
7. Jalkian, R.D.; Denton, M.B. *Appl. Spectrosc.* **1988**, *42*, 1194.
8. Poole, C.F.; Poole, S.K.; Dean, T.A.; Chirco N.M. *J. Planar Chromatogr.* **1989**, *2*, 180.
9. Poole, C.F.; Poole, S.K. *Anal. Chem.* **1989**, *61*, 1257A.
10. Poole, C.F.; Poole, S.K.; Dean, T.A.; Dallas, F.A.A.; Read, H.; Ruane, R.J.; Wilson, I.D., Eds., *Recent Advances in Thin-Layer Chromatography.* Plenum, New York, 1989, pp 11–28.
11. Bertsch, W.; Hara, S.; Kaiser, R.E.; Zlatkis, A., Eds., *Instrumental HPTLC.* Huthig, Heidelberg, 1980, pp 81–95.
12. Prunell, A.; Strauss, F.; Leblanc, B. *Anal. Biochem.* **1977**, *78*, 57.
13. Pollak, V.A.; Schulze-Clewing, J. *J. Chromatogr.* **1988**, *437*, 97.
14. Belchamber, R.M.; Read, H.; Roberts, J.D.M. *J. Chromatogr.* **1987**, *395*, 47.
15. Burns, D.H.; Callis, J.B.; Christian, G.D. *Trends Anal. Chem.* **1986**, *5*, 50.
16. Sutherland, J.C.; Bohai, L.; Monteleone, D.C.; Mugavero, J.; Sutherland, B.M.; Trunk, J. *Anal. Biochem.* **1987**, *163*, 446.
17. Bilhorn, R.B.; Epperson, P.M.; Sweedler, J.V.; Denton, M.B. *Appl. Spectrosc.* **1987**, *41*, 1125.
18. Bilhorn, R.B. Analytical Technologies Division, Eastman Kodak Labs, personal communication.
19. Cosgrove, J.A.; Bilhorn, R.B. *J. Planar Chromatogr.* **1989**, *2*, 362.
20. Jorgenson, J.W. *Anal. Chem.* **1986**, *58*, 743A.
21. Tebo, A., OE Reports, No. 84, Dec. 1990.
22. Mackay, C.D. In *Proc. Meet. Int. Electrophor. Soc., 5th*; Dunn, M.J., Ed. VCH, Weinheim, Germany, 1986.

23. Chan, K.C.; Koutny, L.B.; Yeung, E.S. *Anal. Chem.* **1991**, *63*, 746.
24. Jackson, P.; Urwin, V.E.; Mackay, C.D. *Electrophoresis* **1988**, *9*, 330.
25. Chen-Kiang, S.; Stein, S.; Udenfriend, S. *Anal. Biochem.* **1979**, *95*, 122.
26. Koutny, L.B.; Yeung, E.S. *Anal. Chem.* **1993**, *65*, 183.
27. Sutherland, J.C.; Sutherland, B.M.; Emrick, A.; Monteleone, D.C.; Ribeiro, E.A.; Trunk, J.; Son, M.; Serwer, P.; Poddar, S.K.; Maniloff, J. *Biotechniques* **1991**, *10*, 492.
28. Boniszewski, Z.A.M.; Comely, J.S.; Hughes, B.; Read, C.A. *Electrophoresis* **1990**, *11*, 432.
29. Jackson, P. *Biochem. J.* **1990**, *270*, 705.
30. Karger, A.; Ives, J.T.; Weiss, R.B.; Harris, J.M.; Gesteland, R.F. *Proc. SPIE* **1990**, *1206*, 78.
31. Brown, S.M.; Busch, K.L. *J. Planar Chromatogr.* **1992**, *5*, 338.

7
CCD Array Detectors For Multichannel Raman Spectroscopy

Richard L. McCreery

Department of Chemistry, The Ohio State University
Columbus, Ohio

7.1 Introduction

Raman spectroscopy is based on very weak inelastic scattering of photons from a sample, usually a liquid or solid. The Raman effect was first observed by C. V. Raman and K. S. Krishnan in 1928, when they noted that sunlight was scattered from liquids with a shift in frequency. In 1929 they noted that monochromatic light scattered by CCl_4 consisted of the original wavelength plus several others shifted in frequency by discrete intervals [1]. Although interest in the effect was intense, progress in the field was severely hindered by technological limitations. Typically only one in 10^{10} of the incident photons will undergo Raman scattering, so both an intense monochromatic source and a very sensitive spectrometer are required. For essentially all of the history of Raman scattering, experiments were conducted at the limit of available technology in order to overcome low sensitivity. As a consequence, improvements in applicability and information content of Raman spectroscopy have tracked technological advances. Examples include lasers, photon-counting detectors, intensified photodiode-array (IPDA) multichannel detectors, and operation in the NIR region of the spectrum using Fourier transform techniques. Since Raman scattering is so weak and difficult to observe, technological progress in instrumentation has had and continues to have a significant impact on the utility of Raman spectroscopy.

CCD array detectors represent an important advance for Raman spectroscopy. Their noise characteristics, two-dimensional multichannel capability, and high quantum efficiency all lead to greatly improved performance in terms of signal-to-noise ratio (SNR) and measurement time, and in some cases they permit totally new experimental designs. This chapter describes the value of CCD detectors for Raman spectroscopy, with particular attention to how CCD characteristics affect Raman spectrometer performance. After a brief discussion of the

Raman effect itself, the many variables of spectrometer design will be considered. Then the importance of detector variables to the Raman experiment will be discussed, followed by examples of CCD/Raman applications from the literature.

7.2 Fundamentals

Elastic scattering by gases, liquids, and solids is known as Rayleigh scattering and is exhibited in a variety of familiar effects, including blue sky and red sunsets. Rayleigh scattering is also a weak effect, with roughly one in 10^6 of the incident photons being scattered. Occasionally a scattered photon will gain or lose energy from the scattering molecule, giving rise to the Raman effect. The scattered photon will have higher or lower energy than the incident photon, leading to satellite peaks near the intense Rayleigh peak in the spectrum of scattered light. The energy level diagram in Figure 7.1 shows the most common type of Raman scattering, where the energy gain or loss involves vibrational modes of the scattering material. Raman scattering at longer wavelength (lower energy) than the incident light is more intense than that at higher energy because the ground vibrational state is more populated than the excited vibrational state.

A typical Raman spectrometer is shown in Figure 7.2, configured for collection of scattered light at 90 degrees to the input beam.

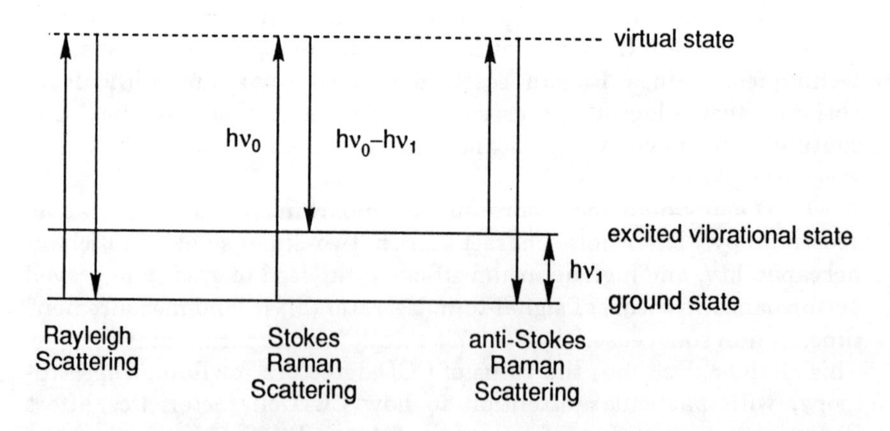

Figure 7.1 Energy levels involved in normal Raman scattering. The virtual state is usually below a real electronic state.

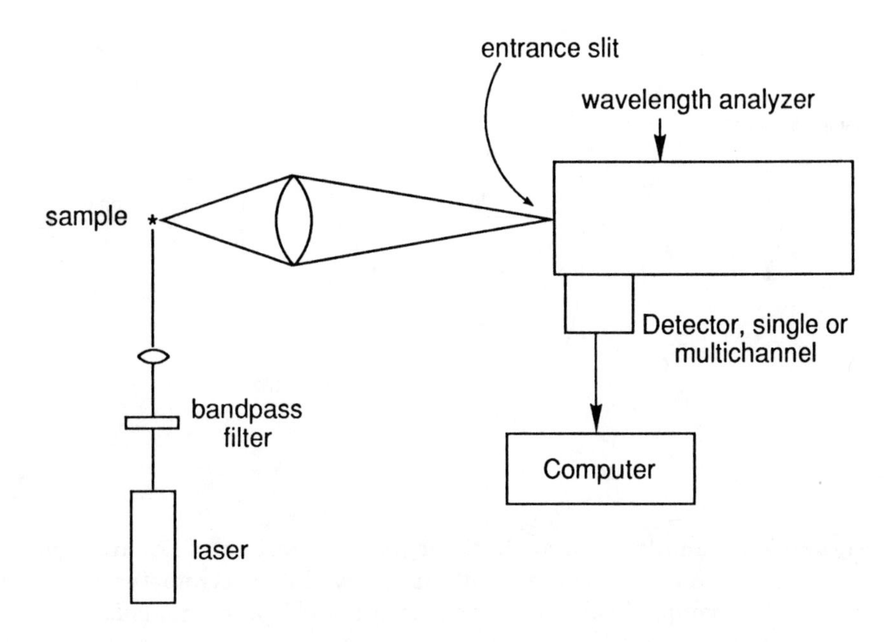

Figure 7.2 Generic Raman spectrometer with 90 degree collection geometry. Alternatives are backscattered geometry and fiber-optic sampling.

Quantitative aspects of Raman scattering are most easily appreciated by considering Figure 7.3, which shows a laser beam passing through a transparent sample. If a is the beam radius and d the beam length monitored by the spectrometer, then $\pi a^2 d$ is the scattering volume and $P_o/\pi a^2$ is the laser irradiance (photons cm^{-2} s^{-1}) with P_o being the laser power (photons s^{-1}). Defining D as the number density of scatterers (molecules cm^{-3}) and β as the differential Raman scattering cross section (cm^{-2} $molecule^{-1}$ sr^{-1}), the total Raman scattering from the scattering volume, I_R (photons s^{-1} sr^{-1}), is given by equation 7.1 [2,3]:

$$I_R = \frac{P_o}{\pi a^2} \beta D \pi a^2 d = P_o \beta D d \qquad (7.1)$$

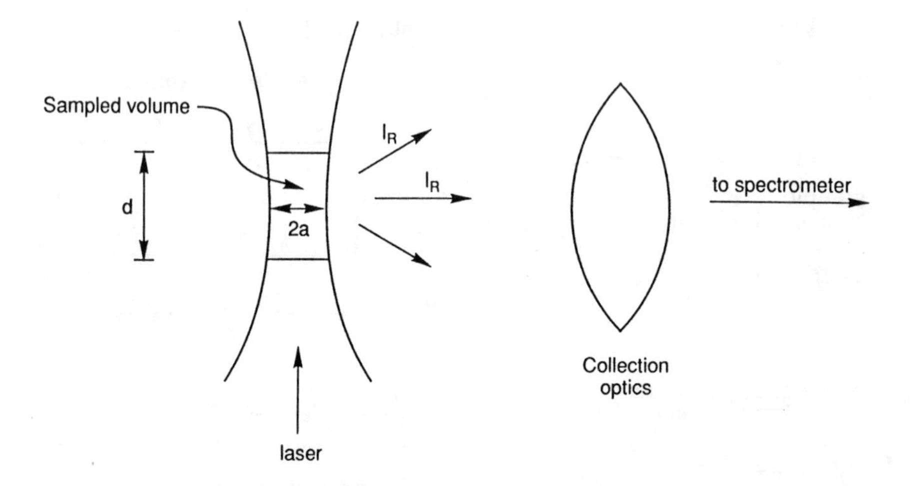

Figure 7.3 Sampled volume for 90 degree observation of liquid or gas samples. The area observed by the spectrometer is the sampled volume viewed from the side and equals *2ad*.

Stated more simply, equation 7.1 relates I_R to the number of incident photons (P_o), the cross section (β), and the number of sample molecules ($D\ \pi a^2 d$).

Several observations relevant to instrument design are useful at this point. First, I_R is proportional to P_o, so higher laser powers yield larger signals. Second, the cross section β (often denoted $d\sigma/d\Omega$) is exceedingly small, with typical values in the region of 10^{-30} cm^{-2} molecule^{-1} sr^{-1}. For example, a 0.02 *M* solution of benzoquinone ($\beta = 3.4 \times 10^{-29}$ cm^{-2} molecule^{-1} sr^{-1}, $D = 1.2 \times 10^{19}$ molecules cm^{-3}) illuminated by a 100-mW laser ($P_o = 2.5 \times 10^{17}$ photons s^{-1}) and monitored with $d = 1$ mm, yields *total* scattering of $I_R = 1.0 \times 10^7$ photons s^{-1} sr^{-1}, or about 1 photon in 10^{10}. Third, Raman scattering is not isotropic. Many discussions of scattering intensity ignore directionality of the cross section β; however, most spectrometer designs use 90 degree or back-scattering geometry under conditions where scattering is maximized. Although anisotropic scattering is very useful in many cases, it will not be considered further here. Fourth, I_R is proportional to number density, D. As detection techniques improve, D, β, or P_o may be decreased and the range of samples increased greatly. Thus spectrometers capable of very-low-light-level operation are very valuable, and it is here that CCDs have major impact.

Unfortunately, a comprehensive treatment of the effect of spectrometer design and detector characteristics on the SNR for Raman spectroscopy is difficult due to the number of variables involved. Sample type, collection geometry, laser-beam parameters, resolution, detector-noise characteristics, and spectrometer type can all profoundly affect SNR, measurement time, and sample-radiation damage [2]. As a result, the application usually dictates the choice of spectrometer design, and no single spectrometer will optimally suit all applications. Although a comprehensive treatment of this issue is not feasible, several specific cases will be considered to illustrate the improvements permitted with CCD technology.

Starting with the 90 degree collection geometry shown in Figure 7.2, it is convenient to define specific intensity, L, as the scattered light intensity per unit of scattering area with units of photons cm^{-2} s^{-1} sr^{-1}. Assuming 90 degree observation and that the beam radius a is small relative to the spectrometer depth of field, L is given by equation 7.2.

$$L = \frac{I_R}{2\pi a d} = \frac{P_o \beta d}{2\pi a} \qquad (7.2)$$

Note that L increases with smaller beam radius (and higher radiation density), since more photons are scattered from a smaller area.

The number of scattered photons reaching the detector will be related to L, but also to the transmission of the collection optics and spectrometer (T) and the solid angle collected (Ω). A fairly subtle but important concept involves the area of scattered light collected, denoted here as A_D. Consider a sample with constant L over some relatively large scattering area, $2\pi ad$. If the spectrometer monitors a large fraction of this area, more molecules are observed and the signal is larger than for a small monitored area. A_D is related to the smallest aperture of the spectrometer, which may be the detector area, the slit area, or even the scattering area if the spectrometer is "underfilled." For now, suppose the detector area is the limiting aperture. In this case, A_D is the *image* of the detector at the sample, possibly magnified or demagnified.

The photon flux reaching the detector, Φ_s (photons/second) is given by equation 7.3,

$$\Phi_s = L \, A_D \, \Omega \, T \tag{7.3}$$

Finally, the signal S (electrons) is easily determined from Φ_s via the quantum efficiency, Q_D (electrons/photon), and measurement time, t. Note that electrons are used as the unit for the signal (as opposed to counts, current, or analog-to-digital-converter units) because ultimately the electronics will detect single electrons. The actual readout of the system is usually proportional to the number of electrons.

$$S = \Phi_s \, Q_D \, t = L \, A_D \, \Omega \, T \, Q_D \, t \tag{7.4}$$

It is useful at this point to classify the variables determining the Raman signal into three groups [2]. First, the signal will increase with L, and therefore with P_o, β, D, and a^{-1}. In principle, L can be increased almost without bound by higher laser power and tighter focus, but in practice L is ultimately limited by radiation damage to the sample. As noted earlier, higher β permits lower D or lower P_o to achieve the same signal. Second, S is proportional to $A_D \Omega T$, which is largely determined by spectrometer variables and is independent of L provided the spectrometer is overfilled. An observation of fundamental importance is the constancy of the $A_D \cdot \Omega$ product through the optical system. An example is shown in Figure 7.4 for a simple case of imaging a scattering area onto a fixed detector area or slit. In Figure 7.4A, the scattering region is imaged without magnification onto the detector, so A_D equals the detector area and Ω is determined by the lens diameter and focal length. In Figure 7.4B, the lens is closer to the sample, thereby increasing Ω, but the scattering region is also magnified as a consequence. The effect is to monitor less scattering area, decreasing A_D. For both Figures 7.4A and 7.4B, the product $A_D \cdot \Omega$ is the same, and the signal is maximized by increasing the $A_D \cdot \Omega$ product.

A common misunderstanding often arises regarding the trade-off of A_D and Ω. At first glance it would seem prudent to increase Ω, for example by using a lower *f*/ collection lens, to increase S. As noted above, however, increasing Ω will decrease A_D, and no gain in signal will occur *if the spectrometer is overfilled.* Under this condition, increasing Ω will collect a larger solid angle of light from a smaller number of scatterers. On the other hand, if the laser beam is tightly focused and the spectrometer is *underfilled*, increasing Ω will increase signal. Since the slit image is larger than the laser spot size for a tightly focused beam, decreasing the slit image will not decrease the signal. For this reason, commercial spectrometers often use a tight

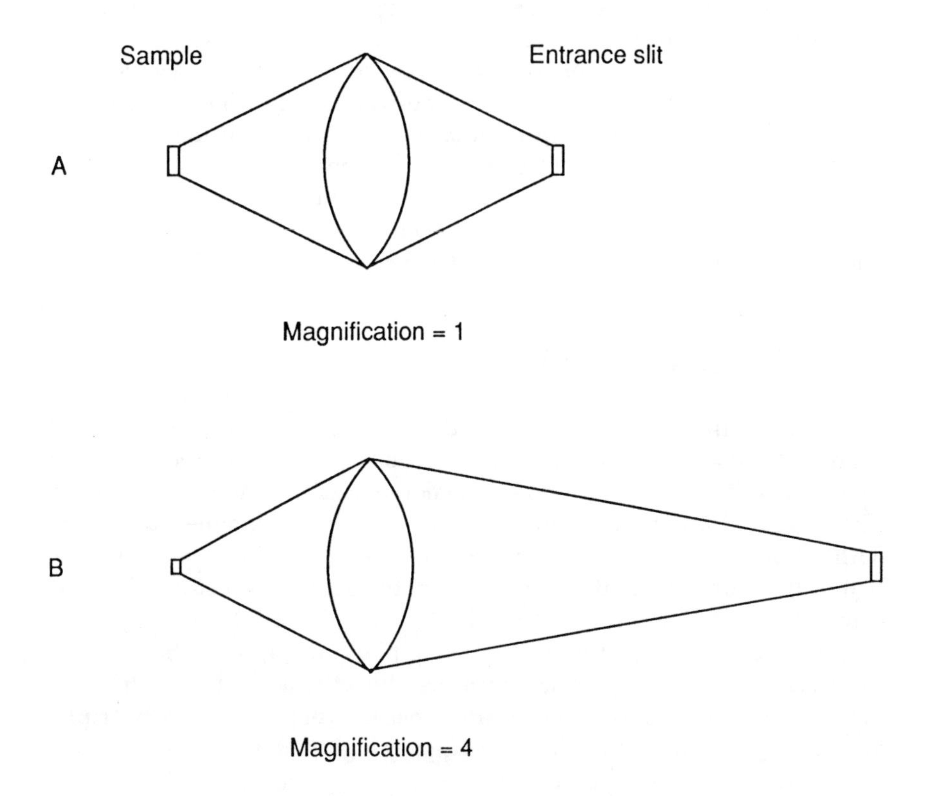

Figure 7.4 Effect of magnification on A_D. Case A: Ω and A_D are the same at the entrance slit and sample; Case B: A_D is smaller and Ω larger at the sample.

beam focus to maximize L and permit a larger Ω. This is an effective practice for increasing signal provided the radiation damage threshold of the sample is not exceeded.

The third group of variables affecting signal includes detector variables such as Q_D and perhaps A_D if the detector area is the limiting aperture. Detector issues will be discussed in detail later, but it is important to note now that both detector area and noise characteristics can be at least as important as Q_D in determining performance. Since Raman spectroscopy involves light collection, the larger the magnitude of A_D (and usually the larger the detector), the more light is collected and the larger the signal. Furthermore, signal magnitude is less important than SNR, and noise characteristics are critical. The factors determining SNR are considered below for several common situations.

To summarize these introductory issues, the Raman signal is determined by laser/sample variables (L), spectrometer variables (A_D, Ω, T), and detector properties (Q_D, area, and noise characteristics). For optimum operation, the spectrometer is usually slightly overfilled so that the Raman scattering originates in an area that is larger than A_D. In this case, S and SNR are determined to a large extent by detector performance. If the spectrometer design is properly optimized, the SNR is ultimately limited by both sample radiation damage and detector characteristics, particularly area, quantum efficiency, and noise.

7.3 Spectrometer Designs

Before considering detectors in detail, we will examine several commonly used spectrometers, with particular attention to the variables relevant to the two-dimensional flat-field format of CCDs. The original Raman spectrometers were single-grating spectrographs with photographic plate detection, but these were replaced with scanning monochromators when photoelectric detectors became available. Until approximately 1980, virtually all Raman spectrometers employed single-photon-counting photomultiplier tubes (PMTs) and a scanning double monochromator. With the advent of multichannel detectors, flat-field spectrographs became popular, usually triple-grating systems. At this writing, single, double, and triple spectrometers are in common use.

Raman spectroscopy puts significant demands on the spectrometer in the area of resolution, stray light rejection, and flat-field performance. Raman spectra cover a relatively narrow wavelength range, particularly when that range is in the blue end of the spectrum. For example, a 0 to 3000 cm^{-1} Raman spectrum is about 96 nm wide when excited by 515-nm laser light, and only 20 nm wide with a 250-nm laser. Thus a spectrometer capable of 1 cm^{-1} resolution for a 515-nm laser must have a wavelength resolution of about 0.03 nm. Relatively high dispersion is required to achieve good resolution with reasonable slit widths. This point will become a significant issue when considering CCD detectors later on.

Excellent stray light rejection is required because the Raman scattering must be discriminated from the relatively intense Rayleigh scatter from the laser. Potential interference from stray light is orders of magnitude worse yet when examining solids, since diffuse scatter can be much stronger than Rayleigh scatter. A typical benchtop monochromator will have a stray light specification of perhaps 10^{-5}, meaning that the monochromator transmission drops to 10^{-5} of the value at its set wavelength when measured at some distance (e.g., 8 bandpasses) away on either side. Since Rayleigh or diffuse scatter can be 10^3 to 10^8 times stronger than Raman scattering, this stray light rejection is often insufficient. Double spectrometers greatly improve stray light rejection at some cost in transmission. A good double monochromator has a stray light specification of about 10^{-12}. The practical consequences of low stray light include low background and the ability to acquire spectra at low Raman shift, near the laser line.

In addition to dispersion and stray light rejection, a third issue of importance to Raman spectroscopy with array detectors is flat-field performance. All spectrometers have minimum aberrations for on-axis light, and near-ideal performance is achieved only for a single point at the entrance slit. In other words, only a "slit" of minimal height and width is ideally imaged on the exit plane of the spectrometer. As the slit height is increased, off-axis light must necessarily be used. Any off-axis aberration in the spectrometer causes some resolution degradation as the slit height is increased. Furthermore, an array detector covering a 1 × 1 cm or 1 × 2.5 cm field on the focal plane requires the use of off-axis light in both dimensions. Without careful attention to aberration correction, significant distortion of the entrance slit image will occur with accompanying loss of resolution. The larger the detector array the more severe these effects, since the light of relevance is imaged further off the optical axis. These aberrations vary greatly with spectrometer design and can be corrected only under certain conditions. The effects are incorporated into a "flat field" specification for the spectrometer. A 10 × 25 mm flat field indicates that the entrance slit is not curved or defocused over a 10 × 25 mm field at the focal plane. The user should be aware that a flat-field specification may be valid only for a particular wavelength range or slit height.

Table 7.1 summarizes the properties of the three spectrometer designs shown in Figure 7.5, with the list being representative rather than exhaustive. Single-grating instruments have good transmission and light-collecting ability (lower *f*/ yields higher Ω) but have relatively poor stray light rejection.

Singles can have fairly large flat fields, currently up to about 10 × 25 mm. Aberration-corrected singles known as imaging spectrographs have particularly low distortion of the slit image and are available with low *f*/. Singles can yield the best SNR due to high T and low *f*/, but some provision to reduce stray light must be made externally. Double monochromators have greatly improved stray light rejection but lower transmission due to the larger number of reflections from mirrors and gratings. Doubles have a relatively small flat field and are rarely used with array detectors. Triple spectrographs are quite common for Raman spectroscopy with array detectors, due

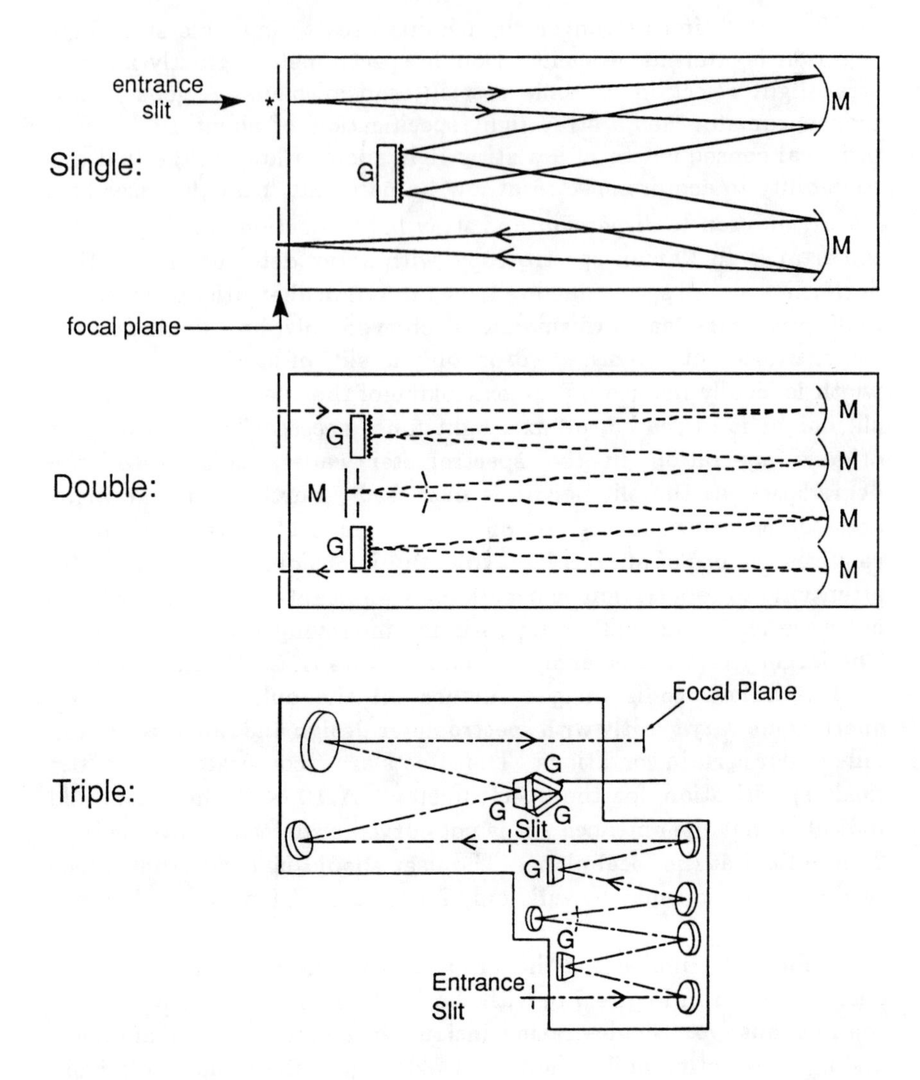

Figure 7.5 Three common spectrograph designs used in Raman spectroscopy. *M* denotes a mirror and *G* denotes a grating.

to their versatility. Most triple designs consist of a subtractive dispersion double monochromator preceding a conventional single spectrograph. The double acts as a tunable prefilter to reduce stray light, and effectively transmits a tunable bandpass of the scattered light to the spectrograph stage. The center and width of the pass band can be selected by the user. Since triples require three gratings and many mirrors, their transmission is low, and aberrations are difficult

Table 7.1 Typical Spectrometer Specifications[a]

Design	f/	Transmission	Reciprocal Linear Dispersion[b] (cm^{-1}/mm)	Stray light rejection[c]
Single[d]	5.2	.32	26	10^{-5}
Double[e]	7.8	.10	12.5	10^{-12}
Triple[f]	9	.11	35	10^{-14}

[a] Data from reference 2.
[b] At 514.5 nm.
[c] Approximate: depends on measurement conditions.
[d] Instruments SA 640, 0.64 meter, 1800 lines/mm grating.
[e] Spex 1403, 0.85 meter, 1800 lines/mm grating.
[f] Spex "Triplemate," 1800 lines/mm final stage.

to correct. The majority of current, multichannel, Raman spectroscopy is performed with triple spectrographs.

Several recent developments have increased the utility of single spectrographs for Raman spectroscopy with array detectors. High transmission, compact size, the availability of rapid, automatic grating changes, and efficient computer interfaces provide excellent performance compared with that of doubles and triples, provided the stray light can be reduced to acceptable levels. The introduction of efficient holographic and dielectric filters to reject the laser and Rayleigh-scattered light before it enters the spectrograph provides the user with high transmission, low *f*/, good stray light rejection, and a compact design.

In summary, spectrometer design and selection involves trade-offs of light collection (determined by T and Ω), dispersion (determined by grating density and focal length), and stray light rejection. As is often the case, the choice is dictated by the application, but all of these variables must usually be considered. When array detectors are used, the added issue of flat-field performance becomes a critical factor determining resolution.

7.4 Detector Characteristics

Before discussing the types of detectors used in Raman spectroscopy, it is useful to consider which detector characteristics are important. Detectors used for dispersive instruments will be discussed in the greatest detail, but comparisons with nondispersive instruments will be made where relevant.

7.4.1 Quantum Efficiency

As noted in equation 7.4, the Raman signal is proportional to the quantum efficiency Q_D, defined as the fraction of photons incident on the detector that produce a detectable particle, usually a photoelectron. As shown in Figure 7.6, Q_D is strongly wavelength dependent, with the typical silicon curve extending significantly deeper into the red than most PMTs. The higher Q_D in the near infrared (NIR) is a significant advantage of CCDs over PMTs, but this issue will be addressed separately.

7.4.2 Detector Area

As noted earlier, the area from which Raman scattering is detected is often determined by the area of the detector, often demagnified by the

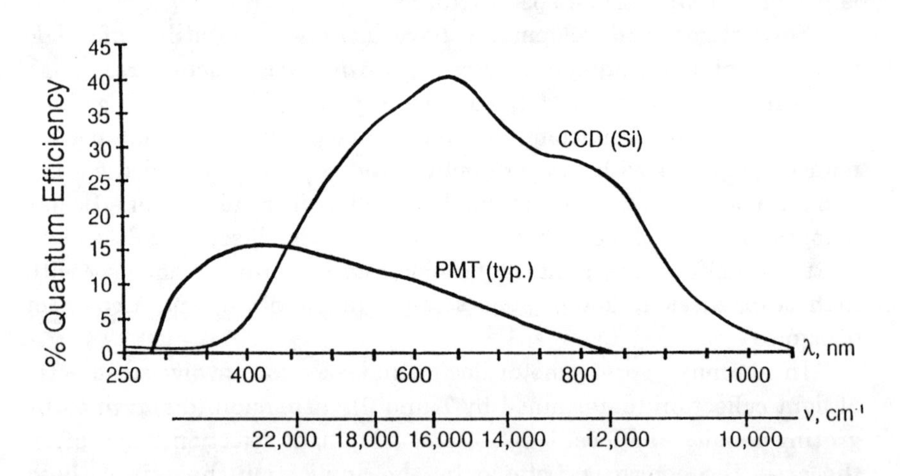

Figure 7.6 Typical quantum efficiency versus wavelength curves for a photomultiplier tube (S-20 photocathode) and a silicon-based CCD. Both curves depend strongly on fabrication and other variables.

collection optics. In most single-channel scanning spectrometers using a PMT, the area A_D equals the entrance-slit image at the sample, assuming this image is smaller than the laser-beam waist. For multichannel detectors, A_D may be limited by either the entrance slit or the area of each detector element. Thus both the height and width of a detector element can be critical in determining A_D and ultimately the signal and SNR magnitudes. Since there is a trade-off between SNR (which is improved with larger elements) and resolution (which is improved with narrower elements), a multichannel Raman spectrometer design must usually compromise these two parameters.

7.4.3 Number of Channels

We will demonstrate later that SNR for a single-channel detector is proportional to the square root of the number of channels, $N_c^{1/2}$, under many common conditions. In addition, when an array detector is operated as a spectrograph with a fixed monochromator position, the number of spectral resolution elements has N_c as an upper limit. In the majority of spectroscopic applications, the number of channels along the axis of wavelength dispersion is most important, and readily available devices contain 384 to 1152 channels. In certain applications, particularly those involving imaging, the axis along the slit dimension becomes important, and the number of channels in two dimensions becomes a design or selection criterion.

7.4.4 Detector Noise Characteristics

Several chapters of this book have dealt with CCD noise sources, and those treatments will not be repeated here. In the context of Raman spectroscopy, however, it is of fundamental importance to understand that the low dark signal and readout noise of CCDs are major benefits of their use. In this section, we will describe those aspects of detector noise that are relevant to Raman spectroscopy; then we will consider the SNR issues for various common experiments.

First, readout noise occurs when a detector signal is converted to some useful quantity, usually a digital number in a computer. For CCDs this noise includes noise introduced by the output amplifier and quantization noise in the A/D converter. For a photon-counting PMT, readout noise is taken as zero, since the electronics either detect a photoelectron or not, and this binary event is readout-noise free. Second, dark counts and noise occur from random counts generated in the absence of light. For PMTs, dark counts are from spontaneous emission of electrons from the photocathode, while for CCDs they come from thermally generated electron/hole pairs. Although dark counts are inherently random, they can be characterized by a flux, ϕ_d, with units of e^-/s. Thus the number of dark electrons (N_d) is given by equation 7.5

$$N_d = \phi_d t \tag{7.5}$$

Since the dark counts follow Poisson statistics, the standard deviation of the dark counts (σ_d) equals their square root [4]:

$$\sigma_d = N_d^{1/2} = (\phi_d t)^{1/2} \tag{7.6}$$

A relevant technical point is the relationship between the electrons generated in the detector (dark or otherwise) and the actual output signal. For a PMT they are the same with each photoelectron being counted. For a CCD, the "gain" (A/D units/e^-) varies with the detector design but is typically 0.25 to 0.008 for systems used for Raman. Since SNR is the inverse of the *relative* standard deviation of S, it can be shown that the gain does not affect the observed SNR.

To look ahead a bit, we will show that Raman experiments fall into two general types with regard to SNR. In the first, the sample signal and background are much larger than readout or dark noise, and detector noise is inconsequential. In the second, the sample is a very weak scatterer, and detector noise can be an issue. We will consider these situations in the abstract initially, and then examine some real detectors and applications.

7.5 Signal-to-Noise Ratio Considerations in Raman Spectroscopy

In most treatments, SNR is defined as the inverse of the relative standard deviation of the signal. For most Raman applications, the signal is the average difference between a scattering peak and the baseline, and the noise is the standard deviation of the peak signal. In some cases the noise is measured for a blank, but it is up to the authors of a given publication to state how SNR is defined. In this discussion SNR will be the average signal relative to a baseline divided by the standard deviation of the signal. The detection limit is the concentration of analyte required for a SNR of 2. For a Raman experiment, noise can arise from background or sample shot noise, laser or background flicker noise, or detector dark or readout noise. We will assume a constant laser intensity and will therefore ignore flicker noise. The remaining noise sources are most easily considered initially as limiting cases where one noise source is dominant.

7.5.1 Analyte Shot-Noise-Limited Raman Scattering

This case occurs for strong scatterers where the Raman scattering is stronger than any background or detector noise sources. We will use "analyte" to denote the scatterer of interest, while "sample" includes both analyte and background scatterers. The signal is given by equation 7.4, and the standard deviation of the signal is the noise (N), as in equation 7.7:

$$N = S^{1/2} = (L\ \Omega\ A_D\ T\ Q_D\ t)^{1/2} \tag{7.7}$$

The SNR for this single-channel case is simply $S^{1/2}$, shown in equation 7.8 after substituting for L:

$$(S/N)_{SC} = \left[\frac{P_o\ \beta_a\ D_a\ \Omega\ A_D\ T\ Q_D\ t}{2\pi a} \right]^{1/2} \tag{7.8}$$

Note that the SNR is proportional to the square root of the sample concentration (via D), and also the quantity $(\Omega A_D T Q_D)^{1/2}$. β_a and D_a refer to the analyte cross section and number density.

Recall that t is the time that the Raman scattering is monitored, so the SNR improves as more photons are collected. For a single-channel spectrometer, t equals the time that each wavelength is monitored, for example by a PMT. To build up an entire spectrum, the total measurement time, T_m, equals the number of desired resolution elements N_R times t;

$$T_m = N_R\ t \tag{7.9}$$

With an array detector, it is possible to monitor many wavelengths simultaneously. If the array detector is configured to have N_R resolution elements (not necessarily equal to the number of detector channels), then the SNR depends on T_m, since all channels collect photons for the entire measurement time:

$$(S/N)_{MC} = \left[\frac{P_o\ \beta_a\ D_a\ \Omega\ A_D\ T\ Q_D\ T_m}{2\pi a} \right]^{1/2} \tag{7.10}$$

or

$$(S/N)_{MC} = \left[\frac{P_o \, \beta_a \, D_a \, \Omega \, A_D \, T \, Q_D \, N_R \, t}{2\pi a} \right]^{1/2} \tag{7.11}$$

So, *if all else is equal*, the multichannel detector SNR will be a factor of $N_R^{1/2}$ higher than the single-channel case for a given total measurement time. This is the so-called multichannel advantage and is a major motivation for the development of array detectors. While $N_R^{1/2}$ is the often-cited SNR gain for multichannel over single-channel systems, it should be noted that more than N_R may vary for the two situations. Taking the ratio of equation 7.11 to equation 7.8 and allowing the detectors to have different values of A_D, Q_D, and of course N_R, equation 7.12 results:

$$\frac{(S/N)_{MC}}{(S/N)_{MC}} = \frac{(A_D \, Q_D \, N_R)_{MC}^{1/2}}{(A_D \, Q_D)_{SC}^{1/2}} \tag{7.12}$$

Thus the $N_R^{1/2}$ advantage of a multichannel system may be increased or decreased by differences in quantum efficiency or detected area, compared with a single-channel system. In some cases the $N_R^{1/2}$ advantage may be lost by smaller A_D or lower Q_D [2,5].

7.5.2 Background Shot-Noise-Limited Raman Spectroscopy

We will define *background* here as any light sampled by the detector (other than Raman scattering from the analyte itself) that is derived from the laser. Assuming linear scattering, the photon flux of background is proportional to input laser power and is caused by elastic scattering, which reaches the detector because of insufficient stray light rejection, inelastic scatter such as fluorescence, and solvent Raman scattering. Despite their distinct origins, these phenomena may be combined for convenience into a single cross section, β_B, and number density, D_B. By analogy to equation 7.4, the background response B will be

$$B = \frac{P_o \beta_B D_B \Omega A_D T Q_D t}{2\pi a} \tag{7.13}$$

If the only sources of noise are analyte and background shot noise, the total noise will be given by equation 7.14, since S and B are additive.

$$N = \sqrt{S + B} \tag{7.14}$$

The signal is now considered to be the mean height of the analyte-signal peak above the background,* which we have defined as S; thus the SNR is

$$\frac{S}{N} = \frac{S}{\sqrt{S + B}} \tag{7.15}$$

Substituting equations 7.4 and 7.13 into 7.15 and simplifying yields:

$$\left[\frac{S}{N}\right]_{SC} = \frac{\beta_a D_A}{(\beta_a D_a + \beta_B D_B)^{1/2}} \left[\frac{P_o \, \Omega \, A_D \, T \, Q_D \, t}{2\pi a}\right]^{1/2} \tag{7.16}$$

Several observations are available from equation 7.16. First, in the common case where background scatter is large relative to analyte scatter, the first term of the denominator is negligible and the SNR is linear with analyte concentration (D_a). Second, large background scatter reduces SNR, since the small signal is superimposed on relatively large background shot noise. Third, if analyte scatter dominates over background scatter, equation 7.16 reduces to equation 7.8. Fourth, the SNR is still proportional to $(P_o \, \Omega \, A_D \, T \, Q_D)^{1/2}$. So if all else is equal, a detector with higher Q_D will have a higher SNR, but only by a factor of the square root of the improvement, $Q_D^{1/2}$. Fifth, the same multichannel advantage, $N_R^{1/2}$, applies for array detectors when the background shot noise contributes or dominates.

7.5.3 Detector Noise Contributions

In the previous two cases, the noise was assumed to be dominated by analyte and background shot noise, both of which are correlated with laser power, collection efficiency, and so on. In contrast, noise sources in the detector such as dark and readout noise have nothing to do with the sample, and lead to quite different equations for SNR [6–8]. Although detector-noise-limited operation is not common when using CCDs in Raman spectroscopy, there are some important situations where it is an issue.

---

* S can also be determined by subtracting a background spectrum from a sample spectrum, rather than measuring the signal peak above a *mean* background response. The former method will have a higher SNR by a factor of $\sqrt{2}$.

With proper temperature control, the dark "current" for a detector can be assumed constant, defined here as ϕ_d, in e⁻/s. The total dark signal will be $\phi_d t$, and the resulting noise if dark current were the only source is $(\phi_d t)^{1/2}$. The readout noise is usually characterized by the standard deviation of a series of readings, σ_R. Thus the total noise, including analyte and background shot noise, dark noise, and readout noise is given by equation 7.17 [9]:

$$N = (S + B + \phi_d t + \sigma_R^2)^{1/2} \tag{7.17}$$

If dark current were the only source of noise, the SNR would be given by dividing equation 7.4 by equation 7.17 and ignoring other terms in the denominator, to yield equation 7.18:

$$\frac{S}{N} = \frac{P_o \beta_s D_s \Omega \, A_D T Q_D t^{1/2}}{2 a \pi \phi_d^{1/2}} \tag{7.18}$$

Note that the SNR degrades as ϕ_d increases, since the signal of interest is on top of larger dark noise. ϕ_d increases rapidly with temperature for most detectors, so all Raman spectrometers have provision for detector cooling. Since SNR increases with $t^{1/2}$, the multichannel advantage applies when more than one wavelength can be monitored. If a multichannel detector is compared with a single-channel system, with all being equal except A_D, Q_D, and N_R, equation 7.19 results for equal total measurement times:

$$\frac{(S/N)_{MC}}{(S/N)_{SC}} = \frac{(A_D Q_D)_{MC} N_R^{1/2}}{(A_D Q_D)_{SC}} \tag{7.19}$$

Note the linear dependence on $A_D Q$, in contrast to equation 7.12 for the sample shot-noise-limited case. These issues have been considered more broadly for spectrochemical applications in general [9].

A useful aside is appropriate here regarding *multiplex* versus *multichannel* spectrometers. A multiplex spectrometer such as a Michelson interferometer directs all wavelengths of interest onto a single detector simultaneously rather than dispersing them onto many detectors. The familiar consequence is an increase in measurement time for each wavelength compared with a single-channel system. Since all wavelengths are monitored at once, each wavelength is monitored for a time $T_m = N_R t$, where t is the time per resolution element for the single-channel case. Considering equation 7.18, a multiplex system would also improve SNR by $N_R^{1/2}$, by increasing the measurement time per wavelength to $N_R t$. This is the usual situation for Fourier-transform-IR and FT-Raman measurements which employ NIR or IR detectors with fairly high dark noise. The $N_R^{1/2}$ is commonly referred to as the multiplex advantage.

The multiplex and multichannel advantages are not equivalent, even though they appear so based on equation 7.18. If sample shot noise contributes to the total noise significantly, the noise will increase as more wavelengths are monitored by a single detector. For example, the relatively high shot noise from a large sample peak adds to the small noise on a small peak. Stated differently, if operating in the sample shot noise limit (equations 8 or 16), a multiplex system will diminish the SNR such that the $N_R^{1/2}$ advantage may be canceled. Depending on the distribution of scattering across the spectral range examined, a multiplex spectrometer can exhibit a disadvantage or advantage relative to a single-channel system; however, if operating in the sample shot-noise limit, a multiplex spectrometer will rarely be more sensitive than a single-channel system, and can often be worse (the so-called "multiplex disadvantage"). Since each element of a multichannel spectrometer monitors only one wavelength at a time, the $N_R^{1/2}$ advantage applies in either the shot-noise-limited case (equation 7.16) or the dark-noise-limited case (equation 7.18), whereas the $N_R^{1/2}$ advantage applies to a multiplex system only in the dark-noise limit (equation 7.18). It should be noted that all currently available detectors for the IR operate in the dark-noise limit, so the multiplex advantage is usually important in the IR spectral region.

Returning to detector noise in array detectors, there remains readout noise to consider. The amplification and analog-to-digital conversion required in most array detectors introduces noise that does not depend on signal size or measurement time, as noted in equation 7.17. If readout noise were dominant (rarely the case), the SNR would be given by equation 7.20.

$$\frac{S}{N} = \frac{P_o \beta_s D_s \Omega \, A_D T Q_D t}{2 a \pi \sigma_r} \qquad (7.20)$$

Since SNR is linear with *t,* both multichannel and multiplex spectrometers would exhibit a N_R advantage when readout noise is the only noise source. This situation is unlikely for multiplex systems but might be encountered for multichannel detectors with very weak scattering.

7.5.4 "Cosmic" Background in CCDs

In addition to sample shot noise, an additional noise source occurs with CCDs that is difficult to treat mathematically: dark and readout noise. Hard radiation of either cosmic or terrestrial origin will generate electron/hole pairs in any silicon device, including CCDs. Since the dark and readout noise of a CCD is so low, the hundreds or even thousands of electrons generated by a gamma photon or other high-energy particle can be much larger than both noise and signal. The result is a spike in the Raman spectrum at the pixel affected by the radiation. Hard radiation rates vary greatly with locality, but in a relatively "quiet" environment there might be 1 to 2 events on a 1-cm^{-2} detector every 10 minutes. If the lab has been used for radiochemistry, the situation might be much worse, to the point of precluding CCD use.

7.5.5 SNR Summary

Table 7.2 summarizes the dependence of SNR on several experimental situations. There is often more than one contribution to noise, but consideration of the limiting cases in Table 7.2 is informative, and these cases apply to a variety of actual experiments. After considering several detectors currently in use for Raman spectroscopy, we will examine how the SNR considerations noted here apply to several real measurements.

7.6 Common Raman Detectors

7.6.1 Photon-Counting Photomultiplier Tube (PMT)

The PMT was the workhorse of Raman spectroscopy through most of its existence until multichannel detectors began to be used in about 1980. PMTs probably still account for the majority of detectors in use but are rapidly being superseded by multichannel systems. The PMTs used for Raman spectroscopy are designed and selected for low dark current and high Q_D and are almost always cooled to reduce dark current to less than 10 e^-/s at the photocathode. A pulse-height discriminator selects PMT output pulses of the appropriate amplitude, which are then counted by a counter or computer. Count rates, and therefore maximum photon flux, are limited by pulse pileup to about 10^6 pe/s. PMTs are fairly easily damaged by overexposure to light and can be destroyed by inadvertent exposure to room light or even scattered laser light.

Referring to equation 7.17, the dark flux for a PMT, ϕ_d, ranges from 2 e^-/s to perhaps 50 e^-/s, at which point the tube should be replaced. The readout noise is essentially zero, since pulse-height discrimination effectively avoids quantization noise. Taking 10 e^-/s as

Table 7.2[a] Signal-to-Noise Ratio Summary

Dominant Noise Source	Dependence of SNR on Concentration	Dependence on Spectrometer & Detector	Dependence on Measurement Time	Multichannel Advantage	Multiplex Advantage
Analyte shot noise	$D_a^{1/2}$	$(\Omega\ T\ A_D\ Q)^{1/2}$	$T_m^{1/2}$	$N_R^{1/2}$	None, or disadvantage
Back ground shot noise	D_a	$(\Omega\ T\ A_D\ Q)^{1/2}$	$T_m^{1/2}$	$N_R^{1/2}$	None, or disadvantage
Dark noise	D_a	$(\Omega\ T\ A_D\ Q)$	$T_m^{1/2}$	$N_R^{1/2}$	$N_R^{1/2}$
Readout noise	D_a	$(\Omega\ T\ A_D\ Q)$	T_m	N_R	$(N_R)^b$

[a] See also reference 10.
[b] Almost never applies to a real instrument.

a typical value for ϕ_d, detector noise for a PMT will not contribute significantly unless the $(S+B)$ signal is less than about 100 pe/s. As noted earlier, the contribution of dark noise to overall SNR depends on the relative importance of the terms in equation 7.17.

The Q_D for a PMT was shown in Figure 7.6 and rarely exceeds 0.2 pe/photon. The active area of a PMT can be large relative to array-detector pixels, about 7 × 23 mm for the commonly used RCA 31034 PMT.

7.6.2 Resistive Anode (RA) Position-Sensitive PMT

The RA detector has not been widely used in Raman spectroscopy but will be mentioned here as a multichannel analog of the PMT. Although multiple PMTs have been used to achieve the multichannel advantage in emission spectroscopy, such an approach would be cumbersome and impractically expensive for Raman. However, the same effect is achieved by preceding a resistive anode with an image intensifier. The heart of the RA is a 1 inch × 1 inch flat surface of a partially conducting material. When electrons strike the surface they migrate toward collection points at the corners of the RA. If arrival time for a pulse of electrons at the four corners is noted, the position where the pulse struck the RA may be calculated. Thus the RA may be effectively divided into approximately 400 × 400 position elements to which arriving pulses of electrons may be assigned. To make the RA into a photon detector, it is preceded by an image intensifier, also known as a microchannel plate (MCP), which amplifies the photoelectrons. For each photoelectron, roughly 3000 electrons emerge from the MCP, at the same position as the input photon. These electrons form the pulse that strikes the RA and is analyzed spatially.

If the RA device is placed at the focal plane of a Raman spectrometer, after many thousands of photons strike the photocathode an image of photon intensity versus position on the focal plane will build up. This image represents a series of entrance slit images for the relevant wavelength range. By summing the resolution elements in the vertical axis, a 400 × 1 array is formed, and effectively 400 wavelengths (or Raman shifts) are monitored simultaneously. Spectral coverage and resolution are determined by the spectrograph dispersion and spatial resolution of the RA.

When used for Raman spectroscopy, the RA detector acts as a 400 × 1 array with about 1 cm × 60 μm resolution elements. Based solely on N_R, a $\sqrt{400}$ = 20-fold gain in SNR would be expected relative to a single-channel PMT; however, the PMT has a larger area than an individual RA resolution element, so all of the $N_R^{1/2}$ gain in SNR may not be realized (see equation 7.12). Although there are examples of RA applications to Raman spectroscopy in the literature [10], RAs have not been used extensively, probably due to the ease of damage and the limited dynamic range. The maximum count rate of 200 to 500 counts/pixel is a fairly serious limitation and constrains the use of the device to low-light-level (and low background) applications. Furthermore, the CCD arrays have a higher Q_D over most of the commonly used spectrum and usually have more than 400 resolution elements.

7.6.3 Intensified Photodiode Array

After a productive but short-lived application of vidicon technology to Raman spectroscopy, PDAs came into use [11,12]. Without an intensifier, the PDA is typically a 1024-element array of 25-μm $\times$ 2.5-mm pixels. Photons striking the photodiode generate current, which depletes a small capacitor of charge. After a suitable integration period (milliseconds to a few minutes, depending on cooling, light flux, etc.), the charge required to recharge the capacitor is measured, with the entire array recharged in about 10 ms. Unlike the PMT or RA detectors, the PDA is integrative, and the signal for each pixel is proportional to the integrated photon flux between readout scans.

The Q_D of a PDA is quite high and follows the silicon response shown in Figure 7.6. A peak Q_D of 0.5 to 0.7 might suggest that the PDA is an attractive alternative to the PMT. However, the recharge process for a PDA is quite noisy, producing a high readout noise of greater than 1000 e^-. This high σ_R severely degrades SNR for weak signals, for which σ_R is large relative to S or B (see equations 7.17). Thus unintensified PDAs have been applied where light levels are high, such as absorption spectroscopy, and are nearly useless for Raman spectroscopy because the readout noise overwhelms the weak scattering signal. Furthermore, the dark signal in PDAs can be fairly large due to capacitor leakage, precluding improvement based on long integration times.

To usefully apply the PDA to Raman spectroscopy, the signal must be boosted well above the readout and dark noise so that the S + B terms dominate over ϕ_dt and σ_R in equation 7.17. To amplify the weak scattering signal, an MCP intensifier very similar to that used in the RA device is employed. The photocathode and microchannel plate are the same but a phosphor layer is added between the MCP and the PDA. The phosphor converts the amplified burst of electrons back to photons, with the overall result being amplification of each photoelectron by a factor of about 3000. In practice, the photocathode, MCP, and phosphor are all contained is a sealed glass disc used as an image intensifier primarily in military applications. The intensifier is placed directly in front of the PDA such that the photon bursts at the phosphor screen are monitored by the array, and the intensifier acts as a photon amplifier with only slight degradation of spatial resolution.

The intensifier increases the $S + B$ terms in equation 7.17 above the σ_R and $\phi_d t$ terms for most situations, the exceptions being very weak scattering and/or long integration times. For many applications with short integration times, the SNR for an IPDA will be governed by equation 7.16, and the SNR will scale with $(\Omega Q_D A_D T)^{1/2}$. For this case, the SNR of the IPDA will exceed that of the PMT according to equation 7.12. As an example, consider a 1024-channel IPDA and a photon-counting PMT. The quantum efficiencies will be comparable, but that of the IPDA may be a bit lower due to the greater difficulty in fabricating a large, uniform photocathode. Because of some resolution degradation in the MCP, the effective number of elements (N_R) is reduced to about 250 to 350, so $N_R^{1/2}$ would be about 18. However, the IPDA pixel is 2.5 mm high, while a PMT active area may be 10 to 20 mm. Assuming a factor-of-two higher Q_D for the PMT and a factor of four larger area, equation 7.12 yields

$$\frac{(S/N)_{IPDA}}{(S/N)_{PMT}} = \frac{1 \times 1 \times 300^{1/2}}{2^{1/2} \times 4^{1/2}} \approx 6 \qquad (7.21)$$

In this example, a multichannel SNR advantage of 6 is predicted for the IPDA, if spectral resolution and total measurement time are equal. Although this advantage is significant, it is not as large as that predicted solely from the $N_R^{1/2}$ factor when the detectors are fully illuminated [2].

As the light level decreases, the readout and dark noise of the IPDA can become significant. To state the conclusion concisely, the SNR scales with $(A_D\ Q_D\ N_R)^{1/2}$ when the experiment is sample shot-noise limited. As the scattering weakens to the point where σ_R and $\phi_d t$ become significant, these variables must be considered. Measurements on an IPDA in our lab (PARC model 1421 IPDA cooled to -20°C, conventional uncooled photocathode) indicate a readout noise of 2 to 3 counts and a dark signal of 38 to 40 counts/s. Based on the manufacturer's specification of 1 count/pe, this IPDA exhibits a σ_R = 3 e^- and ϕ_d = 40 e^-/s. These values permit prediction of the conditions when $S + B$ will exceed $\phi_d t$ and σ_R^2 for a given situation.

Although we will see that CCD performance exceeds that of the IPDA in conventional Raman experiments, IPDAs have a major advantage when time resolution is required. The intensifier may be switched on and off ("gated") quickly, down to 10-ns gate width if desired. Even though the PDA is integrating all the time between readouts, the intensifier gain is so high that only photons arriving when the gate is on are monitored. The "contrast ratio" between gate on and gate off is typically 10^6. Many useful applications of the IPDA have resulted from its time resolution rather than its sensitivity.

Intensified CCDs are available that combine both gating and high CCD performance, and are discussed in Chapter 5.

7.6.4 Charge-Coupled Device (CCD) Detectors for Raman

At first glance, the CCD may not appear to be fundamentally different from the unintensified PDA, since both are based on silicon and they have similar Q_D versus λ curves. The big difference that is so important to Raman is the low dark and readout noise of the CCD. The purpose of the intensifier in an IPDA is to boost the optical signal above the dark and readout noise, but if the nonoptical noise sources are already negligible, an intensifier is unnecessary. Thus from the noise standpoint, the low dark and readout noise of CCD technology make all the difference when the CCD is compared with PDAs. The CCD technology itself is discussed thoroughly in other chapters and will not be repeated here. However, we will consider several CCD properties as they pertain to Raman spectroscopy.

7.6.4.1 Format

Most readily available CCDs are two-dimensional, with several hundred pixels along each axis with a dimension typically of 10 to 30 μm. Several examples are described in Table 7.3. New CCDs are being announced at a rate of roughly one a month, and Table 7.3 is by no means exhaustive; however, the most readily available devices that have been adapted to spectroscopy are listed. In the majority of spectroscopic applications, the detector is operated as a one-dimensional array, with the dimension along the slit being binned in hardware or software. For convenience we will refer to the axis along the slit as Y and the axis in the direction of wavelength dispersion as X. Due to optical aberrations, it is fairly rare to use more than 1 cm of slit height, so a CCD with a Y dimension greater than 1 cm does not provide additional sensitivity. Current spectrometers rarely have a flat field of greater than 25 mm along the X direction, although this specification may improve as the need arises. Consequently a CCD array larger than 1 $\times$ 2.5 cm will be difficult to use with existing Raman spectrometers; however, as such CCDs become available, their advantages of larger detector area and number of pixels may push spectrograph designers to accommodate them effectively.

The number of pixels in the Y direction usually is not critical for nonimaging Raman spectroscopy, but the more pixels along the X axis the better. The resolution and SNR both improve for larger N_R, and N_R controls the resolution/spectral coverage trade-off (see below). N_R can never be too large, since adjacent pixels (or columns) may be binned if the user wants to reduce N_R for some reason. On the other hand, although most spectroscopic applications do not exploit the Y-

Table 7.3 CCD Arrays Used for Raman Spectroscopy.

	Format	Pixel Size (μm)	Dark Current @ Temperature (e^-/pixel/s)[a]	Readout Noise (e^-)	Q at 700 nm
Thompson 7883-PM	384 × 576	23 × 23	6.5 @ −45°C	12	0.46
EEV CCD05-10	298 × 1152	22.5 × 22.5	6 @ −45°C	9	--
Photometrics PM512	512 × 512	20 × 20	3 @ −45°C 0.03 @ −110°C MPP: 0.14 @ −45°C	6	0.44
Tektronix TK512CF	512 × 512	27 × 27	12 @ −40°C	8	0.37
Tektronix TK512 CB/AR (Backthinned)	512 × 512	27 × 27	12 @ −40°C	8	0.80
Kodak KAF-4200	2048 × 2048	9 × 9	MPP: 0.12 @ −35°C	--	0.45
EEV CCD 15-11	1024 × 256	27 × 27	MPP: 3 × 10^{-5} @ −120°C	4	0.40

[a] MPP: multipinned phase.

axis spatial resolution of a CCD, this property can be useful in some cases. When the spectrometer is aligned, a full two-dimensional image can reveal the alignment of the slit image on the detector as well as how many of the Y axis pixels are actually collecting light. In addition, the Y-axis of the CCD may be split in two, resulting in two complete spectra obtained in parallel. If one is contaminated by a "cosmic" event, comparison with the other will reveal the problem and allow objective rejection of these events.

7.6.4.2 Quantum Efficiency

The Q_D versus λ curve for CCDs shown in Figure 7.6 is representative of silicon-based devices but is affected by several aspects of device fabrication. In conventional CCDs, photons must pass through the circuit mask before entering the silicon, so the maximum Q_D is about 0.5 e^-/photon. This value is higher than the best values for PMTs (ca. 0.2). Perhaps more important is the extension of useful Q_D into the NIR, an essential issue for NIR Raman spectroscopy.

If the device is thinned by argon-ion or chemical etching, the photons may enter the silicon from the side opposite the circuit mask. Such "back-thinned" or "backside-illuminated" devices have a maximum Q_D of about 0.8, plus much higher Q_D than conventional CCDs in the near UV. However, the back-thinning process adds another fabrication step, which both reduces device yield and increases cost. Back-thinned devices are just now becoming routinely available. The enhanced Q_D of a typical back-thinned CCD is compared with conventional detectors in Figure 7.7, along with the Raman shift ranges for common lasers.

The increase in maximum Q_D from 0.45 to 0.8 is significant but by itself may not justify the added expense. Recall that SNR scales with $Q_D^{1/2}$ (equation 7.11) in many Raman experiments, so a back-thinned device may improve SNR only by a factor of $\sqrt{1.8}$ or 1.34 at the optimum wavelength. However, the improvement is much larger in the NIR and UV, where the back-thinned devices may have a Q_D 10 times as large as the conventional devices. The only impediment to widespread use of back-thinned CCDs is increased cost due to manufacturing difficulty. On the basis of past experience with CCDs, we expect this impediment to disappear with time.

7.6.4.3 CCD Noise Characteristics

We have already discussed in general terms the critical importance of low detector noise for Raman applications. To repeat, it is important that the readout noise and dark-count flux be small relative to the signal and background photons from the sample. The CCD variables that determine device noise can be broadly classified into two types: device design and operating temperature. Dark-count rates vary greatly for different CCDs, but as a rule of thumb, the dark-count rate

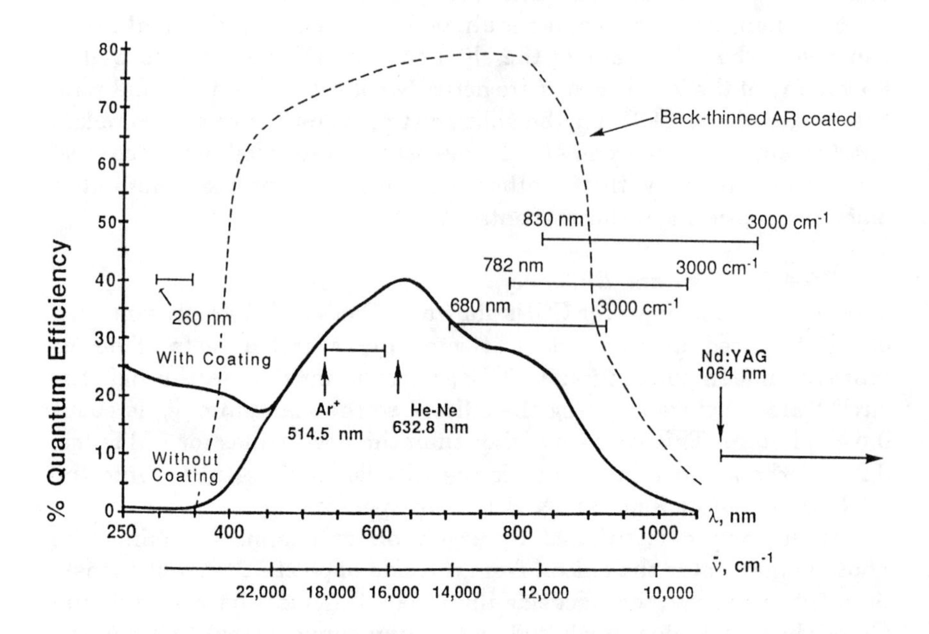

Figure 7.7 Si-based CCD quantum efficiency relative to several laser wavelengths. Bottom scale shows Raman shift relative to a 782-nm diode laser. "Coating": commercial UV scintillator. (Adapted with permission from reference 22.)

decreases by 50% for each 7°C decrease in device temperature. As an example, liquid nitrogen cooling of the Photometrics PM512 CCD to -110°C produces a dark rate of about 1.6 e^-/pixel/min. Impressive as this number is, it still amounts to 14 e^-/column/s when 512 pixels are binned into a 1-pixel-wide column.

A further dramatic reduction in dark flux can be achieved with "multipinned phase" (MPP) operation [13]. Without going into details, MPP devices have dark rates in the region of 1 e^-/pixel/hr at −110°C. This value is sufficiently low that dark rate is unlikely to be an issue in any current Raman experiment, since other noise terms in equation 7.17 are likely to be much larger. MPP does have a price, however, in the form of reduced dynamic range. The full-well capacity (the maximum number of electrons stored in a pixel before readout) is reduced by a factor of at least two, from typically 200,000 e^- to less than 100,000. For exceedingly weak signals where very low dark rate is important, the smaller full-well capacity is unlikely to be an issue; however, the trade-off of dynamic range and dark rate may be a consideration during spectrometer design. More information about MPP operation is found in Chapter 2.

A brief technical note is appropriate here due to the inconsistency in the way CCD results are reported in the literature. As pointed out earlier, CCD systems have different gains (expressed as analog-to-digital units, ADUs, per photoelectron) depending on the choice of A/D converter and other factors. The shot noise limit, the readout noise, and equations presented so far are based on *photoelectrons*, before any analog-to-digital conversion. However, spectra are often presented as ADU, directly from the CCD software. As long as the gain is constant for a given system, it has no effect on the observed SNR. The relative standard deviation of the ADU reading (and therefore the SNR) will be the same as that of the number of photoelectrons, provided both are large relative to the readout noise. This situation applies to the vast majority of CCD experiments, with the exceptions involving very low light levels.

7.6.4.4 NIR Detectors for Raman

The 1.1-μm Q_D limit for silicon-based CCD detectors can be a limitation when the objective is reduced fluorescence interference. As the laser wavelength is extended into the NIR, the likelihood of exciting electronic transitions that yield fluorescence decreases, as amply demonstrated by FT-Raman techniques with 1064 nm lasers [14]. This issue has driven the evaluation of detectors with useful Q_D in the 1- to 5-μm range for use in Raman spectroscopy. Examples are Pt:Si arrays, in which Pt is implanted in silicon to extend sensitivity to about 5 μm, and InGaAs arrays. Although these detectors are currently noisier than silicon-based CCDs, they are quite new and may improve with time. Both Pt:Si and InGaAs array detectors are available commercially, but have not yet been widely used in Raman spectroscopy. An additional factor for wavelength longer than 1 μm is the thermal background. Blackbody radiation, even at room temperature, can be large enough to contribute shot noise to the observed signal, adding a noise term in the same way as dark noise is added. As the detector noise characteristics in the 1- to 5-μm region improve, the thermal background from the sample or the environment will become the new lower limit on noise. Chapter 5 of this volume focuses on the near-IR performance of nonsilicon CTDs.

Table 7.4 compares several detectors of current interest in Raman spectroscopy in a format initially used by Murray and Dierker [15]. Although the choice of detector will depend on the application, there are many cases where CCDs outperform the alternatives.

Table 7.4 Detectors of Current Interest for Raman Spectroscopy

	PMT	IPDA	Conventional CCD	Back-thinned CCD	Intensified Resistive Anode
Number of channels	1	700–1024	384–1152	512–1024	400
Max. Q	0.20	0.20	0.50	0.80	0.20
Useful Q range, nm	<200–800	<200–800	400–1100	300–1100	<200–800
Readout noise	None	~3 counts	~5 e^-	~5 e^-	none
Dark signal	~5 counts/sec	~40 counts/sec	~1e^-/pixel/min[a]	~1 e^-/pixel/min	0.005 counts/pixel/s[b]
Time resolution	~10 nsec	<10 ns[c]	~0.1 s[d]	~0.1 s[d]	~1 ms
System cost, $K	~5–10	~50	~20–35	~25–45	~50

[a] Varies greatly; this value is −110°C. MPP can reduce to 1e^-/pixel/hr.
[b] Depends strongly on how pixels are binned.
[c] Best case.
[d] Determined by shutter, readout time may be longer.

7.7 Application of CCDs in Raman Spectroscopy

The following examples of CCD applications illustrate the unusual properties of CCD detectors. Although these examples are presented in roughly chronological order, they are classified into several types of experiments: low-light-level detection, high-sensitivity spectroscopy of liquids, NIR Raman spectroscopy, and Raman imaging.

7.7.1 Low Light Level Raman Spectroscopy

Although Raman scattering is nearly always a very weak effect, some experiments produce signal levels much lower than others. Early Raman experiments with CCDs were directed toward surface films on metals and dielectrics, in part because of the active history of surface-enhanced Raman spectroscopy (SERS). Starting with Campion in 1982 [11], several investigators sought to observe single layers of molecules on surfaces *without* any surface enhancement, with the objective of determining interfacial structure. Since the predicted Raman signal for a surface monolayer under typical conditions is in the region of 1 photon/s [16], the spectrometer is severally taxed with respect to SNR. The high Q_D, low noise, and multichannel properties of CCDs are critical to these applications.

Murray and Dierker [15] discussed detector characteristics for monolayer films of cadmium stearate on a silica surface. The spectra shown in Figure 7.8 were obtained with an RCA CCD and triple spectrograph [15,17,18]. Each monolayer is about 25 Å thick, and it was possible to obtain a spectrum of the C-H stretch region of a single layer with a 6-min integration time. The changes in the spectrum with layer thickness were interpreted in terms of layer structure.

Campion compared a Photometrics PM512 CCD with an IPDA for examining silicon surfaces both with and without an adsorbed monolayer of oxydianiline (ODA) or pyromellitic dianhydride (PMDA) [19]. Figure 7.9 illustrates two advantages of CCDs over the IPDA. The lower spectrum is of the third-order phonon scattering of silicon obtained using an IPDA at an integration time of 250 s. The middle spectrum is for the same conditions but substituting the CCD. The improved SNR (about a factor of two) is due to the higher Q_D of the CCD. The top spectrum was obtained after elongating the laser spot to completely fill the CCD height but keeping laser power density constant. Since the CCD surface is higher than the IPDA (1 cm versus 2.5 mm), it should yield a higher A_D and better SNR. Figure 7.9 verifies this prediction, confirming our expectations based on applying equation 7.16 to each pixel. For sample shot-noise-limited experiments, the SNR will scale with $(Q_D A_D)^{1/2}$, all else being equal. In the

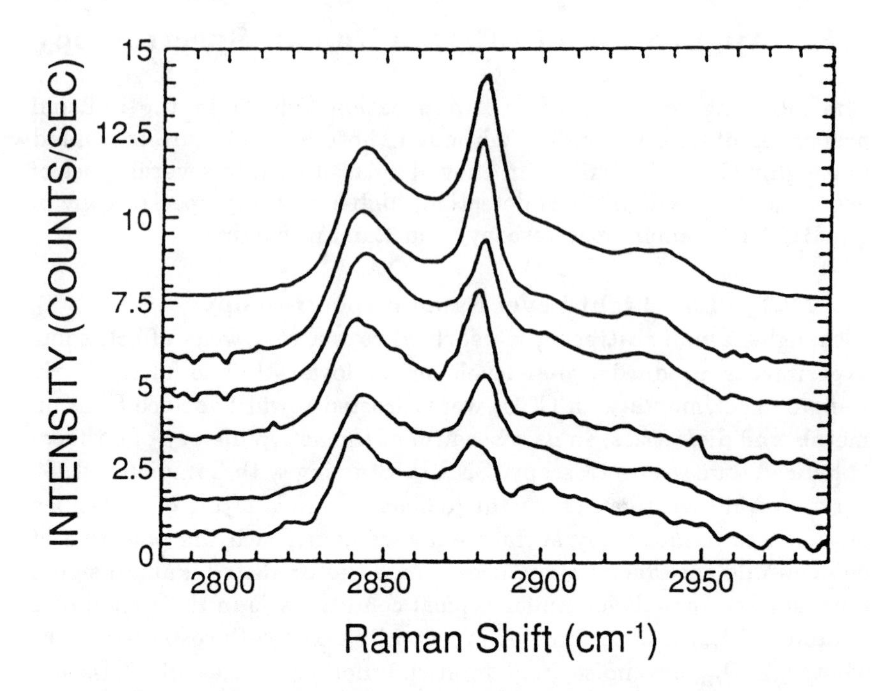

Figure 7.8 Raman spectra of cadmium stearate films on silica surfaces. From top to bottom, film thicknesses are 27, 19, 9, 5, 3, and 1 molecular layer (normalized for layer thickness). (Adapted with permission from reference 17.)

case of Figure 7.9, this amounts to a factor of four improvement in SNR for the completely filled CCD compared with the IPDA.

The stability and high SNR possible with the CCD permits spectral subtraction; in Campion's case the silicon spectrum is removed to reveal the adsorbed monolayer. Figure 7.10 shows CCD spectra of silicon with and without a monolayer of ODA, together with the difference spectrum. The monolayer spectrum is weak but discernible, even though the silicon-substrate spectrum is much stronger. It should be noted that the substrate contributes a type of background scatter and will lead to an SNR governed by equation 7.16. Any time a large background occurs, the higher Q_D and A_D are particularly important to help compensate.

7.7.2 High Sensitivity Raman Spectroscopy of Solution Components

Solution samples consist of dissolved analytes in a solvent, often water, and differ significantly from monolayer films in terms of background

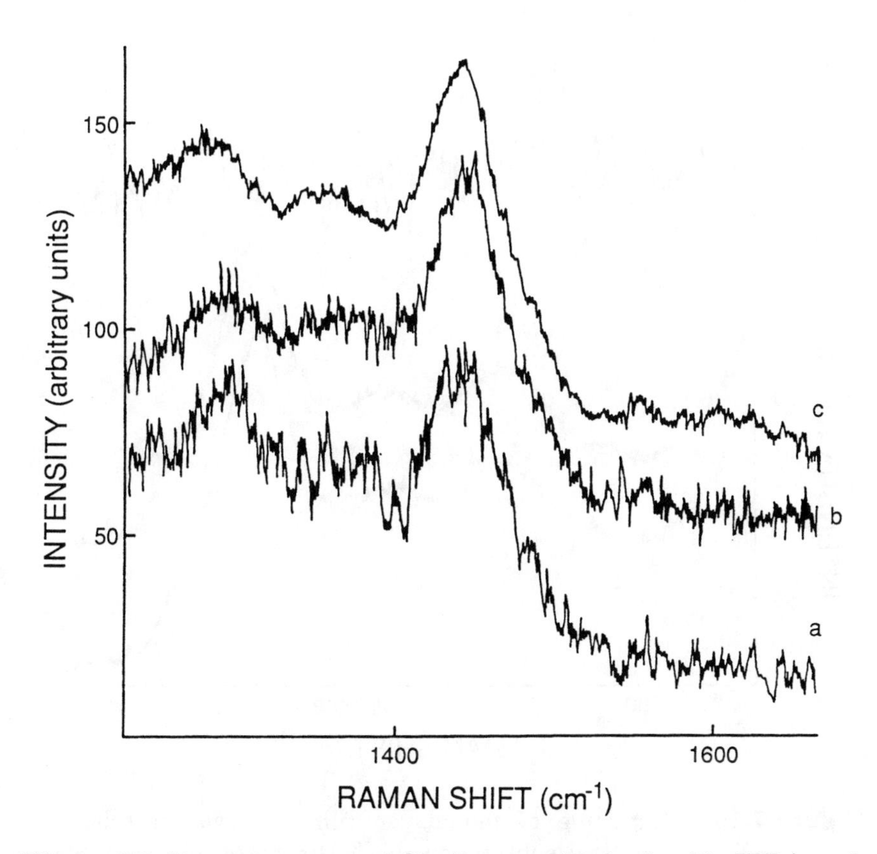

Figure 7.9 Raman spectra of Si phonons obtained with IPDA and CCD. (a) IPDA; (b) CCD with equal resolution and measurement time; (c) CCD with an elongated focal spot that filled the detector. (Adapted with permission from reference 19.)

scatter. Both solvent Raman and other inelastic scatter adds to stray light to produce a higher background than is observed for thin-film samples, so the problem is not so much one of detecting weak signals but rather of distinguishing weak signals from a relatively large background. It is often the case that background scatter is the dominant noise source, leading to a SNR predicted by equation 7.16. For this limiting (but common) case, the CCD's Q_D and A_D are important, but the detector noise terms do not affect SNR provided

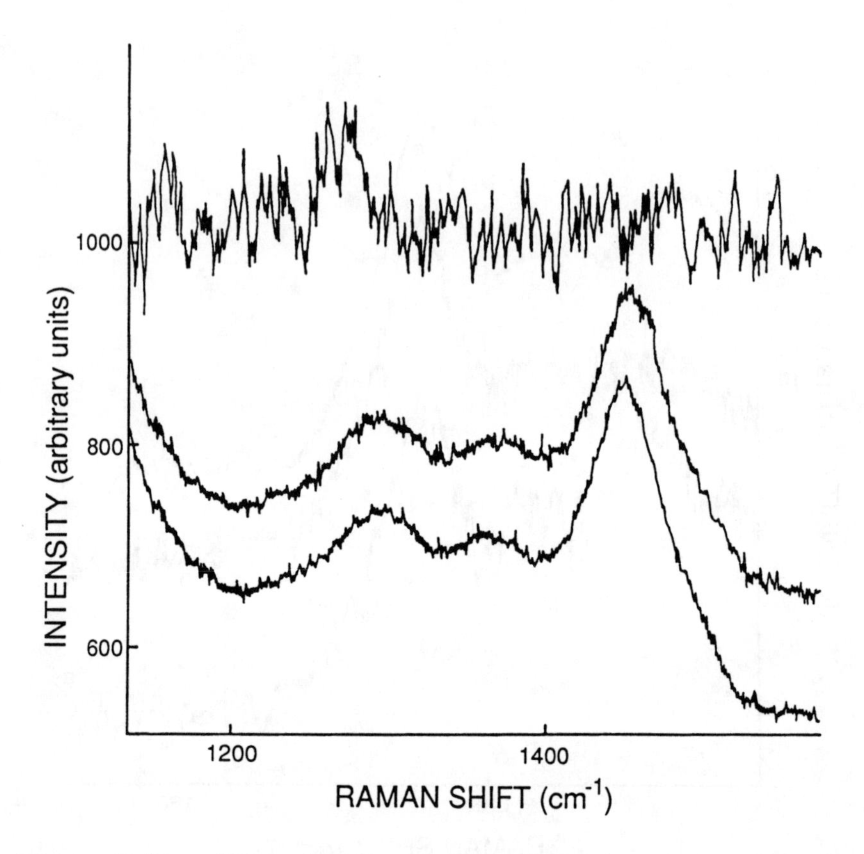

Figure 7.10 Example of monolayer film observation following spectral subtraction of CCD spectra. Lower: clean silicon; middle: Si with oxydianiline monolayer; upper: difference. (Adapted with permission from reference 19.)

they are small relative to background scatter. In some cases the low CCD dark flux permits longer integration time and higher SNR.

CCD detectors have not yet been pushed to the sensitivity limits for solution scatterers, yet some systematic comparisons are available. Pemberton and Sobocinski [20,21] used a CCD with a conventional double monochromator to acquire useful spectra with a low-power visible laser. The spectrum of pyridine shown in Figure 7.11 was obtained with a 10-mW He-Ne (632.8 nm) laser and 5-min integration time, and exhibited a SNR of about 60.

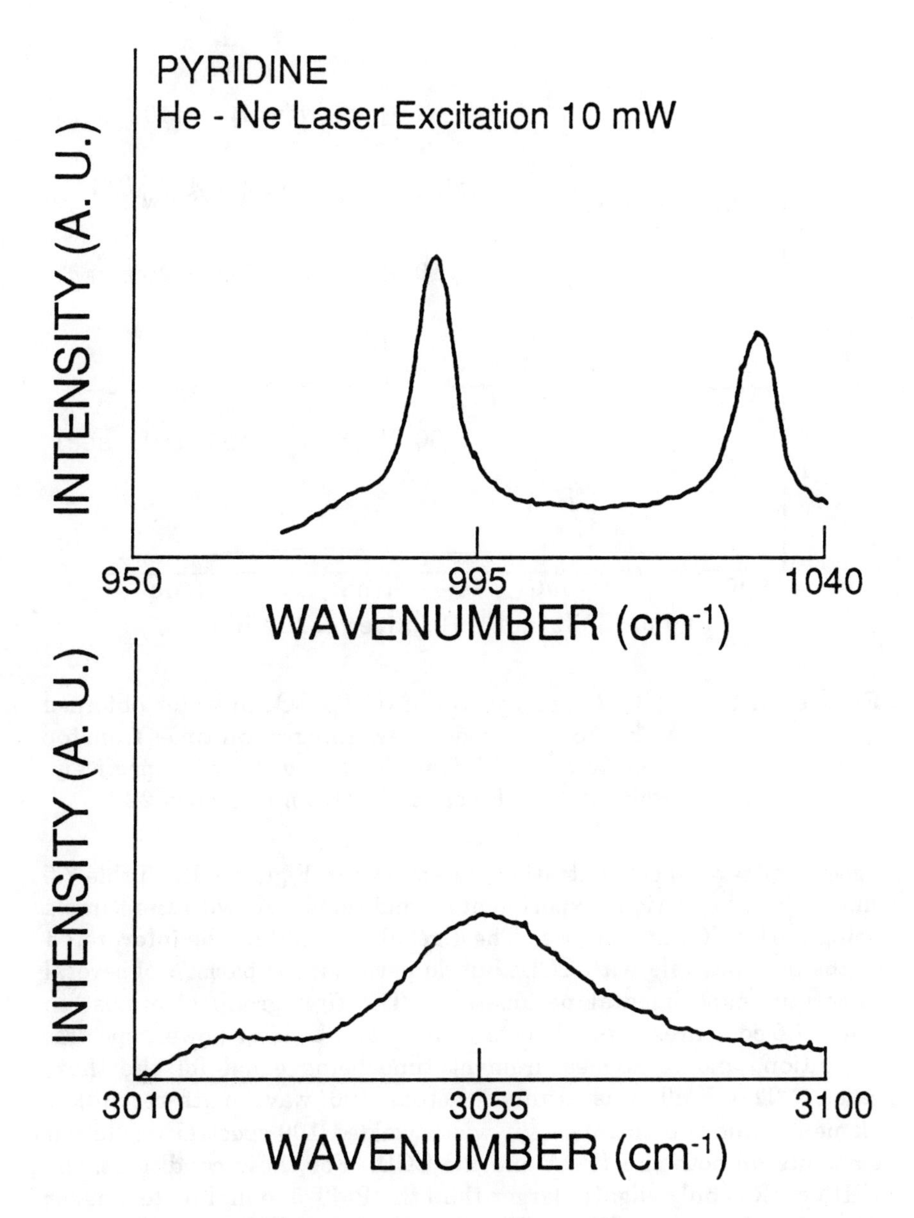

Figure 7.11 Raman spectra of neat pyridine obtained with a CCD and He-Ne laser. A.U., arbitrary units. (Adapted with permission from reference 20.)

Our laboratory has used $(NH_4)_2SO_4$ in water as a standard sample for comparison [2,22] due to some work by previous investigators [23].

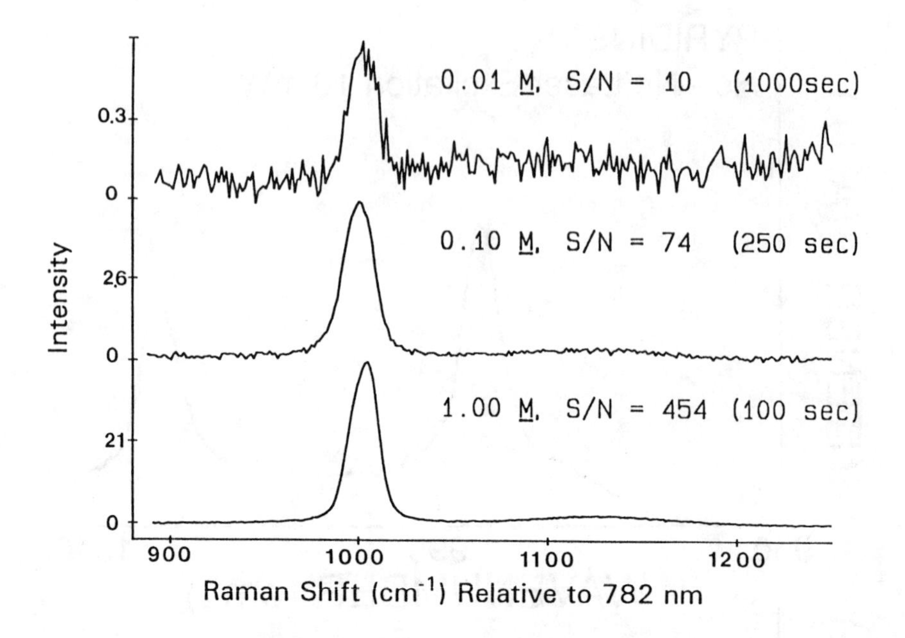

Figure 7.12 CCD Raman spectra of $(NH_4)_2SO_4$ in water obtained with a 782-nm diode laser. Integration times from top to bottom are 1000 s, 250 s, and 100 s, respectively. (Reprinted with permission from reference 22.)

Spectra of several concentrations are shown in Figure 7.12. Table 7.5 shows SNR for various experimental conditions, but with the sample being $(NH_4)_2SO_4$ in all cases. These results should not be interpreted as the best possible with CCDs but do permit a comparison of several important experimental parameters. The first group of entries in Table 7.5 compares three detectors with sample, laser power, spectral resolution, and total measurement time being equal for the three cases. The PMT was scanned across 100 wavelength-resolution elements, and the entrance slit width yielded 100 spectral resolution elements on both the IPDA and the CCD. For these conditions, the IPDA SNR is only slightly larger than the PMT due mainly to smaller A_D and higher noise. The CCD yields a factor of four to five improvement over the IPDA due both to greater pixel height (therefore A_D) and larger Q_D. The second group of entries illustrates the effect of $(NH_4)_2SO_4$ concentration on SNR. The longer laser wavelength decreases cross section and CCD Q_D compared with 514.5 nm, but 782 nm is an attractive wavelength for reasons discussed later. For now, note that SNR approximately scales with $D_a t^{1/2}$, with D_a proportional to concentration.

Figure 7.13 shows that relatively low concentrations of pyridine in water may be observed even with a 782-nm laser. These spectra illustrate a submillimolar detection limit for the conditions employed. PMT-based Raman with 514.5-nm lasers and special sampling techniques have yielded micromolar detection limits for normal Raman scatterers, and approximately 50 nM for resonance Raman scatterers [3,24]. The higher Q_D and multichannel advantage of CCDs should further lower these detection limits, but such experiments have not been reported to our knowledge. It is clear that low micromolar detection limits should be routine with CCDs and visible lasers, provided fluorescence interference is low. In any experiment, the high Q_D, N_R, and A_D possible with CCDs can be exploited to yield useful SNR for cases when β, D_a, P_o, or t are small. Whenever the detector characteristics are improved, the flexibility of the application increases.

Table 7.5 SNR[a] for $(NH_4)_2SO_4$ Solution Samples

$[SO_4^{-2}]$ *M*	Detector	Laser Wavelength (nm)	Time (s)	Spectral Resolution (cm^{-1})	SNR[a]
0.01	PMT	514.5	150	3	8.5
0.01	IPDA	514.5	150	3	12.4
0.01	CCD	514.5	150	3	61
0.01	CCD	782	1000	10	10
0.1	CCD	782	250	10	74
1.0	CCD	782	100	10	454

[a] Total measurement time.

[b] Defined as signal (relative to baseline) divided by the standard deviation of the baseline.

The reader may have noted that the CCD Raman spectra in Figures 7.8 through 7.13 cover a fraction of the total Raman shift range of interest, typically 800 cm^{-1} or less from a total range of 0 to 4000 cm^{-1}. While the Raman shift range covered by a spectrograph/CCD system is a function of CCD size and spectrograph dispersion, it is ultimately determined by the trade-off between resolution and spectral range noted earlier. For a finite number of CCD pixels along the wavelength axis (typically a maximum of 1152 with readily available CCDs), one can increase spectral range only by decreasing resolution (and therefore increasing cm^{-1}/pixel). For example, an 800-cm^{-1} spectrum imaged onto a 512 × 512 pixel CCD yields at best 1.6 cm^{-1}/pixel resolution with the usual practice of binning the vertical pixels to make a 512 × 1 detector. By decreasing the spectrograph dispersion, the same detector could monitor 3000 cm^{-1} but with a maximum resolution of 5.9 cm^{-1}/pixel. A particular application may require only modest resolution or spectral range, but

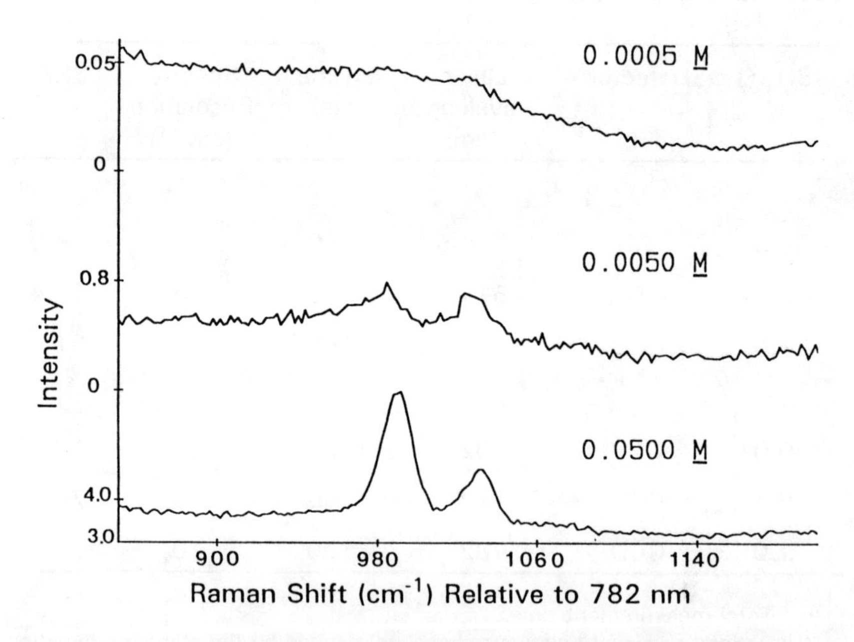

Figure 7.13 CCD/diode laser Raman spectra of three concentrations of pyridine in water. Integration times (top to bottom) were 1200 s, 1200 s, and 100 s. Reprinted with permission from reference 22.)

the simple spectrograph and one-dimensional CCD system will continue to entail a compromise until the number of available CCD pixels along the wavelength axis increases substantially.

A simple solution to the problem is acquisition of several spectral segments following rotation of the grating. For example, four 800-cm^{-1} CCD exposures can be joined to obtain a complete 3200 cm^{-1} spectrum. The cost of this approach is longer measurement time compared with a single exposure, plus some inaccuracy in the wavelength axis. Grating drives are not perfect, and there will always be some small variation in grating position. Nevertheless, the ability to arbitrarily select single or multiple wavelength ranges adds useful versatility to the instrument.

A more elegant but more complex approach has been presented by Pelletier [25], based on an echelle spectrograph similar to those used in atomic emission spectroscopy [26 and Chapter 3]. An echelle grating has high dispersion but also severely overlapping orders. High dispersion is achieved by a coarse grating (79 grooves/mm) configured so that a 4500-cm^{-1} Raman shift range is covered by orders 31 through 40. To separate the overlapping orders, a cross disperser (either a prism or a second, conventional, grating) is used, as shown in Figure 7.14. The end result is two-dimensional dispersion, so that the orders are separated along the Y axis of the CCD, and the wavelength is separated along the X axis. Thus the two-dimensional format of the CCD is exploited to defeat the spectral range/resolution trade-off. The price is a moderate but often acceptable decrease in sensitivity due to smaller A_D.

Figure 7.15 shows a Raman spectrum of cyclohexane covering 3300 cm^{-1}, obtained with the echelle spectrometer. No scanning or multiple exposures were required, and the entire spectrum was obtained with a single 1-s CCD exposure. The resolution exhibited in the spectrum was limited by the finite CCD pixel size to 2 cm^{-1}. Note that the instrument can cover nearly 5000 cm^{-1} of spectral range with 2 cm^{-1} resolution, corresponding to a one-dimensional CCD detector with at least 2500 elements. Thus 2500 resolution elements were achieved with the 512 × 512 CCD detector.

Although direct comparisons are not yet available, the CCD echelle system should be less sensitive than a conventional system due to lower throughput and smaller A_D [25]. If a wide-range spectrum is required, however, the conventional system would have to be scanned and multiple exposures obtained. Provided dark and readout noise are negligible compared with sample shot noise (i.e., equation 7.16 applies), multiple exposures with a one-dimensional spectrometer produce the same SNR as a single two-dimensional exposure covering the same range but with reduced slit height. For example, suppose eight 512-cm^{-1}, one-dimensional spectra were combined to form a 4096

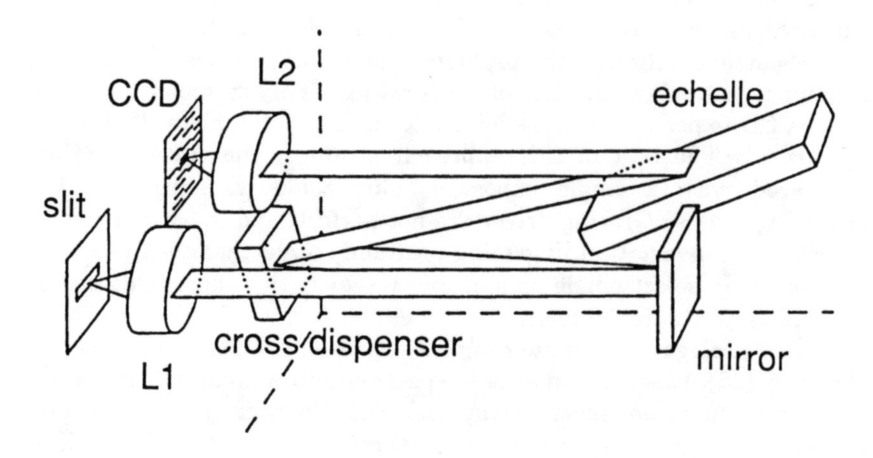

Figure 7.14 Optical arrangement for echelle/CCD Raman spectrometer, L1 and L2 are lenses, and the dashed line indicates the outline of spectrograph housing. (Adapted with permission from reference 25.)

cm^{-1} spectrum, with each obtained with a 1-cm slit height to fully cover a 512 × 512 pixel CCD. Alternatively, an echelle might be configured to project eight spectra on the same detector, each one being 64 pixels high, thus reducing A_D. Referring to equation 7.16, the factor-of-eight decrease in A_D is compensated by the factor-of-eight increase in exposure time, since the two-dimensional system does not require multiple exposures. For the case of a wide-range spectrum, therefore, the echelle system should be as sensitive as multiple exposures from a one-dimensional system, all else being equal. In practice, the echelle system will be slightly less sensitive due to lower throughput and incomplete use of the Y CCD axis. For small spectral ranges, however, a single exposure on a conventional system will have a higher SNR than the echelle system due mainly to larger A_D. The echelle will always cover a wide spectral range unless the optics are reconfigured.

On a practical note, echelle systems have historically been difficult to design and align, due to two-dimensional dispersion and sometimes difficult optical aberrations. Once aligned, however, there are no moving parts and the system should be quite rugged. Until quite large CCDs with 2000 to 4000 pixels along one axis become readily available, the echelle offers an attractive alternative to multiple exposures from conventional spectrographs.

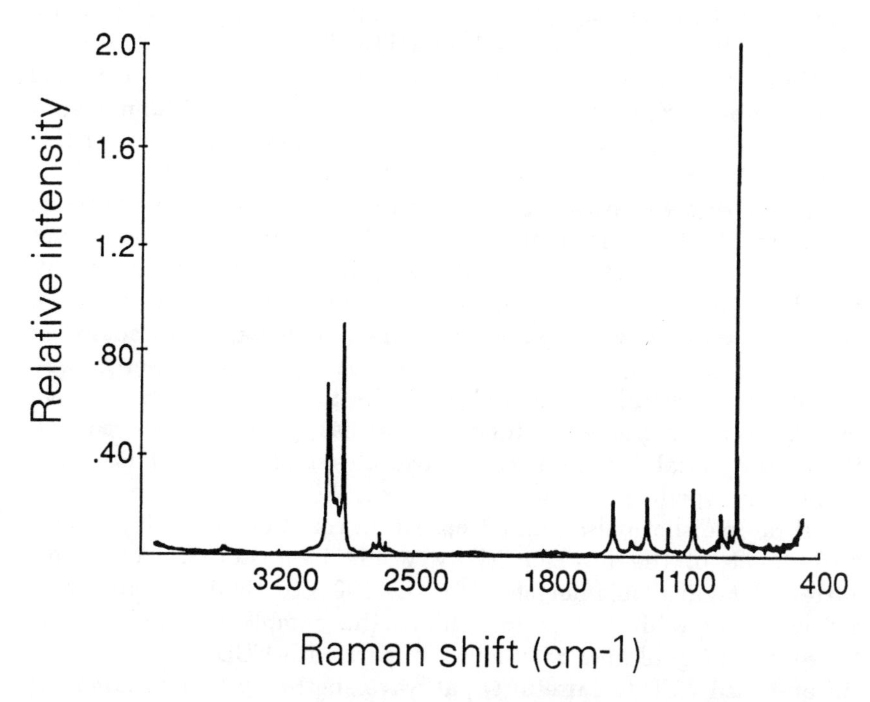

Figure 7.15 Raman spectrum of cyclohexane obtained with the echelle/CCD spectrograph of Figure 7.14, covering 3500 cm^{-1}. (Adapted with permission from reference 25.)

7.7.3 CCD/Raman Spectroscopy with Near-Infrared Lasers

The rapid growth of FT-Raman spectroscopy using a 1064-nm Nd:YAG laser since its introduction in 1985 clearly illustrates the advantages of NIR Raman spectroscopy [14,27–29]. In many potential Raman applications, it is not the weakness of the Raman scattering but rather interference by fluorescence and other background luminescence that prevents acquisition of useful Raman spectra. Trace levels of fluorescent impurities or weak fluorescence of the analyte itself can easily overwhelm the weak Raman scattering, thus increasing the *B* term of equation 7.15 and decreasing the SNR toward zero. When a 1064-nm laser is used, however, the excited electronic states giving rise to fluorescence are rarely populated, and the fluorescence is much weaker or nonexistent. Even though the Raman scattering is weaker in the NIR due to the ν^4 dependence of the cross section, the fluorescence background is so low that excellent Raman spectra may be obtained. Additional advantages of FT-Raman include wide spectral range, excellent wavelength precision, and the availability of data-processing techniques developed for FTIR [28].

These benefits carry a price, however, in the form of increased detector noise. For a 1064-nm laser, the 0- to 4000-cm^{-1} Raman shift range covers the spectrum from 1064 to 1852 nm. Detectors for this range have significant dark flux, so that equation 7.18 applies. Since ϕ_d exceeds sample scattering in most cases, the system never operates in the sample shot-noise limit. The interferometer and FT techniques provide a multiplex advantage that greatly improves the measurement, but the SNR is still determined by detector noise. The result is decreased sensitivity compared with the visible Raman discussed so far, which is often partially compensated by higher laser power. Despite this shortfall in sensitivity relative to visible Raman, the near immunity to fluorescence interference has greatly enhanced FT-Raman's applicability to a wide range of samples, and at least five commercial vendors have emerged since 1986.

If sample shot-noise-limited operation could be maintained in the NIR, much of the lost sensitivity would be restored. There are currently no available detectors for the 1000- to 1800-nm range with sufficiently low dark flux to achieve the sample shot-noise limit. However, the goal may be reached in part with CCDs. Recall that a silicon-based CCD has useful Q_D at wavelengths up to approximately 1.1 μm. This Q_D range is sufficient to permit the use of lasers in the 600- to 800-nm region while still observing most of the Raman shift range of interest [23,31]. For example, a 780-nm laser will generate a 0- to 3000-cm^{-1} Raman shift range from 780 to 1018 nm, which is still within the Q_D range of the CCD. Figure 7.7 shows the position of several common lasers with respect to the CCD Q_D curve. Note that the silicon-based CCD would be useless with a 1064-nm laser, but that it would work with Ar^+ (514 nm), He-Ne (633 nm), Kr^+ (752 nm), or a 782-nm diode laser. Thus by decreasing the laser wavelength from 1064 to the 600- to 800-nm range, the Raman experiment can be conducted with a CCD detector and benefit from its attractive noise characteristics; however, the likelihood of fluorescence is higher at 600 to 800 nm than at 1064 nm, so the end result is a trade-off of detector characteristics and immunity from fluorescence. A 1064-nm experiment will have lower fluorescence but a noisier detector and lower SNR. A 780-nm (for example) experiment with a CCD should have higher SNR, but also more likelihood of fluorescence. The optimum choice is sample dependent. In general, it is desirable to use a laser wavelength that is red enough to avoid fluorescence but not so red as to lose sample shot-noise-limited operation.

Pemberton pointed out in 1989 that a spectrum of anthracene obtained with a He-Ne laser (632.8 nm) exhibited much lower

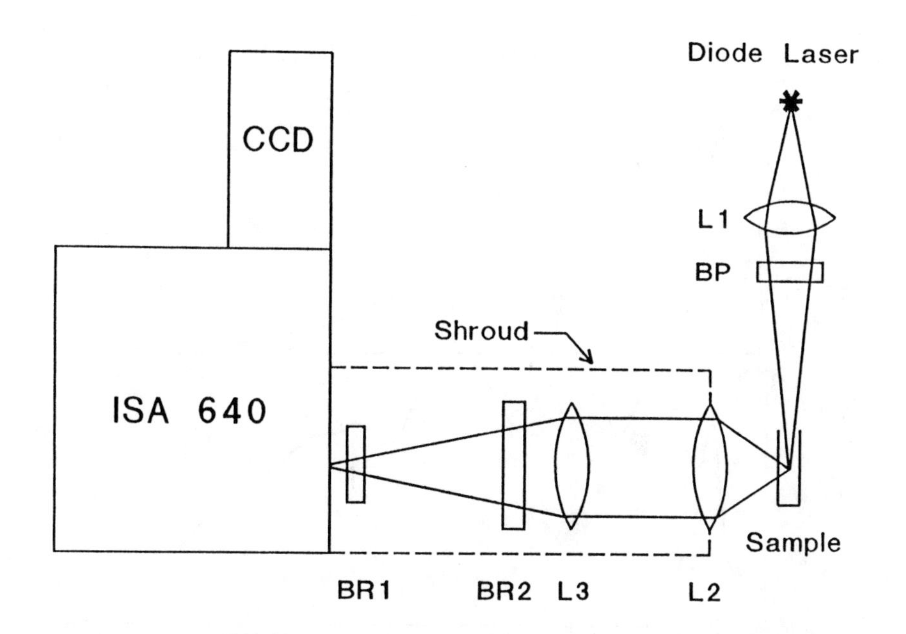

Figure 7.16 Schematic of CCD/diode laser spectrometer. BP and BR refer to bandpass and band reject filters, respectively. Reprinted with permission from reference 22.)

fluorescence than one obtained at 514.5 nm [20]. Our group took the approach further into the NIR, with diode lasers operating at 782 nm and 830 nm [22,30,31]. The apparatus is shown in Figure 7.16. Band reject filters were essential for decreasing elastically scattered laser light and were based on multilayer dielectric or holographic designs. The system response of the filters, spectrograph, and CCD was estimated by obtaining spectra of a tungsten bulb, which has a fairly flat output in the 800- 1000-nm region. Figure 7.17 shows the detector output plotted as a function of Raman shift relative to two common diode-laser wavelengths. This output is proportional to the $T \cdot Q_D$ product, incorporating filter transmission, spectrograph throughput, and detector quantum efficiency. The gradual decrease as Raman shift (and wavelength) increases is due to the Q_D roll-off of the detector, while the oscillations are due to interference effects in the band reject filters. These response nonuniformities can be mathematically corrected, but clearly the SNR will suffer at high Raman shift. Provided a silicon-based CCD is used, use of a laser wavelength above about 780 nm will result in loss of the C-H stretch region above 2800 cm^{-1}.

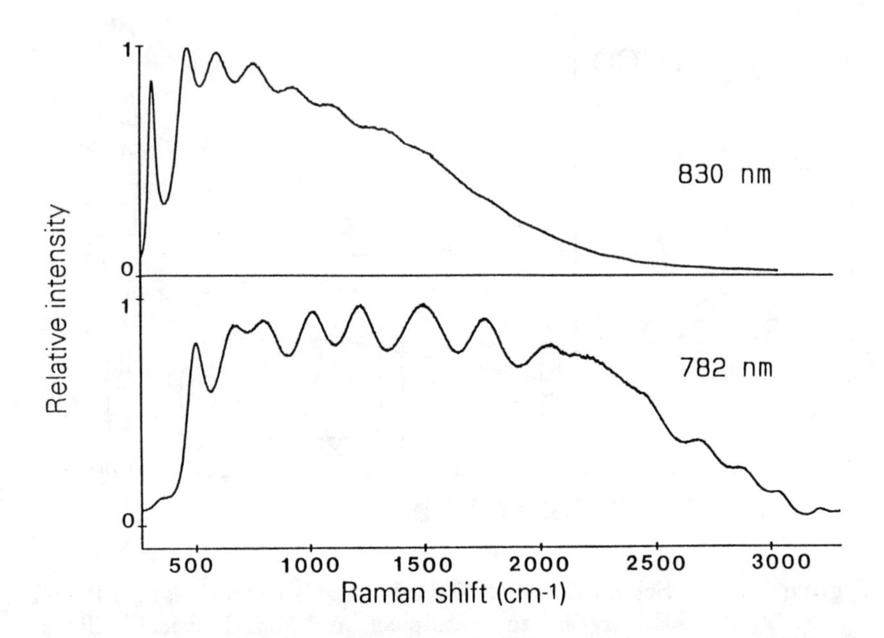

Figure 7.17 System response curves for CCD spectrometer with BR filters relative to two excitation wavelengths using a tungsten lamp source. The CCD response is normalized to maximum value. Reproduced with permission from reference 22.

The sensitivity of the diode-laser/CCD approach was demonstrated in Figures 7.12 and 7.13. An additional issue is the ability to observe minor components in the presence of much stronger scatterers, shown in Figure 7.18. The 992-cm^{-1} band of 0.01% benzene in CCl_4 is observable next to a strong CCl_4 band at ~780 cm^{-1}. This is a case where a multichannel detector outperforms a multiplex system, because noise from the large peak would obscure the small peak in a multiplex spectrometer. The reduced fluorescence of the NIR/CCD system is illustrated in Figure 7.19, showing spectra of a fluorescent laser dye, Rhodamine 6G, obtained with two different lasers. The fluorescence tail diminishes further with longer laser wavelength, but of course one is losing CCD quantum efficiency at the long wavelengths. Until the CCD Q_D curve is extended further into the NIR, the trade-off of fluorescence immunity and high Q_D is hard to avoid.

One is certainly not constrained to diode lasers for NIR/CCD Raman experiments. Dye lasers [33], Kr^+ (752 nm), and Ti:sapphire

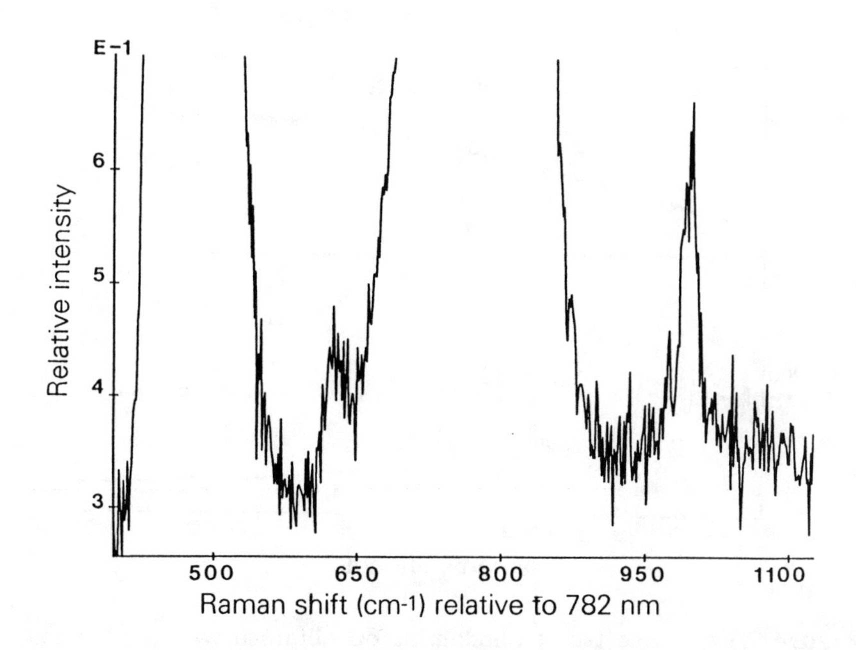

Figure 7.18 Spectrum of 0.01 *M* benzene (992^{-1}) in CCl_4 (459 and ~750 cm^{-1}), obtained with 782-nm diode and CCD, 30-min integration time.

are certainly possible and in several cases have been evaluated. As is the case with spectrometer configuration, there is no "best" NIR laser, but the choice will depend on the application. However, there is reason to believe that a general-purpose NIR/CCD Raman spectrometer suitable to a variety of applications will emerge in the near future.

7.7.4 Integrated CCD Raman Spectrometers

The technical developments involving CCD detectors, laser filters, NIR lasers, and inexpensive computers have stimulated interest in developing commercial CCD spectrometers. Dilor (Princeton, N.J), Instruments SA (Metucheon, N.J.), and Spex (Edison, N.J.) offer CCD attachments to their top-of-the-line spectrometers. More recently, several vendors have designed compact spectrometers based on single-stage spectrographs, holographic laser rejection filters, and CCD detectors. These systems incorporate many of the innovations discussed earlier, and in some cases permit rapid grating changes to vary spectral coverage. Spex offers a system with two electronically

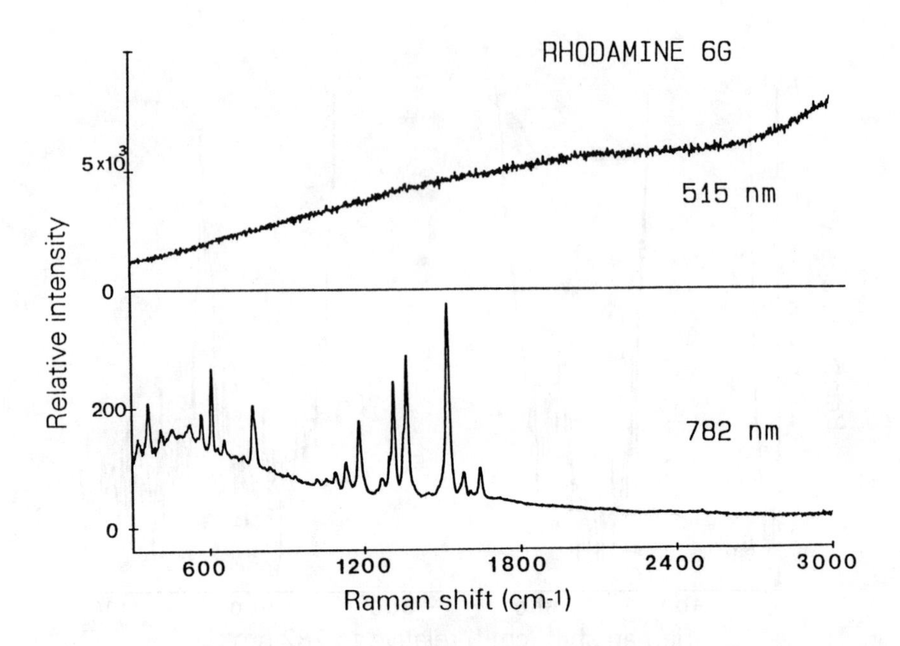

Figure 7.19 Spectra of Rhodamine 6G obtained with a PMT and argon laser (upper curve) and CCD/diode laser (lower curve). 60-s integration time for CCD. From reference 22.

selectable gratings in a 270-mm single spectrograph. Renishaw (Wotton-under-Edge, Gloucestershire, U.K.) offers a CCD spectrometer with a provision for Raman imaging.

The system developed by Chromex (Albuquerque) is shown schematically in Figure 7.20 [33]. A fiber-optic sampling head greatly simplifies operation, such that the fiber probe need only be immersed in the sample, or in some cases directed into the sample through the glass wall of a container. Representative spectra acquired within 2 min of four strong Raman scatterers are shown in Figure 7.21. Fiber-optic sampling has also been developed for remote sensing of hazardous samples, and the issue of fiber-generated background has been addressed [34,35]. The availability of compact, efficient CCD spectrometers plus fiber-optic probes should significantly increase the utility of Raman spectroscopy for routine applications. Such instruments became available only in late 1992, so it is difficult to ascertain the magnitude of their impact on Raman applications.

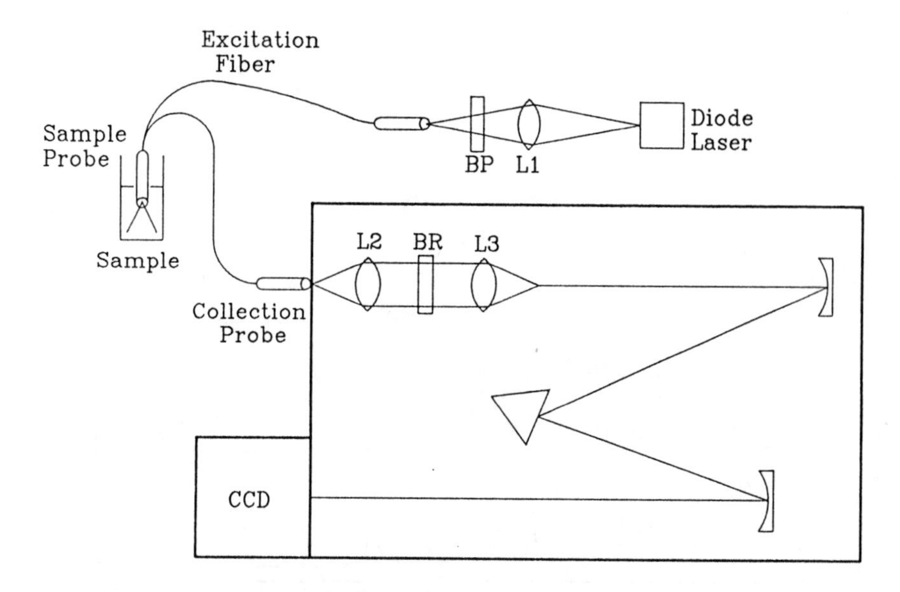

Figure 7.20 Integrated CCD/fiber-optics/diode laser Raman spectrometer based on Chromex imaging spectrograph. (Adapted with permission from reference 33.)

7.7.5 Raman Imaging with CCDs

The Raman microprobe has been available commercially for about 6 years and has proven useful for many applications. The incident laser light is focused through a microscope objective down to a spot $< 5\ \mu m$ in diameter; the scattered light is collected by the same objective and analyzed in a conventional spectrometer. Thus Raman spectra of quite small regions in a heterogeneous sample may be obtained. A significant improvement in information content is achieved with a Raman-imaging experiment, in which an image of a particular Raman feature is constructed by some means. For example, defects on graphite surfaces have a characteristic 1360-cm^{-1} mode that is distinct from the graphite lattice vibration at 1582 cm^{-1} [36]. If an image were constructed with only 1360 cm^{-1} Raman-shifted light, the image would reveal the size and distribution of defects. The desired spatial resolution of such an experiment would extend down to the limit of light microscopy, about 1 μm.

The simplest approach to constructing a Raman image is to combine a conventional microprobe with an automated X-Y translation stage under the sample (X_s and Y_s denote sample coordinates as distinct from CCD coordinates). By rastering the sample with the

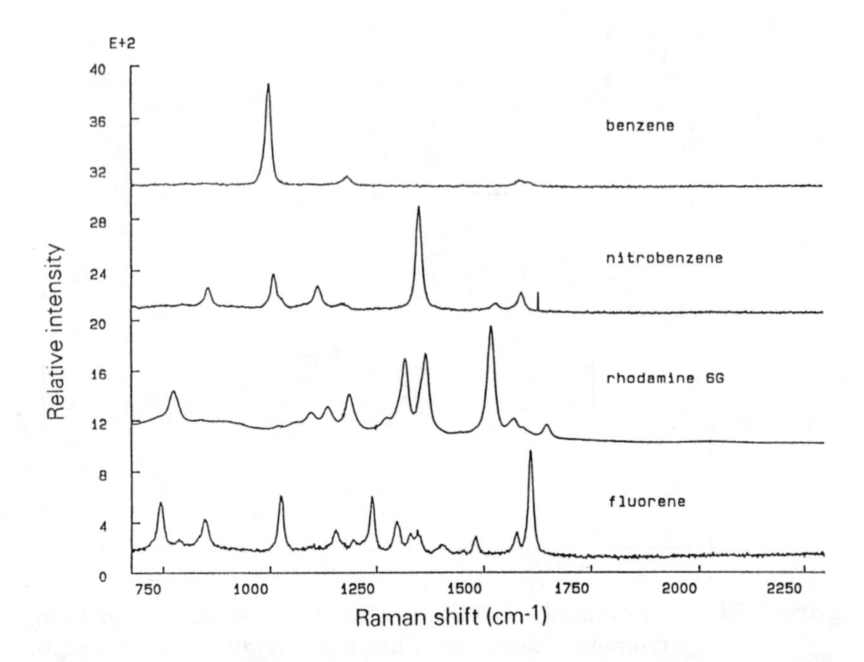

Figure 7.21 Raman spectra obtained in rapid succession (500 ms each) with the apparatus of Figure 7.20. The fiber probe was immersed in benzene and nitrobenzene and positioned above rhodamine and fluorene.

spectrometer set at a particular $\Delta\nu$, an image of intensity versus X_s and Y_s could be constructed. However, this approach is very slow and subjects the sample to possibly high-power density on a small focal spot. This situation could be improved by exploiting the multichannel advantage of a CCD to obtain complete spectra of each spatial resolution element, but one will still have to raster scan the sample to obtain four-dimensional intensity versus X_s, Y_s, and the Raman-shift data set.

Three more promising approaches exploit the two-dimensional nature of the CCD to greatly accelerate the image formation process. The conceptually simplest approach combines a microscope with a CCD camera and was recently introduced commercially by Renishaw Transducer Systems [37]. The sample is diffusely illuminated by an Ar^+ or He-Ne laser, and the scattered light is collected by the microscope objective. The scattered light is then passed through a narrow bandpass filter that is selected to transmit only a limited Raman-shift range (typically 20 cm^{-1}). The image is then formed on

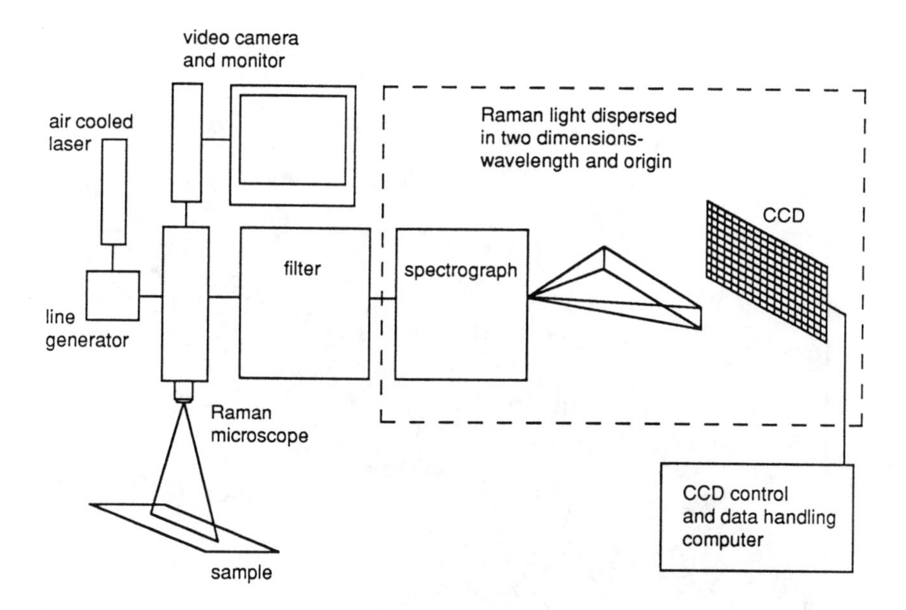

Figure 7.22 Line-scanned Raman imaging apparatus. (Adapted with permission from reference 37.)

the CCD using only the Raman-shifted light, so that the observer can visualize which regions of the sample give rise to the Raman shift of interest. To vary the Raman shift, the filters are angle tuned or switched mechanically. The second approach to Raman imaging involves a laser focused into a line that is scanned along the sample in one direction with the apparatus of Figure 7.22 [38]. The line was imaged into a spectrometer that projected the wavelength axis along one CCD axis and the position along the line along the other. Thus the Y_s and Raman-shift axis were analyzed by the CCD, while the X_s axis of the sample was scanned with the line-shaped laser beam. The spatial resolution was stated as <10 μm, and the spectral resolution was < 4 cm^{-1}. An intensity map versus X and Y for an integrated circuit is shown in Figure 7.23. By imaging the 520-cm^{-1} phonon of silicon, a map of the exposed silicon on the circuit was obtained.

An elegant but more conceptually formidable approach to Raman imaging has been presented by Treado and Morris [39–40]. The objective was to avoid using a focused beam to define spatial resolution at the sample, thereby to decrease the radiation density compared with either a point or a line focus. The sample is washed with a weakly focused laser, covering an ~500-μm-diameter spot. The scattered light is magnified by a modified optical microscope, then passed through a

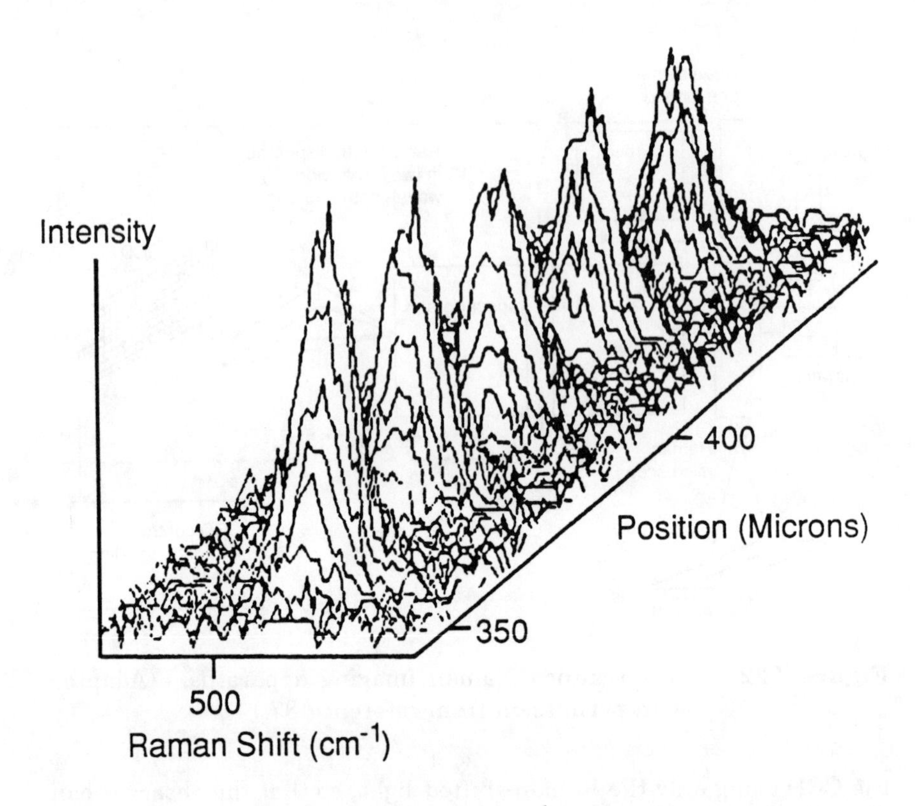

Figure 7.23 Raman image of 520 cm^{-1} Si phonon on an integrated circuit. Adapted from reference 39.

one-dimensional Hadamard mask, shown in Figure 7.24. The mask multiplexes the X_s dimension of the sample in time, so that the spectrograph monitors different regions of the sample as the mask is moved laterally. At the entrance slit of the spectrograph, the Y_s axis of the sample is projected along the slit axis; the X_s axis has been compressed to the width of the slit but multiplexed in time. The Hadamard mask has the major advantage over the line scan of monitoring half the sample at any one time, thereby providing a significant spatial multiplex advantage.

Once the light enters the spectrograph, it is dispersed as usual, then imaged onto the CCD. However, the spectrograph maintains the spatial information along the slit axis so that the vertical axis of the CCD corresponds to the magnified Y_s axis of the sample. The end result is a four-dimensional data set of intensity versus wavelength and versus X_s,Y_s position on the sample. The CCD provides two of the independent variables (Y_s and wavelength), while the Hadamard mask provides the third (X_s). The original references should be consulted for

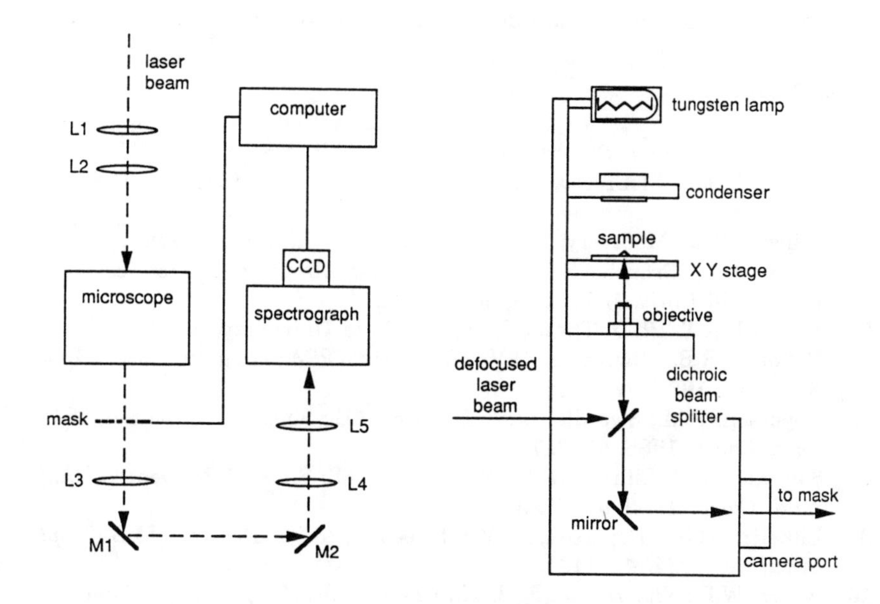

Figure 7.24 Apparatus for Hadamard/CCD Raman imaging. Hadamard mask is moved laterally to multiplex one spatial axis of the sample. (Adapted with permission from reference 40.)

the nontrivial process of mathematically converting the raw CCD intensity data to an image.

The technique was used to record images of laser-induced defects on graphite surfaces in which the damaged region has a distinctive 1360-cm^{-1} Raman band. The damaged areas were clearly observable with a spatial resolution approaching the diffraction limit [41].

7.8 Conclusions

It is clear that CCD detectors have already had a major beneficial effect on our ability to use Raman spectroscopy for chemical analysis. It is likely that the low noise, high Q_D, and two-dimensional multichannel advantages of CCDs will lead to the replacement of most conventional detectors for Raman spectroscopy. In additional, a series of technical developments, including compact spectrographs, laser-rejection filters, NIR lasers, and fiber-optic sampling, have added significant driving force for broader applications of Raman spectroscopy. At this writing, integrated commercial spectrometers combining all or most of these developments are beginning to become available.

Time will tell if the major advantages of these developments will lead to widespread use of Raman spectroscopy outside the research lab.

References

1. Long, D.L. *Raman Spectroscopy.* McGraw-Hill, New York, 1977.
2. Packard, R.T.; McCreery, R.L. *Anal. Chem.* **1987**, *59*, 2631, and **1989**, *61*, 744a, and references therein.
3. Schwab, S.D.; McCreery, R.L.; Gamble, F.T. *Anal. Chem.* **1986**, *58*, 2486.
4. Ingle, J.D.; Crouch, S.R *Spectrochemical Analysis.* Prentice Hall, Englewood Cliffs, N.J., 1988, pp 141–142.
5. Packard, R.T. Ph.D. Thesis, The Ohio State University, 1988.
6. Bilhorn, R.B.; Sweedler, J.V.; Epperson, P.M.; Denton, M.B. *Appl. Spectrosc..* **1987**, *41*, 1114.
7. Epperson, P.M.; Sweedler, J.V.; Bilhorn, R.B.; Sims, R.G.; Denton, M.B.; *Anal. Chem.* **1988**, *60*, 327A.
8. Sweedler, J.V; Bilhorn, R.B.; Epperson, P.M.; Sims, G.R.; Denton, M.B. *Anal. Chem.* **1988**, *60*, 282A.
9. Bilhorn, R.B.; Epperson, P.M.; Sweedler, J.V.; Denton, M.B. *Appl. Spectrosc.* **1987**, *41*, 1125.
10. Acker, W.P.; Yip, B.; Leach, D.H.; Chang, P.K. *J. Appl. Phys.* **1988**, *64*, 2263.
11. Campion, A.; Brown, J.; Grizzle, W.H. *Surf. Sci.* **1982**, *115*, L153.
12. Campion, A.; Woodruff, W.H. *Anal. Chem.* **1987**, *59*, 1301A.
13. Janesick, et al., *Proc. SPIE* **1989**, *1071*, 15.
14. Chase, D.B. *J. Am. Chem. Soc.* **1986**, *108*, 7485.
15. Murray, C.A.; Dierker, S.B. *J. Opt. Soc. Am.* **1986**, *3*, 2151.
16. Van Duyne, R. In *Chemical and Biological Applications of Lasers*, Moore, L.B., Ed. Academic, New York, 1979, Vol. 4, Chapter 4.
17. Dierker, S.B.; Murray, C.A.; LeGrange, J.D.; Schlotter, N.E. *Chem. Phys. Lett.* **1987**, *137*, 453.
18. Schlotter, N.E.; Schaertel, S.A.; Kelty, S.P.; Howard, R. *Appl. Spectrosc.* **1988**, *42*, 746.
19. Campion, A.; Perry, S. *Laser Focus World* **1990**, August, 113.
20. Pemberton, J.; Sobocinski, R. *J. Am. Chem. Soc.*, **1989**, *111*, 432.
21. Pemberton, J.; Sobocinski, R. *Appl. Spectrosc.* **1990**, *44*, 328.
22. Wang, Y.; McCreery, R. *Anal. Chem.* **1989**, *62*, 2647.
23. Freeman, J.J., et al. *Appl. Spectrosc.* **1981**, *35*, 196.
24. Schwab, S.D.; McCreery, R.L. *Appl. Spectrosc.* **1987**, *41*, 126.
25. Pelletier, M.J. *Appl. Spectrosc.* **1990**, *44*, 1699.
26. Pilon, M.J.; Denton, M.B., et al. *Appl. Spectrosc.* **1990**, *44*, 1613.
27. Chase, D.B. *Anal. Chem.* **1987**, *59*, 881A.
28. Chase, D.B.; Parkinson, B.A. *Appl. Spectrosc.* **1988**, *42*, 1186.
29. Lewis, N.; Kalasinsky, V.F.; Levin, I.W. *Anal. Chem.* **1988**, *60*, 2658.
30. Williamson, J.M.; Bowling, R.J.; McCreery, R.L. *Appl. Spectrosc.* **1989**, *43*, 372.
31. Allred, C.D.; McCreery, R.L. *Appl. Spectrosc.* **1990**, *44*, 1229.
32. Harris, T.D.; Schnoes, M.L.; Seibles, L. *Anal. Chem.* **1989**, *61*, 994.
33. Newman, C.D.; Bret, G.G.; McCreery, R.L. *Appl. Spectrosc.* **1992**, *46*, 262.

34. Carrabba, M.M.; Rauh, R.D. U.S. Patent no. 5,112,127, May 1992.
35. Myrick, M.; Angel, S.M.; Desiderio, R. *Appl. Optics.* **1990**, *29*, 1333.
36. Bowling, R.J.; Packard, R.T.; McCreery, R.L. *J. Am. Chem. Soc.* **1989**, *111*, 1217.
37. Batchelder, D.; Cheng, C.; Pitt, G.D. *Advanced Materials* **1991**, *3*, 566.
38. Bowden, M.; Gardiner, D.J.; Rice, G.; Gerrard, D.L. *J. Raman Spectrosc.* **1990**, *21 37*.
39. Treado, P.J.; Morris, M.D. *Anal. Chem.* **1989**, *61*, 723A.
40. Treado, P.J.; Morris, M.D. *Appl. Spectrosc.* **1989**, *43*, 190.
41. Treado, P.J.; Morris, M.D., et al., *Appl. Spectrosc.* **1990**, *44*, 1270.

8
CTD Detectors in Analytical Luminescence Spectroscopy

Robert S. Pomeroy

Department of Chemistry, U. S. Naval Academy
Annapolis, Maryland

8.1 Introduction to Luminescence

The roots of the study of luminescence can be traced back to the observed luminescence from the extract of *Lignium nephiticium* in water reported by Monardes in 1565. The element phosphorus was named in 1669 from the Greek for light bearing, based on the observation that phosphorus glowed in a dark room. Emission from a solution of chlorophyll exposed to sunlight was reported in 1833 by Sir David Brewster, and luminescence from fluorspar was noted in 1852 by Sir G. G. Stokes. Stokes is credited with generating the term *fluorescence*, named after the blue emission seen in fluorspar; he correctly concluded that this phenomenon was different from simple light scattering and postulated that the emitted radiation was derived from the absorption of light of one wavelength and the emission of light of a longer wavelength. Despite such an extensive and rich history, luminescence techniques are still among the most active areas of current research, extending into demanding applications such as probing the contents of single cells [1].

Hot bodies that are luminous solely because of their high temperature are referred to as incandescent. All other forms of emitted radiation are referred to as luminescence. The term *luminescence* was coined by Eilhard in 1888 to describe the emission of radiation not caused solely by an increase in temperature. This distinction also explains the origin of the term cold light when describing luminescence. Hence, a system exhibiting sustained luminescence is losing energy, requiring that energy be supplied from an outside source. The mode in which this outside energy is supplied delineates the classification of the different forms of luminescence: photoluminescence (including fluorescence and phosphorescence), stimulated by IR, visible, or UV light; radioluminescence, excited by radioactive species; sonoluminescence, caused by intense sound waves in fluids; chemiluminescence, utilizing the energy generated by a chemical reaction; electroluminescence, produced by the passage of an electrical current;

triboluminescence, stimulated when certain crystals are crushed; and thermoluminescence, caused by the release on heating of chemically reactive species trapped in a rigid matrix.

One of the most important attributes of luminescence phenomena is that the emitted luminescence can be spectroscopically separated from the excitation. For the forms of luminescence where electromagnetic radiation is not the source of excitation, such as chemiluminescence and triboluminescence, this is self-explanatory. For photoluminescence, the fact that the emission spectrum is always at longer wavelengths than the excitation spectrum (Stokes' law) allows the excitation radiation to be removed, and only the emission energy is monitored. The result is that the emitted radiation is detected against virtually no background. This makes the task of detecting the luminescence radiation much easier and accounts for the exquisite sensitivity achievable. By comparison with absorption techniques, where the detector has to resolve the small difference between two large signals, luminescence techniques have a sensitivity that is 1 to 6 orders of magnitude greater. The range of sensitivities is a function of the luminescence quantum efficiency (QE) and the output of the source at the excitation wavelength; however, even for compounds that exhibit poor luminescence QE, luminescence can still be the method of choice over absorption with the proper optics, detector, and source stability. Another aspect of Stokes' law is that luminescence can be applied to turbid solutions, when absorption techniques cannot, since scattered excitation radiation can be filtered.

Luminescence also offers a degree of selectivity, since many more molecules absorb radiation than re-emit radiation. One can argue that this is a drawback, since luminescence spectroscopy can only be applied to those molecules that luminesce; however, one can attempt to selectively derivatize the analyte of interest so as to make it luminesce. The luminescence phenomenon also provides an additional degree of flexibility over absorption; the analyst can vary both the excitation and the emission wavelengths to increase the selectivity. These attributes account for the wide applicability of luminescence techniques and why it is such an active research topic.

Recent technological advances in solid-state imaging detectors offer the analytical chemist new alternatives for the detection of luminescence emission. Charge-transfer device (CTD) detectors are sensitive multichannel detectors with large dynamic ranges. One particular class of CTD, the charge-coupled device (CCD), is particularly well suited to low-light-level spectroscopies due to its high QE and low noise characteristics.

This chapter is divided into four sections. The first section provides background material on luminescence processes so as to introduce the operational necessities and factors that influence chemical analysis and the measurement. The second section addresses instrumental needs and the present instrumentation. The third section introduces CTD detectors and their capabilities for low-light-level spectroscopy. The final section presents sample applications of CTD systems to analytical luminescence spectroscopy.

8.2 Background

8.2.1 Luminescence Processes

At room temperature most molecules of an analyte are in the lowest vibrational level of their electronic ground state. In the ground electronic state, the electron spins of the electrons occupying the same orbital are opposed (antiparallel). The resulting net spin, S, is zero and the multiplicity of this configuration is determined from the relation $2S+1$. The quantity $2S+1$ is called the multiplicity. For a molecule whose electron spins are opposed, the multiplicity is 1. This configuration is referred to as a singlet. If the spins are aligned or parallel, then the value of S is 1 and the multiplicity is 3. This configuration of this less common arrangement is referred to as a triplet. The absorption of light does not cause a change in the orientation of the electron spin; consequently the transition from a singlet ground state is to a singlet excited state, $S_0 \rightarrow S^*$. The time frame for the absorption of light by a molecule is on the order of 10^{-15} s. Excitation is not limited to the lowest vibrational level of the excited state and can give rise to a vibrationally excited molecule in an excited electronic state. (In solutions, the rotational levels are so closed spaced that they are spectroscopically indistinguishable so that the vibrational levels are typically represented as bands which contain the rotational levels.) The electron returns to the ground state via vibrational relaxation to the lowest vibrational level of the excited state. This process takes place in a time frame of 10^{-14} to 10^{-11} s. From the lowest vibrational level of the excited state the molecule can return to the ground electronic state by a radiationless decay process termed internal conversion. Internal conversion converts the remaining excitation energy to heat through kinetic, vibrational, or rotational modes. The time frame for this process is on the order of 10^{-8} s.

The lifetime of the excited state is long enough that it also may return to the electronic ground state by the emission of a photon. This process is fluorescence and takes place on a time scale of 10^{-9} to 10^{-7} s, directly competing with internal conversion. Although the direct excitation from the electronic ground state to an excited triplet is virtually zero, an excited singlet-state molecule can undergo intersystem crossing, in which a spin flip occurs converting the system to a triplet. The lifetime of these transitions is 10^{-8} s which competes with both internal conversion and fluorescence. Once in the excited triplet state, the molecule usually undergoes vibrational relaxation to the lowest vibrational level of the triplet state. From the lowest vibrational state of the triplet, the excited molecule can return to the ground state by emission of a photon. This form of luminescence is *phosphorescence*. The lifetime of the triplet state can be approximately 10^{-4} to > 10 s. Internal conversion of the excited triplet in solution is fast enough that phosphorescence is rarely observed in solutions at room temperature. Competing with these processes are energy transfers and/or chemical reactions with species in close proximity to the excited molecule. Generally, alternative means of energy transfer must occur on the same time scale in order to compete effectively. Table 8.1 lists the lifetimes of the various processes and Figure 8.1 summarizes the possible fates of an excited molecule.

Two other factors that impact the practice of analytical luminescence spectroscopy are the excitation and the polarization of the emitted radiation. The excitation wavelength is to a certain degree nonspecific; that is, the quantum yield is independent of the wavelength up to a specific limiting wavelength. Past the limiting wavelength there is a sharp decline in the quantum yield. The nonspecific nature of the excitation wavelength eases the restriction on the excitation source, and hence a single source such as a mercury vapor lamp can be a suitable source for many analytes and applications.

Molecules whose fluorescence lifetimes are relatively short compared with their rotational mobility often exhibit polarized fluorescence when the emission is observed perpendicular to the plane

Table 8.1 Lifetimes for Luminescence Processes

Process	Lifetime
Absorption	10^{-15} s
Vibrational relaxation	10^{-14} - 10^{-11} s
Fluorescence	10^{-9} - 10^{-7} s
Intersystem crossing	10^{-11} s
Phosphorescence	10^{-4} - 10 s
Internal conversion	
- Initial deactivation	10^{-11} s
- From excited singlet to ground state	10^{-8} s
- From excited triplet to ground state	10^{-2} s

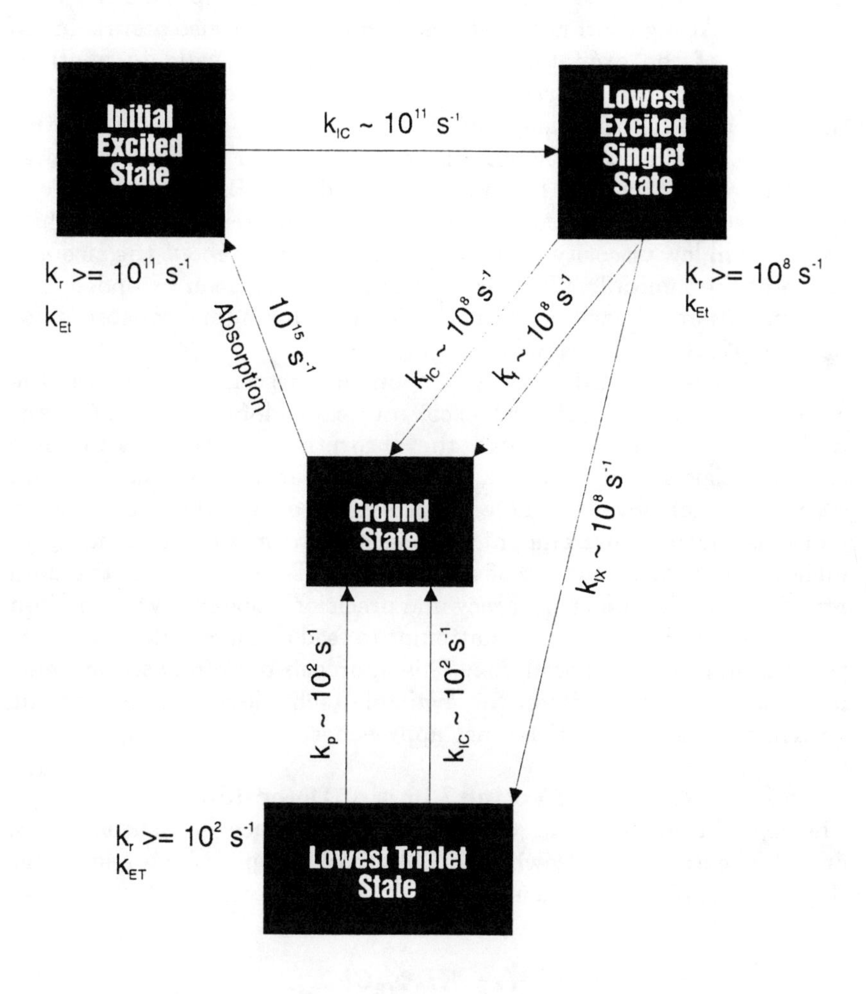

k_r = rate constant for chemical reaction
k_{Et} = rate constant for energy transfer
k_{IC} = rate constant for internal conversion
k_{IX} = rate constant for intersystem crossing
k_f = rate constant for fluorescence
k_p = rate constant for phosphorescence

Figure 8.1 Diagram of possible fates of an excited molecule.

of the electric vector of the excitation. Fluorescence depolarization of dilute samples in high viscosity matrices is due to the difference in angle between the absorption and emission oscillators. The polarization spectrum can thus be used in the determination of the number and character of the electronic transitions that make up the absorption spectrum. At higher concentrations, depolarization also occurs due to migration of the excitation energy from one molecule to another. Examination of the polarization spectrum at high analyte concentration in a high-viscosity medium gives information about the electromagnetic coupling of neighboring molecules. In solvents of lower viscosity, depolarization has a component due to Brownian rotations of the molecule. Measurements of the depolarization for dilute solutions in low viscosity solvents give information about the size and shape of the molecule. Polarization studies have been extended into determination of excited state lifetimes and binding constants for proteins [2,3].

For more detailed information on luminescence processes, the reader is referred to several excellent texts [4–6]. The important message here is that although the absorption spectrum is to some extent nonspecific, there are subtleties in the absorption and emission spectrum that give valuable information about the luminescing molecule and its environment. Modern instruments are tending to higher and higher degrees of automation. To ensure that the data reported are of a known accuracy and precision, modern systems often incorporate rich data sets and attempt to feed as much information as possible into the final analysis. This approach to automated systems is placing new demands on the available technology and instrument design, specifically multichannel approaches.

8.2.2 Factors Affecting Limit of Detection

The signal-to-noise ratio (SNR) in luminescence measurement is described extensively elsewhere [7]. A greatly simplified expression for the luminescence signal component of that ratio is

$$I_f = K I_0 \Phi_f \epsilon C b \tag{8.1}$$

where

I_f = fluorescence intensity
K = instrumental constant relating the transfer efficiency of the light to the sample and subsequent collection of emission
I_0 = source intensity
Φ_f = fluorescence quantum yield

ϵ = analyte absorptivity (from Beer's law)
C = molar concentration of analyte
b = path length

The quantum yield is a key factor in determining the intensity of the luminescence. The quantum yield is the ratio of the number of emitted luminescence photons to the number of absorbed photons. This is a simple statement conceptually, but the practical measure of absolute quantum yields is difficult to perform with precision because it requires consideration of the emission in all directions. More typically quantum yield is measured on a relative basis, the luminescence emission being compared with that of a reference compound. Measurement of the quantum yield also requires attention to quenching of the fluorescence, photodecomposition, the fluorescence blank, inner filter effects, temperature, and adsorption. The ultimate limit to the practical application of luminescence spectroscopy often requires special care and attention to the above-mentioned factors rather than simply use of a luminescent species with a superior quantum yield.

As important as the quantum yield of the luminescent molecule is the number of times that molecule can be cycled through the process of excitation and emission before it undergoes photodecomposition. Photodecomposition is the destruction of the analyte due to the absorption of light. UV and visible photons carry sufficient energy that at high fluxes the problem of photodecomposition becomes an issue. Consequently, molecules that exhibit poor quantum yield but are resistant to photodegradation can still be sensitively detected, the ultimate limit being the luminescence blank, as discussed above. Photodecomposition can be minimized by employing lower photon fluxes, which will in turn lower the luminescence signal. This situation indicates a move away from the application of high-power sources toward increasing the light collection and detection efficiency of spectroscopic systems. Photodecomposition can also be reduced by using a shutter to shield the sample from the source between analyses or by employing short exposure times. This approach is constrained by the speed and precision of mechanical shutters, the transparency of electrooptic shutters, and/or the response time of the detector. Photodecomposition can also be used to advantage in chemical analysis if it is possible to selectively photodegrade an interference, or to create a fluorescent by product from a nonfluorescent species of interest [8,9].

Quenching is the reduction in fluorescence intensity due to the transfer of energy from the excited molecule to the solvent or surrounding molecules; it directly competes with fluorescence and phosphorescence as a means of deactivation of the excited state. Although the processes of internal conversion and vibrational relaxation are internal processes within the molecule, they can be markedly influenced by the environment. These competing means of energy transfer have placed limitations on the application of luminescence spectroscopy to chemical analysis of complex mixtures. For this reason separation and purification techniques such as high-performance liquid chromatography have employed luminescence analysis to advantage as a means of quantitative detection.

The concentration of the analyte also needs to be a consideration. Ideally the excitation should uniformly illuminate the sample cell with all analyte species exposed to the same excitation. If the concentration of the absorbing species gets high enough, the fluorescence intensity may be diminished in two ways: either emitted fluorescence is absorbed before it reaches the detector or the absorption of the source excitation is so high that the intensity of the excitation is drastically altered throughout the sample cell. The result typically is a nonlinear calibration curve and may include distortion of the excitation and emission spectra. These two means of reducing the fluorescence yield are termed inner filter effects and should not be confused with a quenching process. At high concentrations, a chemical effect may come into play: dimer formation can also become a factor, giving rise to altered absorption from the dimer and characteristically different emission from the eximer.

Temperature has a large effect on the many of the processes that compete with fluorescence as a means of deactivation. Molecules that are only weakly fluorescent at room temperatures display strong fluorescence at lower temperatures. In the small volumes of the sample containers, exposure to the source for significant periods of time can lead to heating of the sample and alteration of the observed fluorescence. Shuttering the source eliminates some heating; however, for prolonged exposures of dilute samples, analysis of temperature-sensitive fluorescent species requires some means of temperature control.

Adsorption also plays a role in the emitted fluorescence. Adsorption can be both useful and problematic. It is possible to use the alteration of fluorescence intensity due to adsorption of the fluorescing species as an indicator for precipitation reactions and in the analysis of olefins in gasolines. Adsorption of fluorescent species to the walls of the sample cell is detrimental to quantitative analysis and is a particular concern in dilute solutions, the typical situation in luminescence analysis. Many of these problems can be overcome by proper pretreatment of the cell, judicious choice of solvent, and minimization of cell surfaces.

The nature of the chemical environment, the matrix or solvent, can impact in additional ways. Hydrogen bonding can affect the luminescence, and it is important to avoid changes in the matrix between the sample and standard that could make differences in the hydrogen bonding of the solvent. The hydrogen-bonding abilities of many molecules in the excited state are markedly different from those in the ground state, and the direction of the spectral shift is often difficult to predict. The solvent may also interact in chemical equilibrium with the excited state, thereby affecting the quantum yield. A prime example of this is the change in the pK_a of many luminescent acid compounds upon excitation. The excited compound can undergo proton exchange, a fast reaction at about 10^{11} L/mol-s, giving rise to "anomalous" fluorescence behavior as a function of pH. This phenomenon has been used to advantage in the development of fluorescent indicators for acid/base reactions. The dielectric constant of the solvent also influences the emission energy. The changes in the luminescence spectra due to the changing dielectric constant of the solvent are referred to as polarization shifts. They result from changes in the dipole moment of excited molecules, typically becoming more polar. The resulting changes in the interaction between the solvent and the excited analyte molecule give rise to changes in the intensity and structure of the luminescence emission spectrum.

In almost all luminescence measurements the ultimate limit of detection is dependent on the background luminescence, the luminescence blank. Sources of blank luminescence include the Raman emission from the solvent, luminescence from the sample cells, scattered light not removed by the instrument, impurities in the solvent or reagents, and fluorescence from other species in the sample. Obviously the most desirable situation is to be analyte photon shot noise limited. In practice, however, even with careful preparation of blanks and background subtraction, the limitation is due to the uncertainty in the measure of the blank. Oddly, in comparisons between absorption and luminescence, luminescence is typically characterized as measuring a small signal against a very low background, whereas absorption is the measure of small differences in large signals. However, at the limit of detection, the challenge is to measure small analyte signals against relatively large background signals. Epperson [10] demonstrated in studies on the fluorescence limit of detection that for a 10^{-12} *M* anthracene solution, the signal from the analyte was 2800 e^- and that of the background was 260,000 e^-. The end result is that luminescence limits of detection are typically background photon shot noise limited, that is, limited by uncertainties in determining the blank.

8.3 Instrumentation

The ideal instrument for luminescence analysis collects complete excitation and emission spectra and uses a source that produces sufficient excitation but not deleterious effects such as photodecomposition or sample heating. The optical system would discriminate efficiently against unwanted radiation, and the detector would demonstrate sensitive low-noise response over a broad spectral range and over a large linear dynamic range. The instrument would also be capable of time- and polarization-resolved spectroscopy and would be able to accommodate a variety of sample types. With these ideas in mind, this section will provide a brief description of current luminescence instrumentation.

The instrumentation for monitoring luminescence emission basically consists of a source of exciting radiation and a means of measuring the luminescence emission intensity. Fluorometers are filter instruments used in luminescence analysis, and spectrofluorometers employ gratings or prisms to select radiation instead of filters. Spectrofluorometers are more flexible and can be used either for routine work or research. For routine work, fluorometers can be very effective at anywhere from one-fifth to one-tenth the price.

8.3.1 Source Excitation

Ideally, the power in the excitation bandwidth should constitute a large portion of the overall source power. Many continuous sources have two problems: (1) Only a small fraction of the overall power is present in any particular wavelength, and (2) the removal of the excitation from the desired luminescence is difficult. Use of continuum sources like a xenon-arc lamp or a tungsten lamp virtually demands the use of an excitation monochromator to select the excitation radiation. Line sources such as low-pressure mercury vapor lamps and lasers can often be operated without an excitation monochromator and simply use a filter to remove the unwanted source radiation from the sample and then another filter to remove scattered source radiation from the luminescence.

Major advantages of lasers are (1) the concentration of the power within a narrow spectral range and (2) the fact that the beam can be focused into a small region. The latter is an important practical implication for chromatographic applications that demand small detector volumes. The nearly monochromatic nature of lasers removes the need for an excitation monochromator, and the resulting Raman emission from the solvent will be in a well-defined band where it can be more easily separated from the luminescence emission. Line sources have significant advantages in that they ease the demand on systems for wavelength discrimination; however, the trade-off is a loss in flexibility, since optimal excitation may not be obtainable with a line source. Tunable dye lasers offer both concentrated monochromatic excitation and decreased restriction on the choice of excitation wavelength; however, tunable dye lasers are presently too large, complex, and expensive for general use in analytical luminescence instrumentation.

8.3.2 Cell Configuration

The cell configuration is the relative position of the excitation beam, the cell, and the detector. Among several possibilities [7,8,11], the right-angle and frontal arrangements are most important. The right-angle configuration places the detection system at a right angle from the excitation beam and is best for discrimination against the excitation radiation and collection of maximum radiation. If inner filter effects are significant or if the sample has poor optical properties (e.g., a cracked solid), the frontal geometry is preferred; the fluorescence radiation is collected through the same surface as that through which the excitation beam enters the cell, thereby minimizing reabsorption or scattering of the emitted radiation.

8.3.3 Spectrographs

Instruments for luminescence analysis vary chiefly in the degree of control they exert over the excitation and emission wavelengths. Filter fluorometers employ filters to restrict excitation to a specific wavelength in a line source or to a band of radiation in a continuous source. Filters are also placed between the sample compartment and the detector to select the emission band to be analyzed. Spectrofluorometers utilize one or more monochromators to restrict either the excitation or the emission or both. Use of monochromators allows greater flexibility in the range of excitation and emission wavelengths that can be monitored and has the added advantage of allowing measurement of both the excitation and emission spectra.

There is usually a trade-off between spectral resolution and throughput. So much of the light is attenuated in the monochromator systems of spectrofluorometers that high sensitivity is compromised. Optimum performance requires that entrance slits are at the same width as the exit slits. The amount of light that can be passed through a spectrograph is proportional to the square of the slit width. Increasing the slit width directly increases the band of radiation. This can lead to nonlinearities in the absorption; the Beer's law term in equation 8.1 is based on the change in ϵ over the bandpass being insignificant.

One also must be cognizant of multiple orders with grating instruments, stray-light rejection, the variation of the system efficiency with wavelength, and the variation in detector response with wavelength. The latter two points are particularly important in the correction of the raw data to yield the luminescence spectrum corrected for instrumental artifacts. Obviously this places a premium on the use of powerful sources and sensitive detectors.

8.3.4 Detection

Attempts to increase the sensitivity of luminescence systems have focused on the development of intense excitation sources such as lasers. Lasers have been successfully employed in many systems and have enjoyed great success in research instruments; however, laser systems, especially the flexible tunable dye laser, are expensive, bulky, and clearly impractical for routine analytical instruments. Lasers also run the risk of photodecomposition and temperature effects (see Section 8.2.2). An alternative approach to using high-power sources is to develop better spectrographs, more efficient optical systems, and more sensitive detectors.

The most common detector in luminescence instrumentation is the photomultiplier tube (PMT). The PMT provides high sensitivity, wide dynamic range, and good spectral response over a large wavelength region. These characteristics have held the PMT in high esteem for many years. However, the PMT is a single-channel detector, and there is a great deal of interest in replacing the single-channel PMT with multichannel detectors. Charge-transfer device (CTD) array detectors represent a class of multichannel detector that offers the capability of multichannel detection, and these devices exhibit sensitivity that exceeds that of the PMT on a single-detector-element basis [12]. CTD detectors, when combined with appropriate optics, give rise to multichannel spectroscopic systems that exhibit superior speed, sensitivity, and flexibility over conventional PMT-based luminescence systems [13]. The balance of this chapter will focus on the use of CTD detectors, specifically the charge-coupled device (CCD) detector, for application to analytical luminescence spectroscopy.

8.4 CCD Characteristics Important to Luminescence Spectroscopy

CTD detectors are a class of solid-state, multichannel, integrating detectors. Photons striking the detector create electron-hole pairs. The amount of photogenerated charge is directly proportional to the photon flux. The photogenerated charge is measured either by shifting the collected charge from detector element to detector element to a readout amplifier (intercell transfer), or by monitoring the voltage change when the charge packet is shifted within a detector element (intracell transfer). The differing modes of charge readout delineate the two subclasses of CTD—the CCD, which uses the intercell transfer, and the charge-injection device (CID), which employs the intracell transfer. Although CIDs have the advantage of nondestructive readout (NDRO), the limited availability of scientific CIDs and their higher read noise make the CCD the detector of choice for high-resolution, low-light-level spectroscopies. CCD detectors come in a large variety of formats ranging from a single element to 4096 × 4096 arrays and exhibit spectral sensitivity from 0.1 to 1100 nm.

By comparison with other types of detectors, such as the PMT, the photodiode array (PDA), and the vidicon, the CCD has several advantages. CCDs have high QE (in excess of 80%) over a broad spectral range by comparison with PMTs and vidicons. CCDs have extremely low dark currents when cooled (as low as 0.001 e^-/s at liquid nitrogen temperatures) and can therefore be integrated for long periods to monitor very low light fluxes. The read noise of CCDs is finite (as low as 3–5 e^-); however, it is small compared with that of PDAs (typically 1200 e^-) and vidicons. Although PMTs exhibit no read noise and are extremely sensitive in the photon-counting mode, CCDs have the advantage of being multichannel devices that are immune to light shock. Given both the multichannel advantage of CCD detectors and an ability to detect low photon fluxes that equals or surpasses PMTs makes the CCD the detector of choice for low-light-level spectroscopies such as luminescence. A detailed evaluation of CTDs for low-light-level spectroscopies is available in previous chapters.

8.4.1 The Multichannel Advantage

The need to monitor several wavelengths simultaneously has been a significant driving force behind the development of multichannel detectors. The alternative with single-channel detectors has been to scan the spectrum or use several detectors. As previously pointed out, subjecting the sample to the source for extended periods can result in changes in the QE with time due to photodegradation and sample heating. Instruments can be designed with several PMTs, but the flexibility is poor since the detectors are laboriously positioned; with more PMTs, the cost begins to escalate. Although the cost of multichannel detectors such as the CCD is high, the price is quite competitive on a per-detector basis. Solid-state-array detectors also offer stable spectral registration for accurate background subtraction since there are typically no moving parts in multichannel spectroscopic instruments. The biggest advantage to multichannel detectors is the simultaneous acquisition of the spectrum.

The detection limit for a luminescence analysis is generally determined by factors other than the photon shot noise of the sample emission. For single component systems, sensitive detection requires discrimination of a weak analyte signal from the background signal. Accurate background subtractions require correction for fluctuations in both source output and detector response with time. Using a multichannel detector allows for the simultaneous acquisition of the analyte signal and the background signal. Other instruments may employ a second detector for monitoring source fluctuations but cannot correct for drift in the detector.

Using a simple optical arrangement as shown in Figure 8.2, the CCD-based system gathers the analyte signal, the scattered radiation signal (Rayleigh, Tyndall, and stray light), and the Raman scattering from the solvent; a spectrum of anthracene that includes the Rayleigh and Raman peaks is shown in Figure 8.3. The importance of accurate background subtraction is readily apparent in extremely dilute solution where the background signal can be 1000 times larger than that of the analyte. The dominant source of noise then is the uncertainty in the background signal.

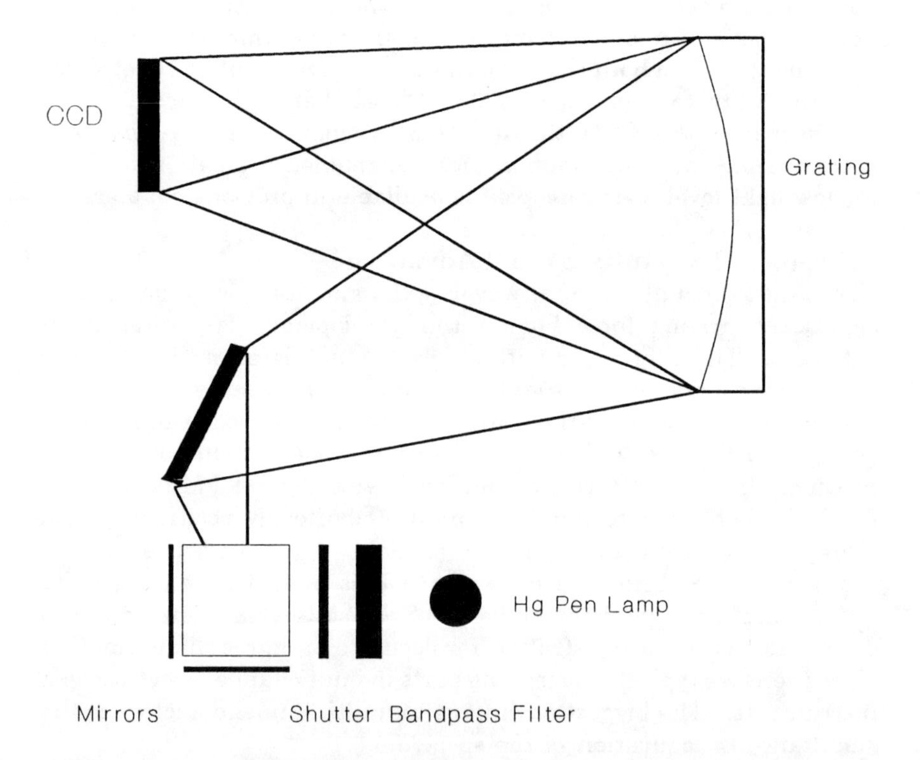

Figure 8.2 Simple configuration for fluorescence measurement requiring only a low-intensity pen lamp as an excitation source. (Reprinted with permission from reference 13.)

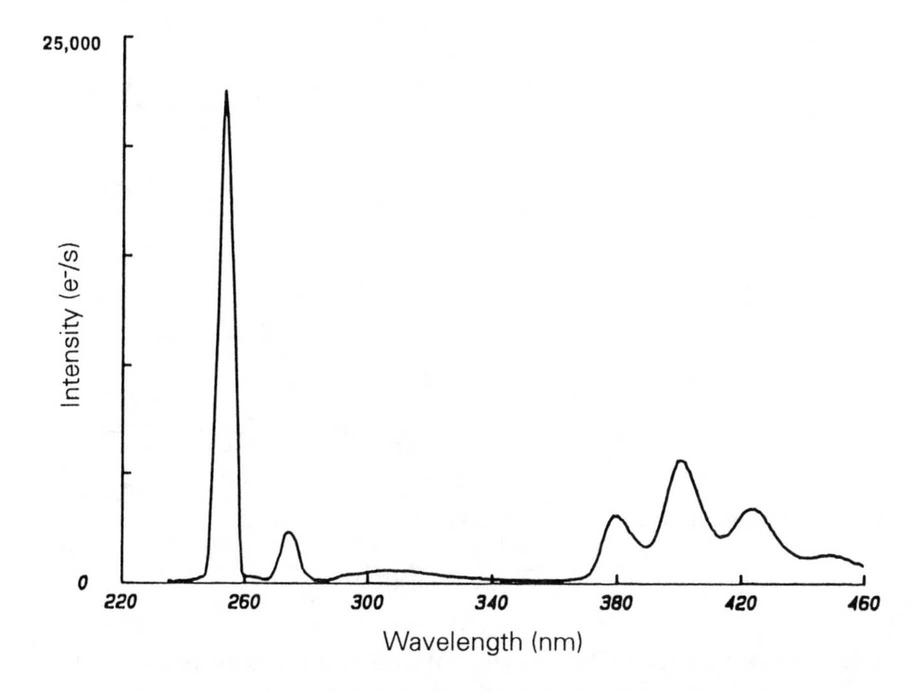

Figure 8.3 Spectrum of 10^{-9} M anthracene acquired with the configuration of Figure 8.2. The Rayleigh- and Raman-scattered peaks are at 254 nm and 274 nm, respectively. Reprinted with permission from reference 14.

By using analyte-independent features in the spectrum such as the scatter from the source or solvent, it is possible to correct for variation in the background and improve the limit of detection significantly. Epperson, Jalkian, and Denton [14] demonstrated that the best results for background subtraction are obtained for the simple optical system in Figure 8.2 when the source and detector drift are corrected using the elastically scattered peak as the reference. The spectra from a series of anthracene solutions is shown in Figure 8.4. Use of the scattered excitation peak to correct for background fluctuations resulted in an improvement in the limit of detection by a factor of 10. This spectroscopic system maintains the high sensitivity achievable with a PMT-based system while offering simultaneous multiwavelength acquisition. This advantage leads to improvements over PMT-based systems since background subtraction methods using scattered radiation cannot be obtained by single-channel PMT-based systems.

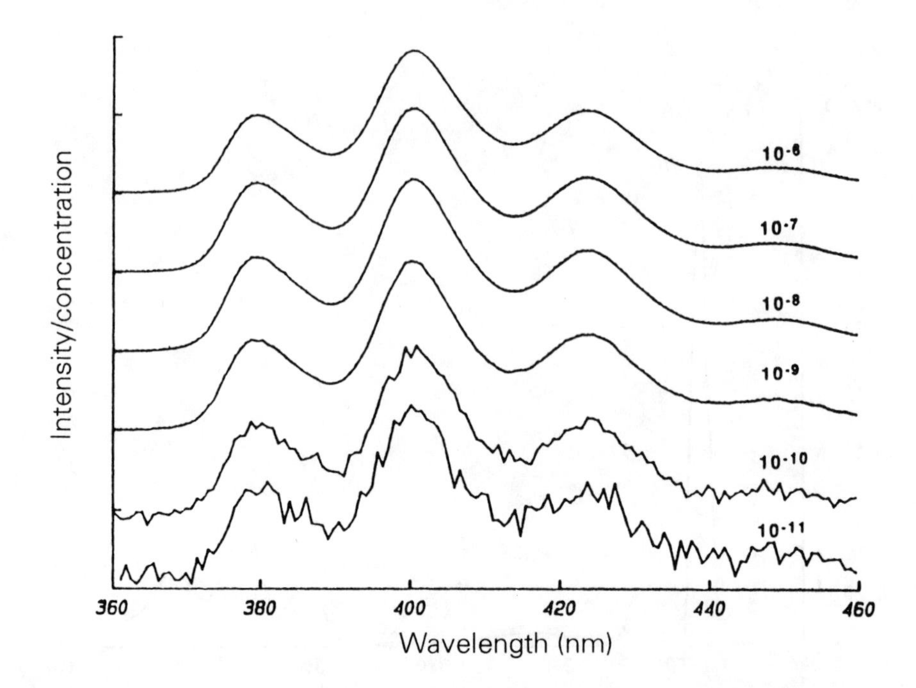

Figure 8.4 Spectra of 10^{-6} to 10^{-11} M anthracene, acquired with the instrument of Figure 8.2, normalized to constant intensity, and offset for clarity. (Reprinted with permission from reference 14).

Where the spectra of all components are accurately known, curve-fitting techniques can be applied to use the information in the entire spectrum, not just the main peak [15]. Applied to the data of Figure 8.4, the limit of detection may well be lowered by yet another decade.

The multichannel advantage also extends to identifying other changes in the spectrum due to factors addressed earlier such as dimer formation, eximer emission, photodegradation, polarization change, and interaction with the solvent. Even though having the entire spectrum will not likely result in correction of these problems, multichannel systems enable these problems to be identified and thereby prevent erroneous analyses.

8.4.2 Readout Modes: Binning

CCDs can selectively combine photogenerated charge from several detector elements into a single charge packet by a process referred to as binning. A detailed description is found in Chapter 3. Binning is used both to improve the sensitivity and to increase the dynamic range of the detector. The usefulness of binning to increase sensitivity may have decreased recently since the read noise associated with newer CCDs has decreased to an extremely low level; however, the use of binning to increase the overall dynamic range of the detector is still a valuable tool in applications to luminescence spectroscopy.

The simple dynamic range of a detector element is the range from the smallest to the largest measurable charge packet. The largest charge packet is limited to the full-well capacity of the detector; exceeding that value causes blooming. The smallest measurable charge packet is approximately twice the detector read noise. The intraspectral dynamic range for an array detector is the ratio of the brightest to the weakest spectral feature that can be measured in a single exposure. Since the full-well capacity of the serial register and readout node is typically larger than the full-well capacity of the parallel register, binning may be possible in the serial dimension even when the pixels of the parallel registers near saturation. By binning the signal from weak spectral features, the charge is accumulated into a single element. The maximum amount of charge that can be binned is determined by the full-well capacity of the serial register or the readout node. By using variable binning—a mixture of normal readouts and binned readouts—it is possible to extend the dynamic range of the device. Additional examples are presented by Epperson and co-workers [16] and in Chapter 3. In their system the dynamic range of the detector could be increased over two orders of magnitude to the point where ultimate effectiveness of variable binning is limited by the range of the analog-to-digital converters.

For some spectroscopic measurements, such as in some chemiluminescence methods, little or no information is contained in the wavelength axis, and the CCD is used simply as a detector with a very large area. The dynamic range can be extended by dividing the CCD into areas whose binning area varies; this technique, reported by Jalkian, Ratzlaff, and Denton [17] is termed simultaneous variable binning. By using a binning organization as shown in Figure 8.5, dynamic range is greatly extended. If the intensity is low, the signal may be difficult to detect in a 2×2 cluster but can be detected in a 70×10 cluster; on the other hand, a larger signal would saturate a 70×10 cluster but be measured readily in a smaller cluster.

A separate but related topic is the orientation of the spectrum across on the CCD. The spectrum can be oriented either parallel or perpendicular to the serial output register of the CCD (see Figure 8.6). Depending on the spectroscopic application, each orientation has certain advantages and disadvantages [16].

Orientation of the spectrum perpendicular to the serial register allows one to use variable binning to increase the dynamic range of the

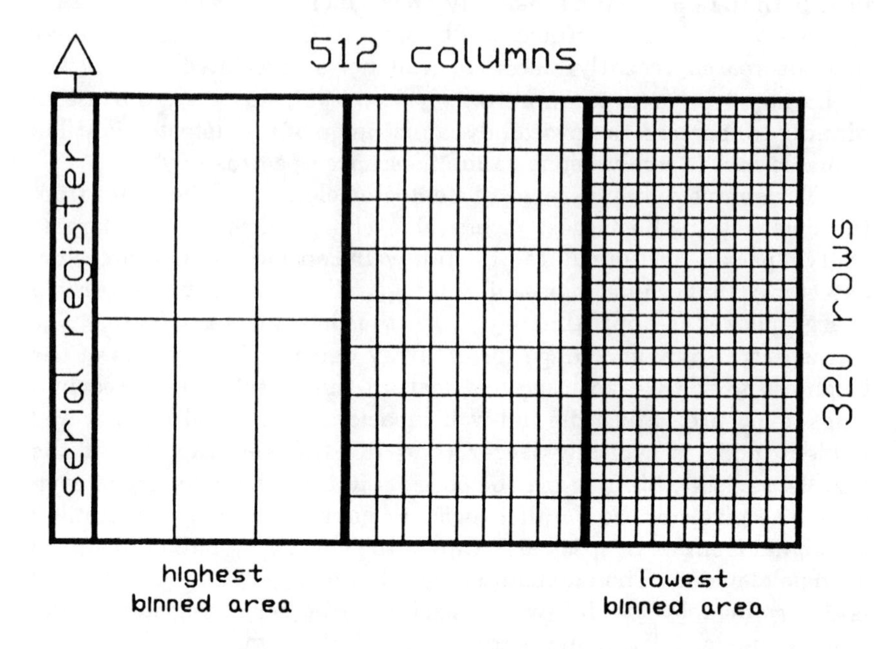

Figure 8.5 CCD surface operated simply as an area detector with simultaneous variable binning with typical cluster sizes of 70 × 10, 10 × 4, and 2 × 2. (Reprinted with permission from reference 17.)

detector. Also, in the case of rectangular format devices, the long dimension is typically the serial direction, so use of this orientation gives rise to increased spectral coverage for equivalent resolution compared with use of an orientation parallel to the serial register. Parallel orientation of the spectrum, however, results in increased readout speed, since fewer serial transfers are required. In addition, channel stops, which are oriented perpendicular to the serial, prevent blooming of charge parallel to the serial direction. The parallel orientation then minimizes the interference from intense spectral features. Consequently, for general analyses where larger dynamic and spectral ranges are useful, orientation of the spectrum perpendicular to the serial register is advantageous. On the other hand, in specific analyses where speed, spectral integrity, and sensitivity are needed, the parallel orientation is best.

8.4.3 Readout Modes: Time-delayed Integration

Time-delayed integration (TDI) is a mode of charge readout in which the movement of the charge packets within the detector is synchro-

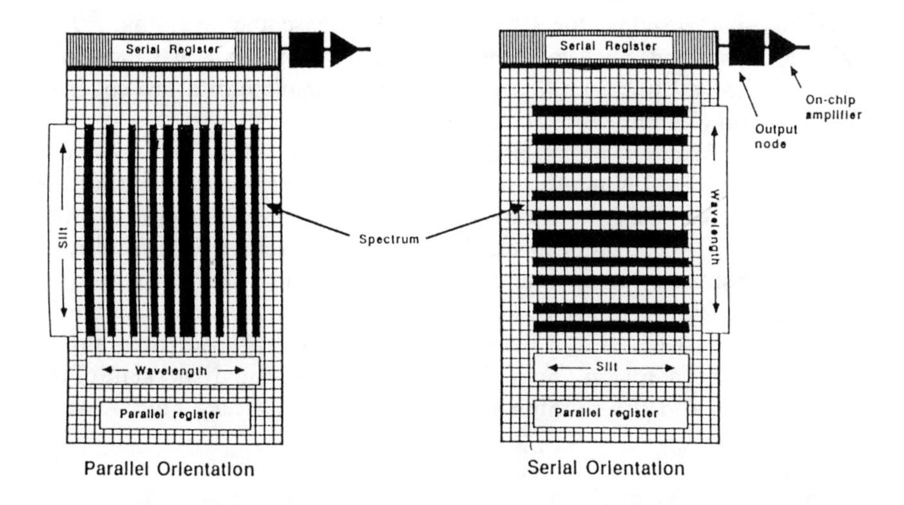

Figure 8.6 Parallel and perpendicular orientations of the array detector. (Reprinted with permission from reference 16.)

nized with movement of the image (see Chapter 3). TDI was first developed for satellite surveillance work. For chromatographic applications we can adopt this mode of readout to monitor the luminescence emission of analyte species in flowing streams by clocking the CCD at the same rate as the analyte flows through a viewing zone. The alternative would be to take a series of shuttered exposures of the analyte and stream them together in much the same way as a cartoon is made. Just as in the cartoon, the number of exposures needs to be large, and the resulting data set can be enormous.

Sweedler and co-workers [18] point out several advantages of TDI-mode luminescence detection for chromatographic analysis. First, only one row of the CCD is read at a time, which vastly reduced the readout rate and the amount of data produced without compromising time resolution. Second, the contents of a single row of the CCD contains the emission from a single band integrated over the entire time the band was in the viewing zone of the detector. The final advantage reported is that the information from the analyte band can be obtained from a single CCD readout instead of summing several CCD readouts from a shuttered experiment. The result is a reduction in the associated read noise by a factor of two to five.

8.4.4 Readout Modes: Fast Spectral Framing

The object of fast spectral framing (FSF) is to provide a means of time-resolved spectroscopy for CCD-based spectroscopic systems. This method, described by Aikens and colleagues [19] and in Chapter 3, requires that all but a small group of detector elements are covered. One then aligns the output from a spectrograph onto the exposed detectors. After a predetermined integration time, the encoded spectrum is shifted beneath the masked region of the detector (but not read out), and a new integration begins. This process is repeated until the entire device contains spectral images. The fastest time resolution by this technique is reported to be on the order of a microsecond; much faster than mechanically shuttered systems that operate on the order of 1 to 10 ms. Each microsecond, the entire spectrum may be recorded, the resolution of which is dependent on the dispersion of the spectrograph and the number of detector elements on a single row of the CCD.

8.4.5 CCD Format and Correlation to Spectroscopy

New advances in optics have produced holographically ruled, aberration-corrected imaging spectrographs [13]. Not only do they provide high-quality dispersion along the wavelength axis, but the vertical (Y) position on the entrance slit is maintained with remarkable integrity. These spectrographs are perfectly suited to two-dimensional array detectors; they provide the capability to utilize the second dimension to perform several independent luminescence analyses simultaneously or to perform multidimensional spectroscopy. The spatial resolution in the Y axis afforded by these spectrographs is quite good, and the spectral resolution depends on the grating element.

For luminescence spectroscopy, where spectral resolution of 2 nm is acceptable, a system can provide spatially resolved spectral images over a broad spectral region. These spectrographs have been applied in several ways to obtain more information from a single exposure. First, an entire excitation-emission matrix (EEM) may be acquired simultaneously by using a predispersing monochromator to excite at several wavelengths; then the entire emission spectrum is acquired simultaneously for all excitation channels. Second, for parallel spectroscopy, fiber-optic inputs have been used to provide several (up to 35) spectra simultaneously from as many independent sources.

With conventional single-channel instruments, acquisition of an EEM is a time-consuming process, since many tens of thousands of data points may be acquired to obtain the entire EEM. Interferometric techniques offer high throughput; however, these systems suffer from a multiplex disadvantage since the signal from strong spectral features is mixed with the signal from weak spectral features, and the shot noise from the strong spectral features can wipe out detection of the weak spectral features. This has led to the development of spectroscopic systems employing television-type detectors to create spectroscopic systems capable of acquiring EEMs simultaneously. Early array detectors such as the silicon intensified target vidicon suffered from a lack of linearity, dynamic range, geometric stability, and sensitivity and they exhibited great variation in response from detector to detector. Consequently, they are inadequate for use in analytical systems. The CCD offers the sensitivity of PMT detection, and the array format is matched to the two-dimensional output of imaging spectrographs. This allows for simultaneous acquisition of EEMs. The time differential is on the order of tenths of seconds for the CCD-based system versus several minutes for the single-channel PMT-based systems.

Although some degree of selectivity can be achieved by variation of the excitation radiation and the luminescence emission detected, the broadband nature of luminescence excitation and emission generally makes selective excitation and detection ineffective for multicomponent analysis. One approach to multicomponent analysis is to use the spectral information from both the excitation and emission spectra to resolve luminescent species. The use of EEMs for the analysis of mixtures is based on the idea that an EEM is the linear sum of the contributions of the luminescence from all individual components. Figure 8.7 shows a high-resolution EEM of a mixture of polycyclic aromatic compounds (PACs) taken with a CCD detector [20]. There may exist spectral regions where components uniquely absorb and/or emit. In other cases it is possible to apply a least squares approach to the analysis of samples with known components. Even in unknown samples it may be possible to determine the number of species present, and EEMs have been successfully used in fingerprinting complex samples such as fuels and oils [21], as well as in pattern recognition [22] and disease diagnosis [23].

The flexibility afforded by a single compact instrument with no moving parts that can simultaneously monitor several distinct luminescent sources speaks for itself.

8.4.6 Spectral Correction

There are slight differences in QE from detector element to detector element, and any spectrograph exhibits variation in light-gathering power with wavelength; however, these effects are constant in a system with no moving parts. The geometric stability of CCD-based systems gives them important advantages for making spectral (or flat-field) corrections; correction of a spectrum is easily accomplished with the computer. The geometric stability is also a necessity for background correction, noted in Section 8.4.1.

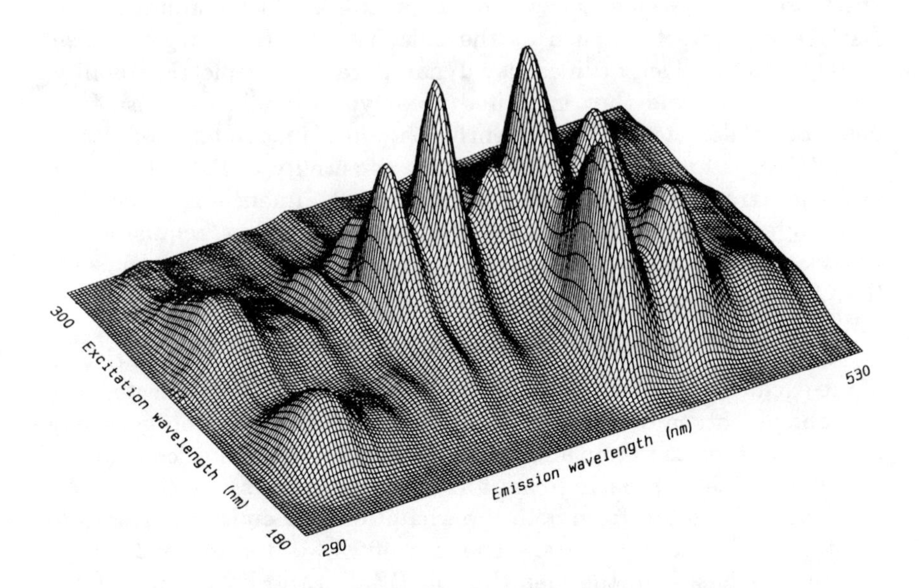

Figure 8.7 A high-resolution excitation-emission matrix of a mixture of 7 polycyclic aromatic compounds, acquired with a CCD detector. (Reprinted with permission from reference 20.)

8.4.7 Summary

CCD detection offers many significant advantages over both single-channel PMT-detection and detection with other forms of array detectors such as photodiode arrays and vidicons. Other array detectors suffer from problems of read noise, sensitivity, linearity, and detector calibration. The typical format of the photodiode array is linear, and although it is a multichannel detector, it lacks the two-dimensional format necessary for imaging or acquisition of EEMs. The CCD matches and in many cases exceeds the performance of PMT-based instruments while having the added advantage of multichannel, solid-state detection. The CCD also offers flexibility in readout modes to enhance the dynamic range and efficiency of data collection, as demonstrated by variable binning and TDI. The advantages of spectroscopic systems with fixed components are realized with precise background subtractions and spectral correction.

One drawback is the time frame over which time-resolved spectra may be obtained. Fluorescence lifetime measurements must be made two to three orders of magnitude faster than can be handled by the read rates for low noise systems. This precludes high-speed spectral analysis without an external gating device such as an image intensifier (described in Chapter 5). Advances in electrooptic shutters may overcome this deficiency; until then short-time-scale spectroscopy may be best done with a PMT.

8.5 Applications

Although the application of CCD detectors to analytical luminescence spectroscopy has been slow, the results obtained from such application have been impressive. The lack of commercially available complete camera systems at a low cost has slowed CCD use. These detectors were originally designed as imagers exhibiting sensitivities that rival those of conventional PMT-based systems. Many of the initial CCD luminescence applications involve their use as imagers in fluorescence microscopy or in the location and quantitative analysis of analytes in planar chromatographic techniques [24–26]. Chapter 6 discusses luminescence application to planar separations. The application of CTDs to luminescence microscopy [27–29] is such a broad and rapidly changing subject area that the interested reader is referred to the above references. The focus for the remainder of this chapter will be on applications that reinforce the important characteristics of CCD detection as they apply to analytical luminescence spectroscopy: simultaneous spectral acquisition, multidimensional analysis, flexibility, and excellent sensitivity over several orders of magnitude.

8.5.1 Spectral Acquisition

Szabo and colleagues [30] used a linear 2048-element CCD to obtain the chlorophyll fluorescence spectrum of leaves as a nondestructive means of studying the mechanism of photosynthesis. The multichannel advantage of the CCD allowed for the rapid acquisition of fluorescence at several wavelengths. In this case the fluorescence ratio at 690 and 735 nm is used to determine the chlorophyll content of the leaves. Complete acquisition of the entire spectrum with time resolution as short as 10 msec allowed these researchers to monitor fluorescence induction kinetics by following the chlorophyll content spectroscopically.

Sweedler and co-workers [31] describe a unique interferometer based on the common path design with which they monitor the fluorescence of anthracene with a two-dimensional CCD. The optical configuration is shown in Chapter 3. Figure 8.8 shows the resulting spectrum for a solution of anthracene. Applications of interferometers in the UV-visible portion of the spectrum tend to suffer from Felgett's disadvantage, the distribution of the noise from bright features throughout the interferogram, decreasing the detectability of weaker

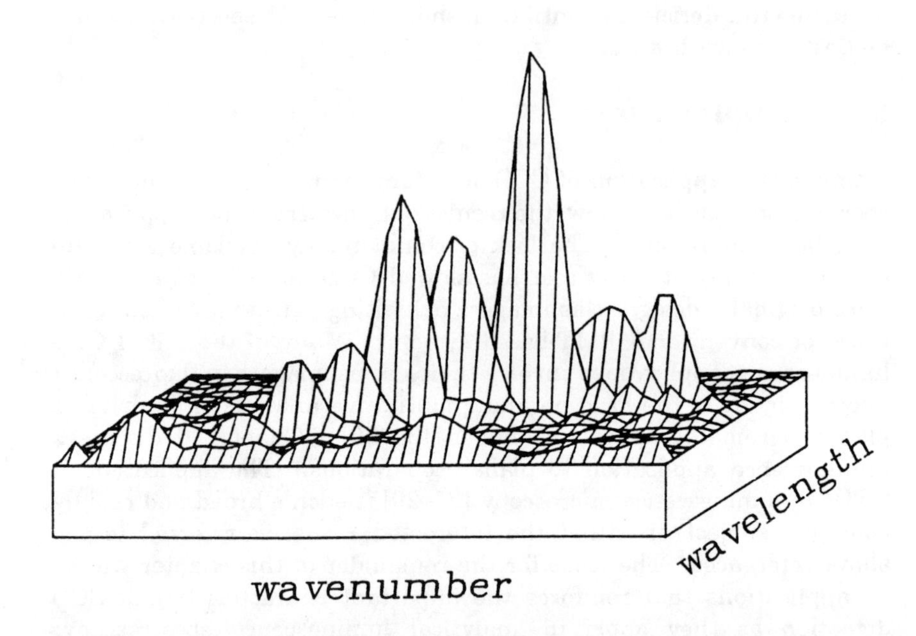

Figure 8.8 Spectrum of anthracene acquired with a common-path interferometer and CCD detector. (Reprinted with permission from reference 31.)

features also present in the interferogram. This common-path interferometer employs a cross dispersive element (a prism or a grating) to create a series of interferograms that are monitored by a CCD array detector. This isolates the noisy, bright spectral features to discrete interferograms and eliminates the distribution of the noise from these features to the other interferograms. The result is a two-dimensional interferometer that gathers the series of interferograms simultaneously without the use of some type of prefilter, and that employs no scanning as in Michelson interferometers. This system can therefore monitor transient sources while at the same time reducing the multiplex effects from bright spectral features.

8.5.2 Total Luminescence

Among the various luminescence techniques, chemiluminescence can be performed with exquisite sensitivity due to extremely low or nonexistent background signals. The excitation energy is derived from chemical sources rather than from excitation by radiant energy.

Experimentally there is no need for filters or dispersing devices since the only source of emission is the chemiluminescent material itself. Femtomole limits of detection have been demonstrated for H_2O_2, Cu^{+2}, Co^{+2}, and Cr^{+3} in solution using a luminol chemiluminescence reaction [17].

Two experimental systems were employed for static chemiluminescence measurements, a wavelength dispersive system and a nondispersive system. The nondispersive system is very simple in design (see Figure 8.9) and provides sensitive, reproducible analysis; it may be used for the quantitative work after the analytical method is developed. On the other hand, the dispersive system finds important application in the development of analytical methodology. The dispersive system is able to monitor for light leaks, sources of background luminescence, and contamination. For example, interference from the luminescence emanating from the borosilicate glass was identified with the dispersive system, and the problem was resolved by storing the sample tubes in the dark for a specified time before analysis.

With the nondispersive system these workers also demonstrated the advantages of simultaneous variable binning as a means of

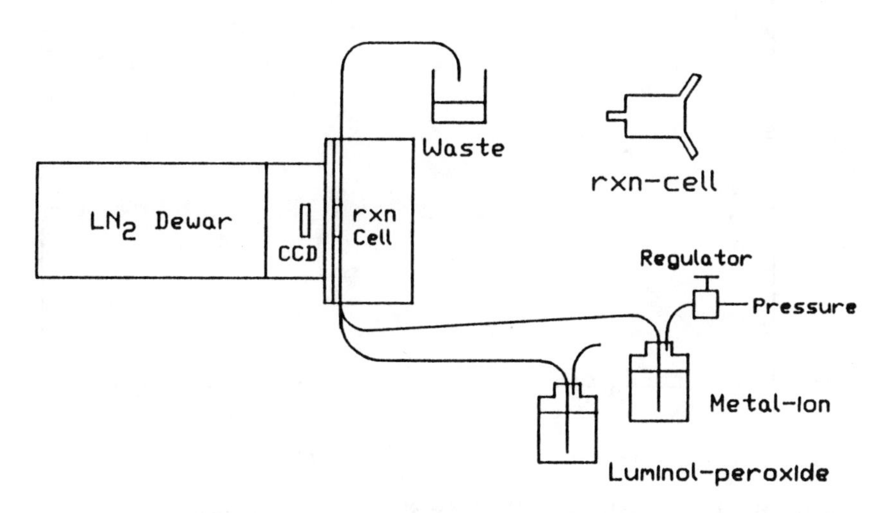

Figure 8.9 A CCD-based system with nondispersive optics for chemiluminescence. Rxn cell, chemiluminescence reaction cell. (Reprinted with permission from reference 17.)

extending the linear dynamic range of the analysis. A linear dynamic range of over seven orders of magnitude was achieved without any prior knowledge of the luminescent intensity; the results are displayed in Figure 8.10. In addition to the static measurements, measurements were made in a flowing stream and coupled to HPLC separation. The limits of detection employing the CCD-based system rivaled those that employed sample preconcentration.

The total luminescence from fluorescent pH indicators as a sensitive spectroscopic means of pH determination is currently being investigated [32]. Spectroscopic determination of solution pH has many advantages over use of potentiometric techniques, especially in complex matrices or when deployed in field applications. The calibration of spectroscopic determination is locked into the thermo-

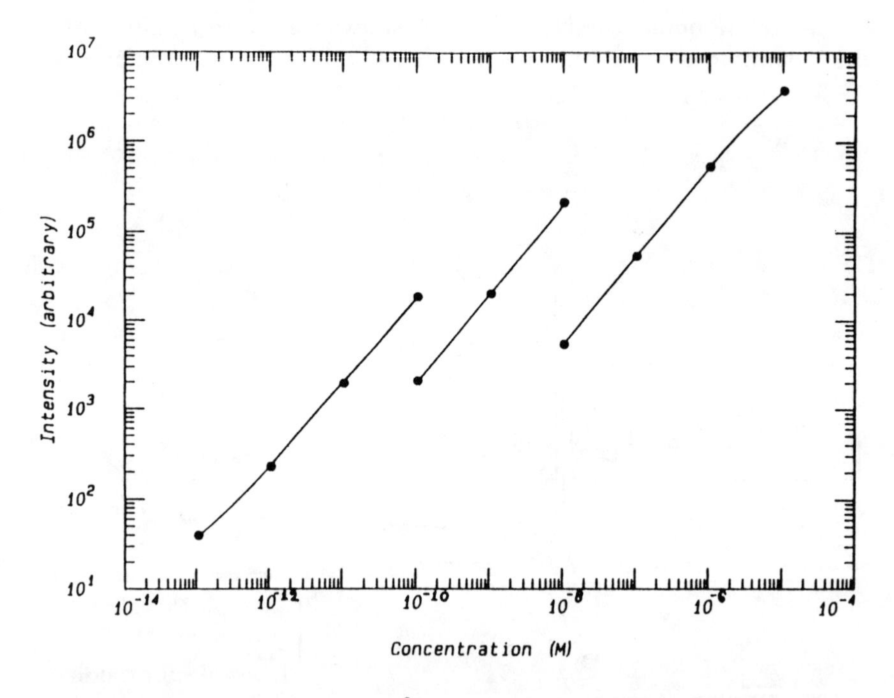

Figure 8.10 Analysis of Cr^{3+} by chemiluminescence using simultaneous variable binning. The curves are for bin sizes defined in Figure 8.5. (Reprinted with permission from reference 17.)

dynamics of the dye material. Proper characterization of the dye allows for prolonged analysis without the problems associated with electrode fouling or drift. The spectrophotometric approach also allows for correction of data sets if at some time in the future more is learned about the sample or the thermodynamic constants. This cannot be done with potentiometric data. In addition, spectroscopic measurement of indicator dyes for pH determinations has shown the potential to be both more accurate and more precise. Fluorescence has advantages over absorption indicators in that the need to add less of the indicator dye minimizes or even eliminates the need to correct for $[H^+]$ derived from the addition of the weak acid indicator dye. The pH range over which selected indicators exhibit a change in fluorescence can be quite narrow, giving rise to spectroscopic resolution of changes in pH of less than 0.001 pH units.

8.5.3 Separation Techniques

Chromatographic separation prior to luminescence analysis has met with great success by several investigators. The chromatographic separation eliminates the problems associated with overlapping spectra from co-components in a sample. Even though chromatographic separation typically results in dilution of the sample, the excellent detection available to luminescence analysis more than compensates. CCDs have been used for thin-layer chromatography [33]; a treatment of applications with planar media is found in Chapter 6. Imaging of luminescent species in flowing streams has found use in several techniques [16,18,20].

8.5.3.1 High-Performance Liquid Chromatography

The added dimension brought about by CCD detection is the ability to obtain luminescence spectra for the various components as they elute from the column. With a single-channel device one would be required to scan a monochromator, requiring that either the flow be stopped or that scanning occur while the sample passed by the detector. There are several advantages to obtaining the spectral output of chromatographic effluents, the most important of which are to help detect co-eluted peaks and to aid in the identification of the component. The application of a CCD to HPLC detection has been investigated by Jalkian and colleagues [20], monitoring the separation and fluorescence spectra of a series of polycyclic aromatic compounds, PACs. Jalkian observed picogram to femtogram detection limits with linear dynamic ranges of four to six orders of magnitude. Figure 8.11 shows a chromatogram obtained employing a CCD detector for a mixture of seven PACs at 10^{-6} *M*. These workers also employed the techniques of binning and derivative spectroscopy to aid in detection of eluted peaks.

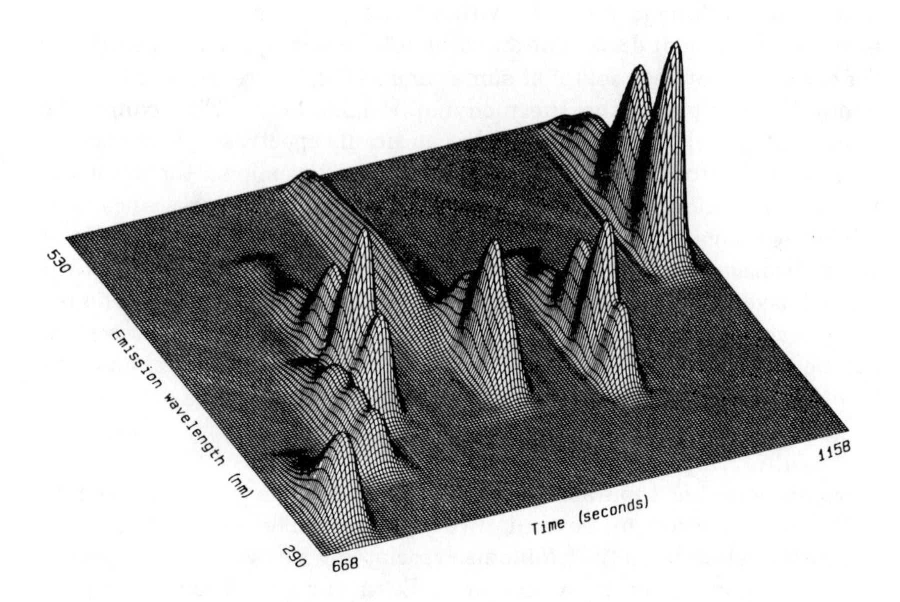

Figure 8.11 A chromatogram of a mixture of polyaromatic hydrocarbons acquired with CCD detection. (Reprinted with permission from reference 20.)

The flexibility of this spectroscopic system derives from the ability to collect complete EEMs, and this information can be used to provide optimal detection and resolution of PACs. In addition, this system exhibits sensitivities that are superior to those obtained using laser excitation or photon-counting PMTs for detection.

8.5.3.2 Capillary Electrophoresis

Capillary electrophoresis (CE) has quickly become one of the most powerful techniques for the separation of complex samples. This technique has found its greatest application in bioanalytical chemistry and has been extended to the sequencing of fluorescence-tagged DNA and the monitoring of the contents of single cells. Laser-induced fluorescence (LIF) coupled with CCD detection for on-line fluorescence analysis has been investigated by several investigators [1,18,34–36].

The need for multichannel detection is extremely important for some applications. An important example is in the use of CE in DNA sequencing. Researchers at DuPont have developed a group of fluorescein derivatives that selectively tag the four different bases [37]. Each base can be distinguished by distinctive fluorescent emission signatures. This requires four channels to resolve the tagged bases.

Use of single-channel detectors like the PMT requires either four separate detectors, a single detector with an emission scan, or a single detector with a filter wheel equipped with four bandpass filters.

It has been shown that it is much easier to accurately identify the tagged base from the entire spectrum [35]. For the single-channel systems, either (1) one must scan the emission of the analyte in a flowing stream and consequently either miss components or slow the flow rate, (2) one must suffer the throughput inefficiencies associated with filters, or (3) the measurement must suffer throughput loss associated with splitting the beam to the four separate channels. Clearly the CCD has important advantages. Despite the need for additional optical elements, the higher QE of the CCD and the efficiency of the reflection optics result in a superior overall efficiency of the CCD system compared with PMT-based systems, while at the same time providing the entire spectrum. Thereby the precision of the base identification is improved. Karger and co-workers [35] found the accuracy of the sequencing to be to comparable to that of slab techniques, and under automation the authors conclude that on-line CE sequencing of DNA fragments provides a rapid and cost-effective alternative to DNA sequencing by slab electrophoresis.

Other researchers have also confirmed LIF-CE with on-line CCD detection to be an effective means of high throughput total emission analysis. Sweedler and colleagues [18], however, added an important innovation to LIF-CE by employing TDI mode integration of the migrating bands in CE. Figure 8.12 displays the optical system used by Sweedler, and the results of a CE run in TDI mode are shown in Figure 8.13 for several amino acids. The previous experimental setup operated in a "snapshot" mode in which the detector is exposed to the luminescence and then the entire frame is transferred to the computer. For scientifically operated CCDs, the read rate is approximately 50 kHz, and so a large array can require a significant amount of time to read and a great deal of memory for a single CE run. Sweedler points out that using a 516 $\times$ 516 CCD array, a read every 3 s over a typical CE run will accumulate over 300 MBytes of data. The CCD is also off-line for a significant amount of time while the device is being read out, so the snapshot mode results in a low-duty-cycle detector that generates a vast amount of data. In addition, all the luminescence intensity can be extracted from a single read as opposed to having to be extracted from several full frame exposures.

8.5.4 Imaging

Imaging for chemical analysis with CTD detectors has quickly grown into a powerful technique. A unique luminescence imaging application not falling into these domains is the detection of latent fingerprints [38]. Fingerprints have proven to be one of the strongest pieces of

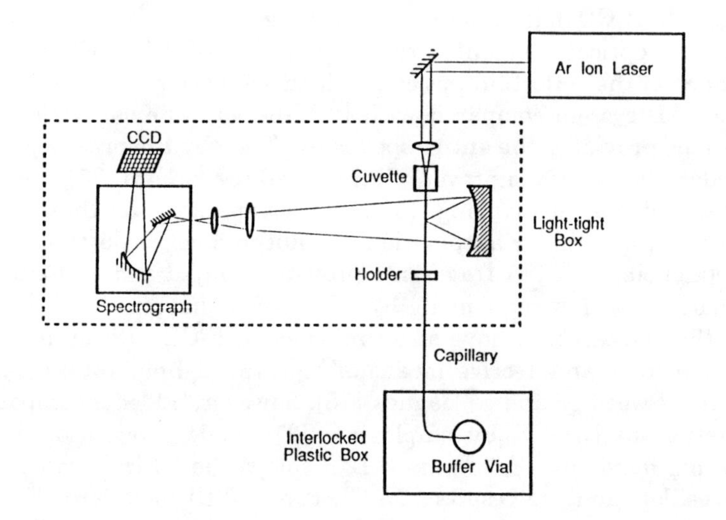

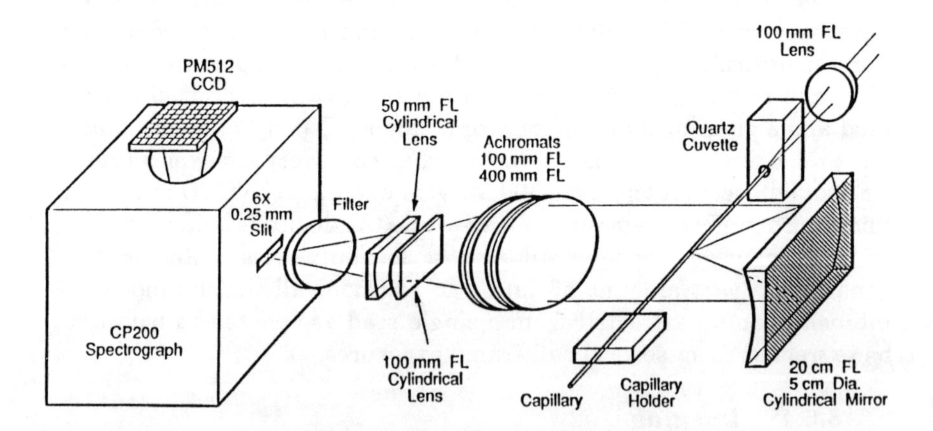

Figure 8.12 Optical configuration for time-delayed-integration-mode CCD detection. (Reprinted with permission from reference 18.)

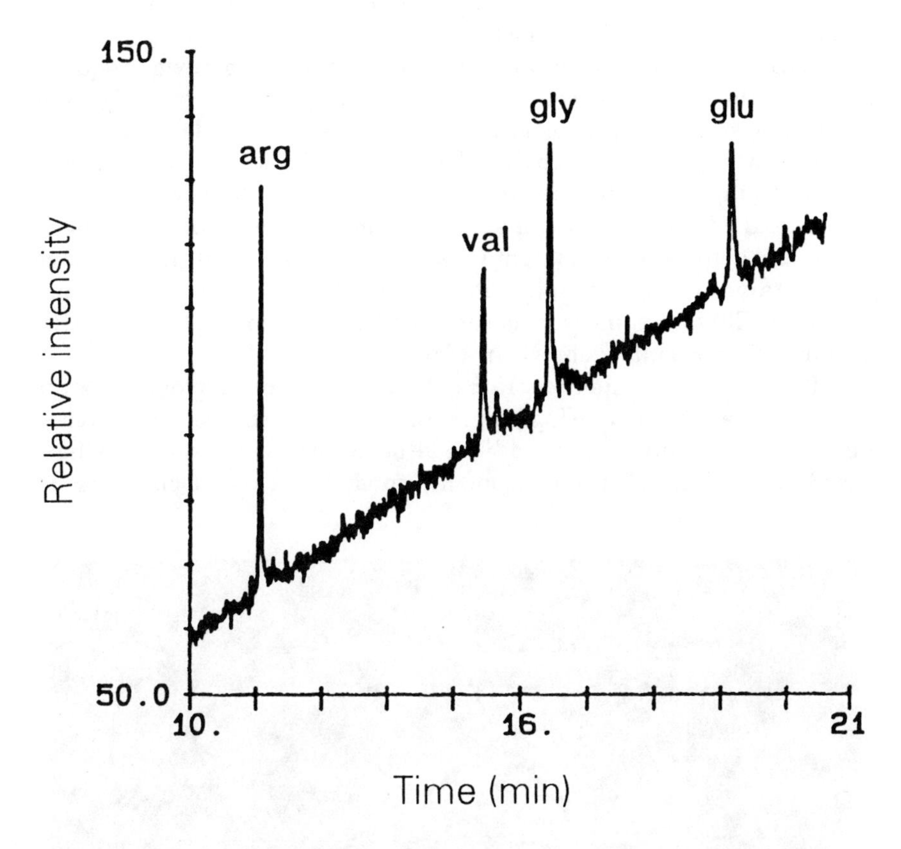

Figure 8.13 The results of a capillary electrophoresis run with CCD detection in time-delayed integration mode. The fluorescence signal is plotted in arbitrary units. (Reprinted with permission from reference 18.)

physical evidence in a court of law. The ability to detect the minute quantities of oils and salts left behind on a multitude of surface types represents a challenging analytical task. Fingerprints were initially distinguished with dusting powders, which simply adhered to the residue left behind. The powder was chosen to contrast with the substrate, and photographs were taken and analyzed. Ninhydrin was developed as a selective reagent that reacted specifically with amino acids to create a colored product. In 1976, Menzel began pioneering the use of laser excitation for luminescence detection of latent fingerprints [39]. The power of the laser enabled the luminescence visualization of naturally fluorescing compounds in fingerprint residue or fluorescence dusting powders; fluorescent derivatizing reagents also were employed to develop the image of the fingerprints. The detector in this scheme was a 35-mm camera.

CCD acquisition of the luminescence reduces the power requirements of the excitation source and automatically records a digitized image for storage, enhancement and analysis. The laser systems required when using 35-mm film detection are bulky, and their power and cooling requirements make them difficult if not impossible to use in the field. Portable instruments often mean a compromise in laser power and detectability of latent fingerprints. Fingerprint images have been obtained using a simple hand flashlight as the source, a 450-nm bandpass filter, a camera lens, and a CCD. An example of a CCD-acquired fingerprint is shown in Figure 8.14.

This application demonstrates the approach of employing better detection versus increasing the power of the source to enhance luminescence analysis. An additional advantage is the time-savings; the 35-mm film, after development, must be scanned in order to

Figure 8.14 Fluorescent image of a fingerprint acquired with a CCD camera. (Reprinted with permission from reference 38.)

perform computer-based enhancement and identification-search routines. The CCD-derived data are already in digital form.

8.6 Conclusions

Luminescence analysis is one of the most sensitive methods of spectroscopic analysis and is an extremely active area of current research. CCD detection for low-light-level spectroscopies such as luminescence has shown tremendous potential. Despite the success of CCD application to analytical luminescence spectroscopy, CCD detectors are not in routine use today. This is due in part to their high cost, although their power and flexibility as detectors actually may prove them quite cost effective. Another drawback is the slow development of spectroscopic systems designed specifically for the CCD so that their full power can be realized. Their potential has been demonstrated in several applications, and on-going research is improving their performance all the time. Charge-transfer devices will eventually replace the PMT in luminescence instrumentation with the possible exception of one case: time-resolved spectroscopy. Currently the time frame for fluorescence time resolution is beyond the capability of mechanical shutters or the read rates of the CCD. Electrooptic shutters have insufficient throughput efficiency and inadequate bandpass over which the electrooptic becomes transparent when "open." Gated CCDs using image intensifiers suffer from the poor QE and noise characteristic of the image intensifiers themselves, and so PMT detection continues to dominate.

Despite these drawbacks, the multichannel advantages, low noise, and high QE characteristics of the CCD and the flexibility of different formats and readout capabilities make them valuable detectors for low-light-level spectroscopy and imaging analysis.

References

1. Sweedler, J.V.; Shear, J.B.; Fishman, H.A.; Zare, R.N. *Proc. SPIE* **1992**, *1439,* 37.
2. Weber, G. *Trans. Faraday Soc.* **1948**, *44*, 185.
3. Laurence, D.J.B. *J. Biochem.* **1960**, *75,* 345.
4. Bolotin, B.M.; Krasovitskii, B.M., Eds. *Organic Luminescent Materials.* VCH, Weinheim, Germany, 1988.
5. Hercules, D.M., Ed. *Fluorescence and Phosphorescence Analysis.* Wiley, New York, 1966.
6. White, C.E.; Argauer, R.J. *Fluorescence Analysis.* Dekker, New York, 1970.
7. Ingle, J.D.; Crouch, S.R. *Spectrochemical Analysis.* Prentice Hall, Englewood Cliffs, N.J., 1988.
8. Parker, C.A. *Photoluminescence of Solutions.* Elsevier, Amsterdam, 1968.

9. Levinson, G.S.; Simpson, W.T.; Curtis, W. *J. Am. Chem. Soc.* **1957**, *79*, 4313.
10. Epperson, P.M. Ph.D. Dissertation, University of Arizona, 1987.
11. McRea, E.G.; Kasha, M. *J. Chem. Phys.* **1958**, *28*, 721.
12. Sweedler, J.V; Bilhorn, R.B.; Epperson, P.M.; Sims, G.R.; Denton, M.B. *Anal. Chem.* **1988**, *60*, 282A.
13. Jalkian, R.D.; Pomeroy, R.S; Kolczynski, J.D.; Denton, M.B.; Lerner, J.M.; Grayzel, R. *Am. Lab.* **1988**, *21(2)*, 80.
14. Epperson, P.M.; Jalkian, R.D.; Denton, M.B. *Anal. Chem.* **1989**, *61*, 282.
15. Ratzlaff, K.L. *Anal. Chem.* **1980**, *52*, 1415.
16. Epperson, P.M.; Denton, M.B. *Anal. Chem.* **1989**, *61*, 1513.
17. Jalkian, R.D.; Ratzlaff, K.L.; Denton, M.B. *Proc. SPIE* **1989**, *1055*, 123.
18. Sweedler, J.V.; Shear, J.B.; Fishman, H.A.; Zare, R.N.; Scheller, R.H. *Anal. Chem.* **1991**, *63*, 496.
19. Aikens, R.S.; Epperson, P.M.; Denton, M.B. *Proc. SPIE* **1984**, *501*, 49.
20. Jalkian, R.D.; Denton, M.B. *Proc. SPIE* **1989**, *1054*, 91.
21. Bentz, A.P. *Anal. Chem.* **1976**, *48*, 455A.
22. Rossi, T.M.; Warner, I.M. *Appl. Spectrosc.* **1985**, *39*, 949.
23. Wolfbeis, O.S.; Leiner, M.J.P. *Anal. Chim. Acta.* **1985**, *167*, 203.
24. Koutny, L.B.; Yeung, E.S. *Anal. Chem.* **1993**, *65*, 183.
25. Karger, A.; Ives, J.T.; Weiss, R.B.; Harris, J.M.; Gesteland, R.F. Report, TR-26-ONR, Gov. Rep. Announce. Index (U.S.) **1991**, *91(18)*, Abstr. # 149,558.
26. Kolbe, W.F.; Turko, B.T. *IEEE Trans. Nucl. Sci.* **1989**, *36*, 731.
27. Silcock, D.J.; Waterhouse, R.N.; Glover, L.A.; Prosser, J.I.; Killham, K. Appl. Environ. Microbiol., **1992**, *58*, 2444.
28. Chan, W.S.; MacRobert, A.J.; Phillips, D.; Hart, I.R. *Photochem. Photobiol.* **1989**, *50*, 617.
29. Denton, M.B., Ed., *Proc. SPIE* **1992**, *1439*.
30. Szabo, K.; Lichtenthaler, H.K.; Kocsanyi, L.; Richter, P. *Radiat. Environ. Biophys.* **1992**, *31*, 153.
31. Sweedler, J.V.; Jalkian, R.D.; Sims, G.R.; Denton, M.B. *Appl. Spec.* **1990**, *44*, 16.
32. Pomeroy, R.S; Baker, M.B.; Dickson, A.G.; Denton, M.B. Pittsburgh Conference, Atlanta GA, March, 1993.
33. Cosgrove, J.A.; Bilhorn, R.B. *J. Planar Chromatogr.* **1989**, *2*, 362.
34. Cheng, Y; Piccard, R.D.; Vo-Dinh, T. *Appl. Spectrosc.* **1990**, *44*, 755.
35. Karger, A.E; Harris, J.M.; Gesteland, R.F *Nuc. Acids Res.* **1991**, *19*, 4955.
36. Taylor, J.A.; Yeung, E.S. *Anal. Chem.* **1992**, *64*, 1741.
37. Prober, J.M.; Trainor, G.L.; Dam, R.J.; Dobbs, F.W.; Robertson, C.R.; Zagursky, R.J.; Couzza, A.J.; Jensen, M.A.; Baumeister, K. *Science* **1987**, *238*, 336.
38. Pomeroy, R.S; Baker, M.E.; Radspinner, D.A.; Denton, M.B. *Proc. SPIE* **1992**, *1439*, 60.
39. Menzel, R.E.; Fox, K.E. *J. Forensic Sci.* **1980**, *25*, 150.

9
Two-Dimensional Array Detectors for Plasma Diagnostics

Cheryl A. Bye and Alexander Scheeline

School of Chemical Sciences, University of Illinois
Urbana, Illinois

9.1 Introduction

Plasma systems at steady state are characterized by their temperature and density of species [1–4]. Typically these systems are not at equilibrium, so that the distribution of energy among excited states of the various species cannot be described by a single temperature parameter [5–8]. To characterize a plasma, one typically wishes to know the concentration of each species *in toto*, the excitation distribution of each species, and the translational energy of neutral atoms, ions, molecules ("heavy particles"), and electrons [1–4]. In many instances, rates of vaporization or condensation of particles or droplets may also be of interest. These properties all vary in space and may well vary in time [9–17].

Until the advent of charge-transfer device (CTD) array detectors, one was frequently faced with balancing trade-offs in light-detection performance that constrained the quantity or quality of data available for characterizing plasmas. One can achieve time resolution (with a photomultiplier) at the expense of being able to observe only a single point in space or a single wavelength [18]. One can obtain two-dimensional spatial resolution (with photographic film or plates or, more recently, with vidicons) at the expense of temporal response and linearity of intensity response. Linear diode arrays permitted some relaxation of these constraints, but dynamic range in intensity was still inadequate in many cases, and an additional measurement dimension was frequently desirable [19]. Various CTDs, usually charge-coupled devices (CCDs) but in some cases charge-injection devices (CIDs), have overcome many of the limitations of competing detectors, providing a convenient means for obtaining optical diagnostic data on plasmas in a form readily processed into useful information [2,19].

In this chapter we describe various diagnostic techniques used to study plasmas. These include characterization of excited state distributions using atomic and ionic line-emission intensity, measurement of rotational and vibrational temperature by observation of *ro-*

vibronic bands, and estimation of free electron density by line-shape analysis (linear Stark broadening of hydrogen-atom line emission or quadratic Stark broadening of noble gas line emission) [2,20–22]. Alternatively, the free electron concentration can be found by Thomson scattering [23–26]. For cylindrical plasmas, the spatial structure of all these diagnosed quantities may be determined through the use of Abel inversion [27–31]. For noncylindrical sources, tomography must be used [32]. Means for time resolution, using the clocking capabilities of the CTDs and external means such as gated image intensifiers and mechanical shutters, will be discussed as appropriate.

Typical applications for plasmas include vaporization and excitation of aerosols for elemental analysis. Similar plasmas are used to make refractory powders, to implant nitride layers on steel, and to deposit thin layers on semiconductors [33–35]. In low-pressure plasmas, such as those used in deposition, wall effects and departure from equilibrium are most pronounced. Only under particular circumstances does equilibrium apply in these systems. Yet, while in plasma systems pressures can range from a few microtorr to several atmospheres, currents from microamperes to kiloamperes, voltages from millivolts to kilovolts, powers from microwatts to megawatts, time variations from nanoseconds to hours, and ionization extent from a few parts per million to nearly 100%, many of the same plasma spectroscopic diagnostic tools can be used throughout [2].

9.2 Detector and Experimental Restrictions

Before we proceed further it is necessary to review briefly the predecessors of CTDs to be able to appreciate the revolutionary impact CTDs have had on analytical plasma spectroscopy. Several types of detectors have been used in the characterization of plasma sources. Of these, the human eye was the first, used for the observation of the optical radiation emitted by a plasma source [19,36]. The eye is a remarkable device in that it is capable of processing information over a wide dynamic range of intensities, automatically compensating (down to photon-counting level, if necessary) for the particular light level present. It is no wonder that spectrometer detector combinations have been patterned after the eye with its integration of both imaging and detection. Yet it is important to consider detector characteristics separately from these combinational systems to deconvolve detector performance from system performance. This section addresses detector characteristics followed by a discussion of overall system design.

9.2.1 Photographic Detection Devices

Photographic detection paved the way for today's solid-state detectors; in fact, it has several attractive characteristics that still make it the method of choice in certain instances. First, photographic emulsions are readily available in a wide range of sizes and sensitivity ranges. As a consequence of their large physical size (4 inches × 10 inches for a typical plate), photographic plates can cover an enormous spectral range at high resolution (typically 200- to 800-nm slices depending on spectrograph resolution). In fact, the information bandwidth of photographic emulsions is far superior to that of presently available CTDs.

Photographic emulsions are not without their problems. Besides the time-consuming and labor-intensive development process, photographic emulsions suffer from several spectral degrading effects, such as band broadening, the Eberhard effect, the intermittency effect, emulsion fogging, and the most severe limitation, the nonlinear nature of the intensity/density ratio [37–41]. Since absolute or relative intensities are commonly required in spectral plasma diagnostic techniques, this last characteristic as well as poor sensitivity makes plasma diagnostics using photographic emulsions extremely difficult. Consequently they have been replaced by more convenient forms of detection, namely photoelectronic detectors such as photomultiplier tubes (PMTs), vidicons, photodiode arrays (PDAs), and recently, CTDs.

9.2.2 Photoemissive Detectors

The photomultiplier tube is by far the most commonly used photoemissive detector. It offers a wide dynamic detection range (over six orders of magnitude) enabling the detection of a variety of signals. In addition, its relatively high quantum efficiency or QE (10% to 20% at the optimum response wavelength), photon-counting capability, wide range of spectral sensitivity (UV to visible), temporal response (typically 10 ns to as short as a few picoseconds), and low detector cost are responsible for its dominance throughout the spectroscopic community [2,19,42–45]. PMTs have been used extensively in plasma characterization studies and for both quantitative and qualitative analysis of a wide range of analyte systems [46,47]. Transient events in plasma systems are difficult to monitor using a single-channel mode of detection. Consequently some form of time gating and signal averaging is typically utilized to acquire both wavelength and temporal information about these sources, with boxcar integration being the most common [48,49]. The single-channel method of measurement is valid only when dealing with systems that are at steady state, reproducible, or extremely stable. In addition this method of detection is complicated by background drift. As a consequence, single-channel methods of detection are not optimal for many of the plasma diagnostic tools; therefore, it is not surprising that multichannel detectors based on the photoemissive technology have been fabricated.

Of these multichannel-tube-based photoemissive detectors, the image dissector has been the most utilized for plasma characterization studies. An image dissector essentially is a scanning PMT; that is, the detector is capable of monitoring a discrete location on the photocathode by means of electron focusing optics. In addition, the photocathode surface can be scanned discretely to build up an image of the incident photosignal [19,50–55]. The image dissector differs from the CTD in that the signal at any discrete location on the photocathode is transient in nature; unlike images provided by photographic emulsions and CTDs which are integrating detectors, the images recorded by the dissector are highly time dependent. Incident photon fluxes must be greater for a comparable response to be registered, due to both the reduced sensing area and the temporal nature of the photosignal [55,56]. However, these detectors are especially useful for monitoring of plasma events that occur on a ≥ 1-ms time scale. Image dissectors have been used to determine the temporal evolution of the electron temperature of high-energy plasma systems (pinches and flashlamps) where high photon fluxes are common [57,58]. They have also been used in conjunction with multiwavelength detection schemes, such as direct reader and echelle spectrometer systems, with success [52,59, 60]. The high tube and support electronics cost combined with the requirement of large incident photon fluxes have limited the use of the image dissector in the plasma-spectroscopy community [19].

In contrast to the tube-type detectors, microchannel photoemissive imagers have found extensive use in astronomical plasma studies. A microchannel plate (MCP) can be thought of as a parallel grouping of individual miniature PMTs bundled into an extremely compact area (see the discussion and figures in Chapter 5 of this volume). They typically consist of a series of extremely small channels (10–100 μm in diameter) that have been optimized for secondary electron emission. An amplification of 100-fold for each individual channel can be achieved [19,61,62]. As a consequence, MCPs are less dependent on incident photon flux than are image dissectors. However, an extra transducer is required to convert the electrons generated in the channels into image form. As a consequence, it is not uncommon to find MCPs coupled with phosphor screens, PDAs, and more recently, CTDs [63,64]. However, with the addition of an anode grid on the backside of the MCP, the electron image of the MCP can be encoded into an electronic signal that can subsequently be decoded to form a digital record of the image. These types of detectors are called multianode microchannel arrays (MAMA) and represent the current state of the art in microchannel systems [65–68]. MAMAs come in a wide range of physical sizes and are particularly attractive from the standpoint of the large detector areas available, that is, arrays with 2048 × 2048 detector elements (pixels) each with a physical dimension of 25 × 25 μm. This results in an overall detector size of 2 inches × 2 inches square [68]. MAMAs and MCPs are likely to find use in characterizing analytical plasma systems, since the ground state transitions of the major plasma species, Ar^+, are in the vacuum ultraviolet [69–71]. MAMAs and MCPs used in conjunction with a CCD transducer are a viable alternative to CTD detection in the UV spectral region where the CTD's performance is poor [72–76].

9.2.3 Solid-state Detectors

Plasma emission spectroscopy experienced a renaissance in the 1970s and 1980s with the commercial availability of solid-state detection devices, such as the PDA, vidicon, and CTD. These semiconductor-based devices are capable of transducing a wide photon-energy range, from the x-ray to near-IR region in the case of CTDs [72–74,76,77]. However, the UV response of CTDs is poor in the absence of coatings or special fabrication [74]. These solid-state devices exploit the high QE of silicon substrates and the ease with which these substrates can be fabricated into a diversity of structures [72–76]. The most primitive of these devices is the photodiode. Photodiodes typically lack the sensitivity required for plasma characterization studies and have found more utility in the laser field for monitoring laser power output [2,19]. In contrast, the multichannel variant of the photodiode and the later generations of these solid-state detectors have properties that make them the detector of choice for plasma characterization studies [77–80].

Unlike the successes of both the PDAs and CTDs, the vidicon has found limited use in the spectroscopic community [19,81–83]. Vidicons come in several different forms and are basically arrays of photodiodes; however, vidicons differ from conventional PDAs in that an electron beam is used for array readout [12,19,52,63,84]. The dynamic range of vidicons is limited to only about 250 (it can be enhanced somewhat by random-access readout techniques), which is a major drawback to using vidicons for spectrochemical plasma diagnostics [19,85]. This is particularly true for low-light-level studies. With a few exceptions, vidicons suffer from charge blooming (it has been compensated for in a few vidicon designs) and image burning, (i.e., they can be overexposed) [19,52,56,57,63]. The fact that vidicons can be time gated with aperture times as short as 40 ns makes them useful when monitoring the temporal evolution of plasma sources such as the high-voltage-spark discharge [12,84,86]. Vidicons are not without their timing problems as they suffer from lag; however, one can compensate to increase the temporal resolution [86]. Both the temporal and spatial resolutions that vidicons offer have made them attractive to high-energy-plasma spectroscopists, where high photon fluxes and pulsed sources are the norm [12]. However, due to the poor dynamic detection range and expensive high-voltage power supply of the vidicon, attention has shifted to more sensitive, rugged, and convenient detection devices, namely the PDA and CTD [87].

Both PDAs and CTDs are finding expanded use in the plasma spectroscopic community. PDAs have been the forerunner in multidimensional spectroscopic investigations and find extensive use in quantitative and qualitative spectrometer systems [88–101]. In contrast to the electron beam utilized for vidicon readout, with the PDA, readout is accomplished through addressing each photodiode using prefabricated connections parallel to the photosensitive junctions [19,64,84,87,102]. This provides greater geometrical accuracy, important for spatial resolution. In addition, PDAs have an approximately fourfold greater dynamic working range than vidicons. Unlike vidicons, PDAs do not suffer from image burning, magnetic field interference, and image lag, but they have reduced intrinsic speed. This limits the temporal resolution of the PDA to the microsecond regime ($\sim 10\ \mu s$) [19,84,87,102]. With additional gating devices, the temporal response can be extended to the nanosecond regime (~ 5 ns). PDAs have been successfully used for Boltzmann distribution analysis, lineshape studies, and the monitoring of spatial emission intensity profiles in the inductively coupled plasma (ICP) source [95,100, 103–115]. However, PDAs are slowly being displaced for the increased sensitivity, wider dynamic range, and better SNRs that the CTDs have to offer.

Charge-transfer devices have been previously reviewed and the reader is referred to Chapters 1 through 3. However, it is important to note that CTD detectors are themselves not without limitations. Table 9.1 lists these. With appropriate precautions, some limitations can be overcome. The sensitivity and the multichannel nature of these devices have made them the spectroscopist's latest panacea, but unless CTDs are coupled with the appropriate dispersing systems, some of their distinguishing features may be inaccessible.

9.3 Spectrometer/Detector Systems

The combination of spectrometer and detector is an important factor to consider in any plasma characterization study. First, the experimentalist must choose the type of measurement to be made. Once this is decided, the appropriate detection scheme can be designed. Since CTDs are generally two-dimensional, they offer a wide range of experimental flexibility. Their response linearity, geometry stability,

Table 9.1 CTD Limitations and Consequences of Those Limitations

Problem	Consequence
Blooming	Cannot observe weak features near bright features
Full-well capacity	Limits dynamic range
Number of pixels	Limits number of resolution elements
Pixel dimensions	Determines minimum space between distinguishable features
Quantum efficiency	Defines ultimate sensitivity limit
Read noise	Excess noise limits light detection
Trapped charges	Need to prefog CCD; apparent loss in low-intensity sensitivity
Variation in quantum efficiency	Requires intensity calibration

extended spectral sensitivity, and high QE give them inherent advantages over TV-type (vidicons and image dissectors) and PMT detectors [19,77–80]. However, for monitoring transient, irreproducible events, a CTD-based system might not be the system of choice. As with any experimental design, trade-offs and choices must be made of which experimental parameters are important. This chapter focuses on the options to consider when using specific CTD-detection-based systems and the types of plasma information that can be obtained using them. Specifically, this section will evaluate the spectrometer/-CTD systems, illustrated in Figure 9.1, that can be used in plasma characterization studies.

The four grids in Figures 9.1A through 9.1D represent the individual pixel wells of the CTD. The horizonal and vertical axes represent the type of experimental information obtainable. Arrows point to increasing values of parameters. Figure 9.1A represents a dispersing/imaging arrangement. Figure 9.1B represents use of the CTD in camera mode. In Figure 9.1C, the CTD is gathering two-dimensional wavelength information. Finally, Figure 9.1D represents the type of information that can be gathered using the CTD in streak camera mode.

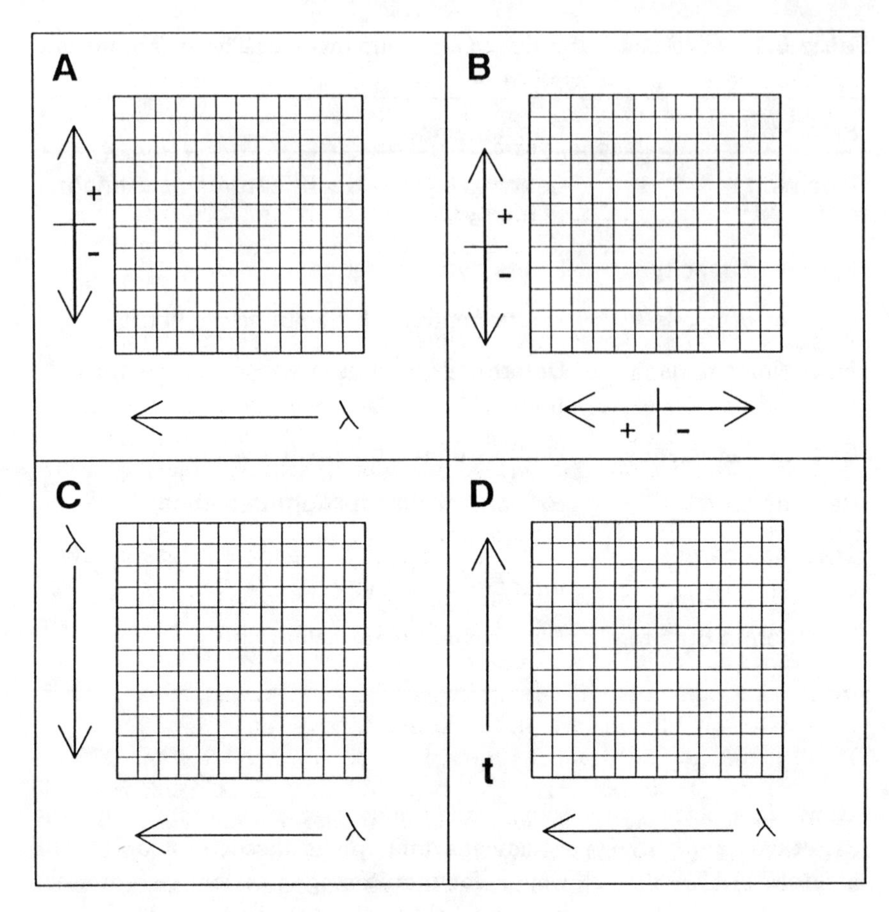

Figure 9.1 A block diagram of possible representative experimental setups using CTD detection. See text for the description.

9.3.1 Imaging Spectrometer Systems

The imaging spectrometer system is represented by Figures 9.1A, 9.1B, and variants of the 9.1D theme. These systems offer some spatial information on the observed source. The simplest form of imaging setup is that shown in Figure 9.1B. The source of interest has been optically transferred (generally by a lens system) and imaged directly onto the CTD, which is acting like a digital piece of film [75,87, 116–22]. By sacrificing spatial resolution in one dimension, the observer can obtain spectral information, as shown in Figure 9.1A. This configuration is by far the most common experimental design currently in use [16,17,77,122–129]. The CTD is placed on the focal plane of a spectrometer. Therefore, one of the spatial dimensions is used for obtaining spectral information on the source of interest.

Figure 9.1D represents how CTDs can be used in streak camera mode [130–35]. The wavelength axis in Figure 9.1D can easily be used to obtained spatial information as well. In addition, CTDs have also been used as detectors in spatial Fourier transform systems [136–38]. This type of setup will not be discussed here. The reader is referred to Chapter 3, which deals specifically with these types of systems. This section will focus primarily on the setup in Figure 9.1A and the experimental difficulties associated with designing such a system, briefly touching on other imaging systems.

In any imaging system, one must be concerned with the aberrations introduced by the collection optics. Optical aberrations, such as chromatic aberration, spherical aberration, astigmatism, coma, distortion, and curvature of field decrease image fidelity [139–141]. Chromatic aberration (the change in focal length as a function of wavelength for refraction-based optics) and spherical aberration (a change in the focal position due to point of optic illumination) can be eliminated by using nonspherical reflective optics. However, in many Czerny-Turner and Fastie-Ebert monochromator systems, spherical aberration is minimized by using higher *f*/ optics [128]. The off-axis aberrations, such as astigmatism and coma, are more difficult to minimize. In the Czerny-Turner system, there is some residual astigmatism (tangential and sagittal focal points occur at different locations) and coma (a blurring of the image, which resembles the shape of a comet's tail, hence the term *coma*) which is completely compensated at only one particular wavelength [142]. Astigmatism can be eliminated by using feed optics in the form of off-axis mirror pairs that are oriented orthogonal to those of the spectrometer [143,144]. In addition to the Siedel aberrations, focal plane flatness and tilt are of concern when adapting a PMT or photographic-detection-based spectrometer for use with CTDs [127].

Adapting flat-field spectrographs for use with CTDs is relatively easy, since the focal plane was originally optimized to be nearly planar [127]. An example of such an adaptation is diagrammed in Figure 9.2. Note the high image fidelity that can be obtained with such a system. The image in Figure 9.2A was obtained using the optical train diagrammed in Figure 9.2B with the ruler at the focal point of the over-under mirror pair; the spectrometer is set to zero order, and the entrance slit width is set to 3.5 mm. The reason for the extra side-by-side mirror pair in Figure 9.2B is twofold. First it serves as a light pipe; but more important, this configuration allows one to perform stroboscopic time gating by spinning the 45 degree folding mirror, which is a six-sided mirrored polygon [17,145]. By narrowing the spectrometer's entrance slit and scanning the spectrometer off zero order, both spatial and wavelength resolution (~0.05 Å/pixel for a 23-μm pixel width) can be achieved, as in Figure 9.1A. In addition to

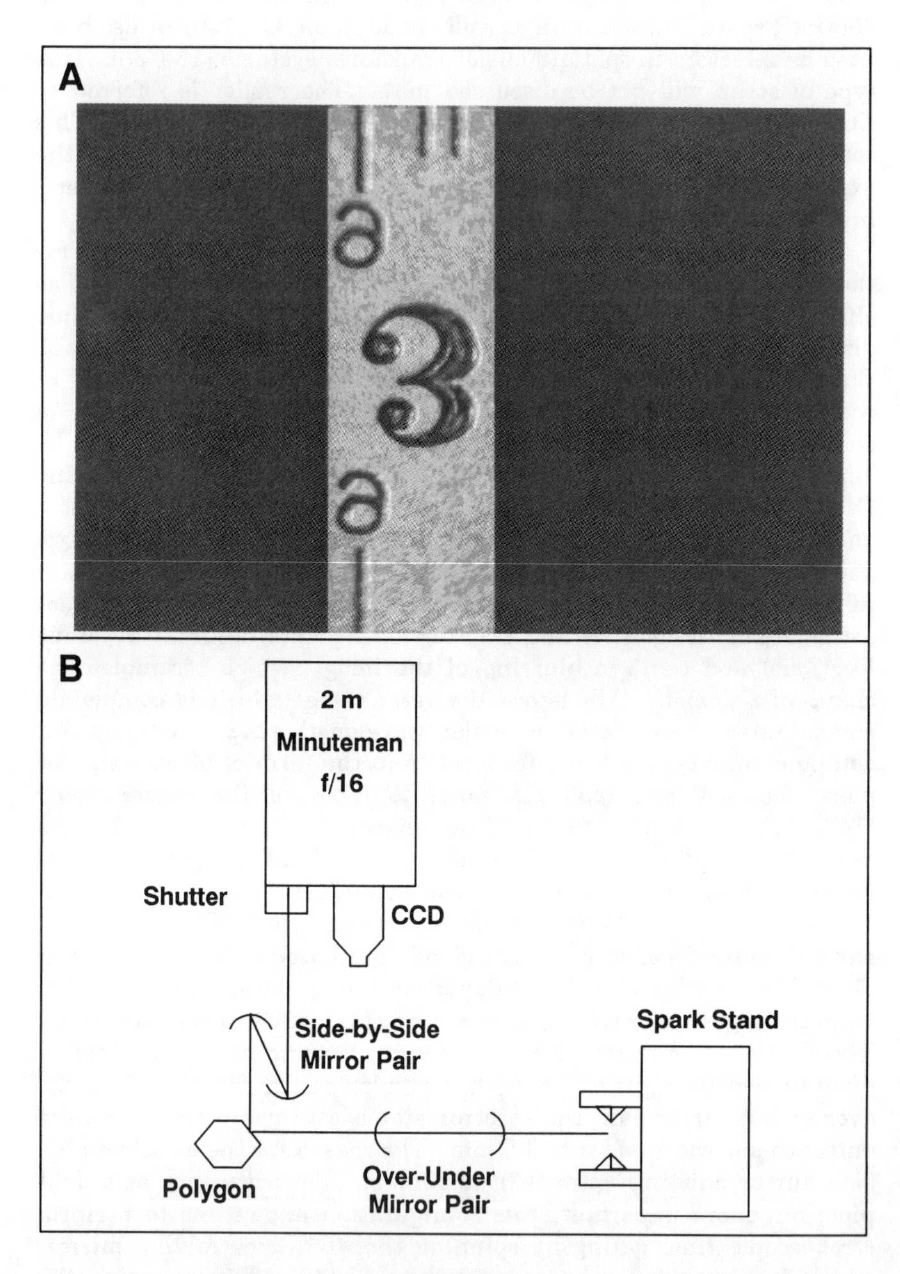

Figure 9.2 An example of a CCD imaging spectrometer system. **A** is an actual CCD zero-order image taken using the optical train block diagram of part **B**.

imaging systems like the one diagrammed in Figure 9.2, commercially available superflat-field *f*/8 dispersing systems using torroidal mirrors have been successfully coupled with CTD detection (Chromex, Inc., Albuquerque, N.M. and Instruments SA, Edison, N.J.) [127,146]. As can be seen in Figure 9.2, not only is the CCD imaging system capable of high image fidelity, but also a high wavelength resolution is obtained, which is extremely important for lineshape studies. Not only are CTD imaging systems useful for lineshape analysis, but they are being used successfully in analyte excitation studies as well [147]. However, the limited bandpass (5 nm in the system described above and shown in Figure 9.2) can be a problem when performing Boltzmann excitation studies on spectrally poor analytes such as copper. By sacrificing the spatial resolution, one can obtain the increased spectral bandpass that is necessary for such studies, which is exactly the trade-off that is made in multidimensional dispersing systems.

9.3.2 Multidimensional Dispersing Systems

Of the multidimensional dispersing systems (Figure 9.1C), the echelle spectrometer is the most common [80,148–160]. Echelle spectrometers are orthogonal dual-dispersion devices usually employing a refractive element (prism) as an order sorter and a high-dispersion element (echelle grating) for wavelength resolution. In contrast to the more common grating designs that rely on a high groove density to achieve the desired resolution, the coarsely ruled echelle grating utilizes a steep grating angle (typically greater than 45 degrees) that probes diffraction order, m, in a high range (anywhere from 28th to 120th+ order) to achieve the same effect [160–163]. Here the grating angle must be set to the blaze angle for the echelle to work, since any m,λ (order, wavelength) combination sufficiently off the blaze angle will lead to severely attenuated spectral intensity. Consequently, echelles are fixed-angle spectrometers, in contrast to Czerny-Turner spectrometers. The orthogonal dispersive element (order sorter) provides the extended wavelength coverage by controlling the position at which wavelengths are imaged in space before being dispersed by the echelle grating. This gives the echelle spectrometer its two-dimensional wavelength dispersion character. The typical echelle is capable of covering the spectral range from 230 nm to 860 nm at a resolution of roughly 0.1 Å using a large-format photographic emulsion [148]. Recently, similar wavelength coverage and resolution have been obtained using CTD-based echelle systems [152,154,157–160]. As a consequence of the enormous information bandwidth of these systems, the experiment actually can be designed after the fact for spatially homogeneous plasmas. Throughout this chapter, depending on context, *bandpass* will have two distinct, but related, meanings:

(1) the range of wavelengths sensed by a single detection element or (2) the range of wavelengths sensed by an array of detector elements.

Not only can plasma-temperature measurements be made using Boltzmann excitation methods, but both gas-kinetic-temperature studies (from molecular Boltzmann studies) and an electron-density measurement (from Stark broadening) can be obtained from a single echellogram [160,164]. The echellogram in Figure 9.3 is that of a plasma-enhanced chemical vapor deposition (PECVD) reactor operating at 500 watts of 13.56-MHz rf energy on a mixture of argon, silane, and methane gases. Wavelength is dispersed horizontally within a given order and wavelength decreases (vertically) as order number increases. A calibrated overlay of the orders and the wavelength dispersion along order 56 has been added to the grayscale image. The darker spots represent areas of high intensity against the white background. A crosshair (+) has also been placed on the 4158-Å Ar(I) emission line in order 56. The band spectrum of SiH is also captured in order 56. In order 60, three distinct CN bands are present, and CH band emission is observed in order 54. The presence of the Stark and Doppler broadened H_β emission can be observed in the left-hand corner of order 48.

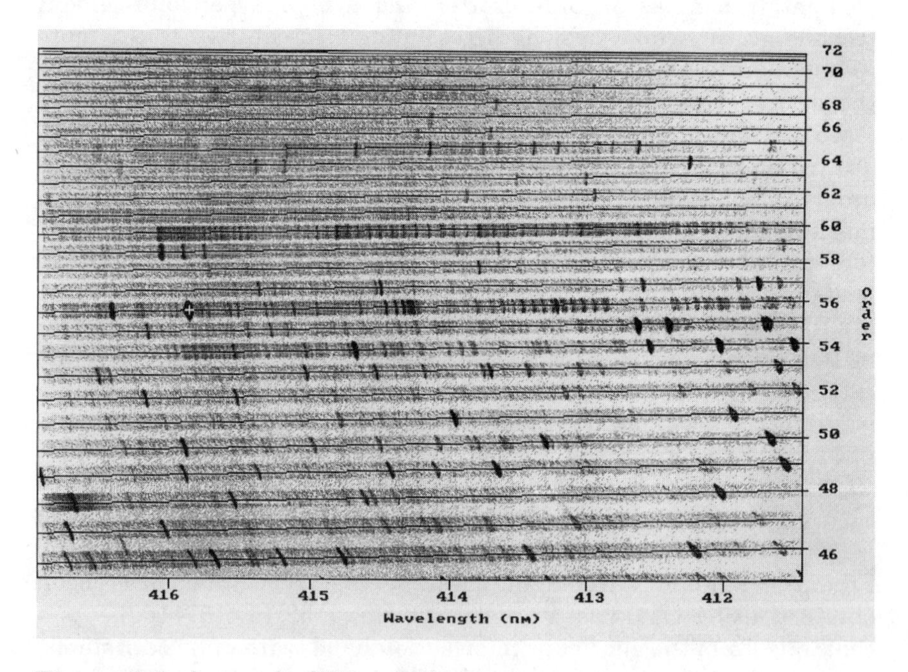

Figure 9.3 A sample CTD echellogram.

Despite all their advantages, echelle systems are not well suited for many types of measurements due to low throughput, high levels of stray light, and the complexity of echelle data analysis [148,154,157, 159,160]. Since echelle systems consist of essentially two spectrometers, their throughput will be lower than a similar *f*/-matched, single dispersion system, due not only to the larger number of optical components, but also to the extra dispersive element, which can be a significant light sink. For example, in an orthogonal Czerny-Turner CTD-optimized echelle design [159,160], the grating order sorter cuts the intensity of the echelle (*f*/8) such that its throughput is worse than an *f*/16 system with a comparable number of reflections (optical train in Figure 9.2A). Other CTD-optimized designs, such as that of Bilhorn and Denton [157], suffer from similar attenuation due to the optics used to reduce the focal plane's size. However, a design recently patented by Baird Corporation [165], shows more promise in improving the CTD-based-echelle throughput. In addition to throughput considerations, echelle gratings have a larger amount of stray light, thereby reducing the dynamic detection range of echelle systems in general [148,159,160]. Likewise with the large spectral coverage, it is more likely that one will run into dynamic range problems, since the intense lines cannot be easily discriminated against as in single-dispersion systems. Both of these factors limit the dynamic detection range; however, some of the dynamic range problems can be overcome using the random-access-read capabilities of CIDs, such as in the design by Denton and co-workers [78,129,157,158,166]. Despite the improvement in dynamic range gained from using CID detection, the added complexity of random-access-read algorithms coupled with the already complex nature of echelle spectra serve to limit the CTD-based echelle system's utility in plasma-characterization studies.

Not only do CTD echelle systems have a large spectral information bandwidth, which is necessary for many of the plasma-diagnostic techniques, but they also have all the problems associated with such high bandwidth systems, namely added data-analysis complexity. This complexity stems from the two-dimensional nature of the wavelength dispersion, which leads to wavelength calibration and spectral-data-extraction difficulties [154,155,157,158,160,167–171]. Since dispersive elements (prisms or gratings) separate spectral information in a continuous fashion, cross-dispersed echelle orders are tilted (see Figure 9.3). The severity of this tilt varies with wavelength, which complicates automated data extraction [154]. When using a grating for the order sorter, this tilt angle can be calculated fairly simply from the grating equation [172]. For refractive predispersion, nonlinear heuristic equations must be used to predict dispersion and tilt variation [155,167,168,171]. In addition, residual aberrations such as coma can further complicate data analysis, especially when lineshape-based plasma-diagnostic techniques are used [160]. Due to the uniqueness of the various echelle designs, data-analysis algorithms that are not instrument specific are either too general, making them of little utility, or require significant modification [155,167,168,171, 172]. As a consequence, echellograms are difficult to analyze, and analysis is further complicated by the sheer amount of data. Eventually, with faster microcomputers, data analysis may become a trivial consideration. Despite these limitations, CTD-based echelle systems offer substantial experimental flexibility and are quite useful in a number of plasma-characterization studies, especially in Boltzmann excitation measurements [147,164,172].

9.4 Plasma Diagnostic Techniques

Plasmas are extremely complex environments that exhibit both gas-like and fluid-like behavior. The gas-like attributes arise from the fact that plasmas are highly collisional systems in which the kinetic energy of the charged particles serves to partially negate the electrostatic repulsive and attractive forces. Hence, the charged particles (electrons and ions) can act somewhat independently of one another as in a gas; however, since the plasma species are charged, electrostatic forces are present as well which serve to unite the electrons and ions, much as in the case of a dense fluid. In other words, one cannot influence the electrons without affecting the ions at the same time, and consequently, plasma behavior is highly nonlinear with a variety of oscillations and instabilities occurring throughout the plasma's lifetime [173–178]. Typically, to deal with these nonlinear features one tries to analyze plasma behavior relative to some form of equilibrium and treat the nonlinearities as perturbations to the equilibrium or steady state. Many of the plasma diagnostic techniques to be discussed later depend on the assumption that the plasma is in a particular form of equilibrium [2,4,22,175,179–183]. Therefore, a brief discussion about different forms of plasma equilibria is helpful before moving to the individual diagnostic techniques.

9.4.1 Plasma Equilibria

Several competing energy-dissipation reactions in a plasma need to be considered in any equilibrium determination. The main plasma processes can be written in mathematical form for analysis and are given by Planck's, Maxwell's, Boltzmann's, and Saha's distributions, which govern the radiative, collisional, excitation, and ionization equilibria, respectively [2,4,179–181]. All of these distributions have a temperature dependence. When each of the plasma processes is in detailed balance with its reverse reaction, the plasma is said to be in complete thermodynamic equilibrium (TE) [2,180], and therefore, the individual temperatures (color, kinetic, excitation, and ionization) are equal. It is fortunate from the analytical chemist's standpoint that complete radiative equilibrium is not achieved, since laboratory plasmas are not blackbody emitters and usually have optical mean free paths greater than the physical dimensions of the plasma. Instead, laboratory plasmas achieve some form of partial equilibrium. Historically, local thermodynamic equilibrium (LTE) is the degree of equilibration on which plasma diagnostics studies have been based most frequently [182,184–202].

LTE can be summed up quite simply as more heat and less light, that is, the radiative processes governed by Planck's distribution are far from equilibrium, while the remaining processes are in equilibrium with one another. A plasma in LTE can be described by a complete TE system that has the same density distribution, temperature, and chemical composition, but in which radiative equilibrium is not achieved [2,22,179–182]. Since many of the plasma diagnostic techniques rely on the existence of some form of LTE, one must take care when invoking the assumption that the plasma source is at LTE. Therefore, the Griem criterion ($N_e \gg 1.6 \times 10^{12} \, \Delta E^3 \, T_e^{1/2} \, \text{cm}^{-3}$) should be used to assess whether LTE will exist for excited-state energy levels of interest [2,182]. A prime example can be found in Boltzmann excitation analysis for plasma temperature determination, where the LTE assumption is at the heart of the measurement.

9.4.2 Boltzmann Excitation Analysis

If a plasma source is at LTE, then an analysis of the population distribution of excited states can be used to obtain the plasma's electron temperature (T_e). Optical techniques are by far the most common method used to measure excited-state distributions, since the emission intensity of a transition from an excited state is directly proportional to the number of atoms in that excited state [2,22,179, 185] and one can indirectly measure the excited state density from an emission line's radiance. In an optically thin plasma, this proportionality relation is governed by equation 9.1 [22].

$$I_{ij} = \frac{1}{4\pi} A_{ij} h \nu N_i \qquad \frac{watts}{m^2 \, srad} \tag{9.1}$$

In the laboratory, one can measure the emission radiance, I_{ij}, of a spectral line. This in turn is related to the Einstein spontaneous emission coefficient (A_{ij}), the emission frequency (ν), the density of excited states (N_i), and the plasma thickness (l). By measuring the emission radiance of a spectral line, the population density of the upper energy level (N_i) can be determined. Subsequently, this measured value of N_i can by substituted for the excited state density in the Boltzmann equation of states:

$$\frac{N_i}{N_j} = \frac{g_i}{g_j} e^{-(E_i - E_j)/(kT_{app})} \tag{9.2}$$

For convenience, the *j*th state is taken to be the ground state, whose population and statistical weight are N_0 and g_0, respectively. Given that the ground state energy (E_0) is zero, one can substitute for N_i in the emission radiance equation (equation 9.1) and after linearization, the familiar form of the Boltzmann plot equation is obtained [2,22,183]:

$$\ln\left[\frac{g_i A_{ij} c}{I_{ij}\lambda_{ij}}\right] = \frac{E_i}{kT_{app}} + \ln\left[\frac{4\pi g_o}{hlN_o}\right] \tag{9.3}$$

Here g_i is the statistical weight of the excited state *i*, E_i is the energy of the excited state, and T_{app} is the apparent excitation temperature. The statistical weight can be calculated from the level's total angular momentum, J ($g_i = 2J + 1$). Since excited-state energy levels, statistical weights, and Einstein spontaneous-emission coefficients are known for some levels, one need only measure the emission frequency and radiance for several transitions to measure the apparent excitation temperature. By plotting the log term on the left-hand side of equation 9.3 against the excited level energy, one can calculate the apparent excitation temperature from the slope [2,22,183]. Note that this will only yield a linear relationship when the source is at or close to some form of LTE. In addition, only when the source is in LTE will the calculated temperature ($T_{app} = T_{ex}$) have some meaning. With this in mind, we can now concentrate on the difficulties associated with making the measurement and how the large information bandwidth of CTDs can be exploited in such measurements.

9.4.2.1 Spectral Bandpass Constraints

Since several emission lines are required for the construction of a Boltzmann plot and the relationship assumes knowledge of the emission radiance, a large spectral instrument bandwidth and an instrumental intensity calibration are required. For spectrally rich elements (the lanthanides, actinides, and some of the transition metals), a relatively small wavelength range is required for adequate upper energy level (E_i) coverage. A spectral bandpass of ~5 nm has proven successful for Fe(I) excitation studies in the high voltage spark and iron hollow-cathode lamp, shown in Figure 9.4. In this case, both limited bandwidth studies (obtained using apparatus diagrammed in Figure 9.2B) and high bandwidth echelle measurements are represented. The Fe(I) Boltzmann plots in Figures 9.4A and 9.4C were obtained using a focused iron hollow-cathode lamp source. The Fe(I) Boltzmann plots in Figures 9.4B and 9.4D were obtained using a focused high-voltage spark source. Parts 9.4A and 9.4B are the results of the limited bandwidth measurements. Likewise, parts 9.4C and 9.4D are the results of the echelle measurements. The echelle spark (Figure 9.4D) measurements were taken on the discharge axis and in the cathode space charge region. The appropriate spatial information was extracted for comparison with the limited bandwidth spark measurements presented in Figure 9.4B. The error bars represent 1 standard deviation.

In the past such excitation measurements typically were made using a single-channel method of detection, namely a PMT [187, 189,193,194,197]. Only if the plasma source under consideration was

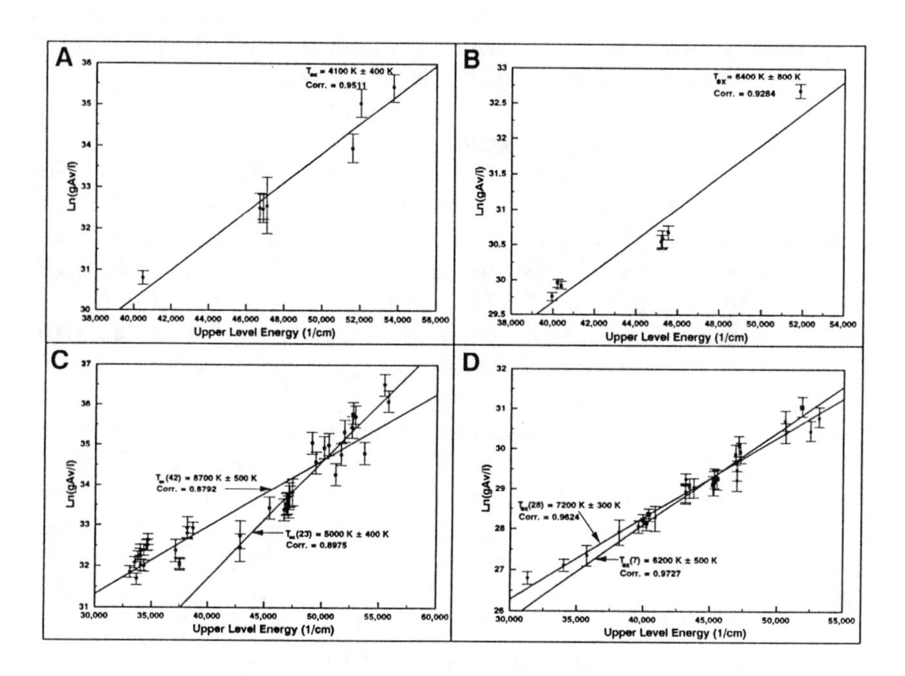

Figure 9.4 A comparison of the Fe(I) excitation distributions in the iron hollow-cathode lamp and the high-voltage spark. (See appendix to define *G, A, ν,* and *I.*)

extremely stable, reproducible, or at steady state did this single-channel method yield meaningful information. This is where the large information bandwidth of CTDs can be exploited. When used in the imaging and multidispersion configurations in Figures 9.1A and 9.1C, a more-than-adequate spectral bandpass can be achieved, and the constraints on the plasma behavior are eased, since all the spectral information is collected at the same time under the same conditions. In addition, if the CTD is used in conjunction with an echelle spectrometer (Figure 9.1C), one can obtain sufficient spectral information to analyze the validity of the LTE assumption [147]. While one may be able to measure T_{app} from a limited number of lines, the number available in a narrow wavelength range may not be sufficient to test the assumption of LTE. Deviations from LTE appear as nonlinearity in the Boltzmann plot. One must note, however, that in this configuration (Figure 9.1C) spatial information has been sacrificed. When making plasma excitation measurements, spatial averages of plasma behavior must be used with care if the plasma source is heterogeneous or has a spatial variation in its density and temperature distributions. To treat such spatial variations appropriately, some form of spatial resolution should be achieved, whether this be Abel inversion for cylindrically symmetric sources or tomography for others [27–32]. Which of these CTD configurations (Figures 9.1A and 9.1C) is best depends on the particular plasma source.

9.4.2.2 The Intensity Calibration Problem

In general, the spectral bandpass issue is negligible in comparison with that of intensity calibration. Fortunately, one need not perform absolute intensity calibration, which is very difficult. Instead, the spectrometer can be relative-intensity calibrated against a standard lamp of known irradiance [2,179]. Typically, the standard lamp is a graybody emitter of known color temperature. For a complete calibration of the instrumental throughput, the diffuse reflection of the standard lamp off a MgO-frosted microscope slide is imaged by the collection optics into the spectrometer and subsequently detected by the CTD [203,204]. MgO is chosen as the reflector since its reflectivity response varies by only 1% across the visible spectrum [205]. Note that care must be taken to minimize the amount of stray light from the standard lamp that is collected by the optical train or enters the spectrometer by direct light paths. This can be accomplished by using a series of light baffles. Light from the standard lamp that does not go completely through the identical experimental optical train (collection, dispersion, and detection) can cause severe errors in the intensity calibration, that is, any stray light entering the spectrometer will over-emphasize the dispersion contribution to system throughput. Due to the diffuse reflectivity of the MgO and low spectral irradiance of the standard lamp, exposure times can be on the order of minutes. (For proper calibration, one must expose the CTD until the maximum pixel charge is at least two-thirds the saturation level.) Also, if short experimental exposure times are used, the shutter speed can be a significant factor, since various portions of the array will be exposed for differing amounts of time. For example, note the pinwheel closing pattern of our shutter in Figure 9.5. In addition, if the CTD suffers from charge trapping, the need for a preexposure using a flash ring or diffuse source is important for both intensity calibration and spectral data acquisition [158,206]. The calibration procedure outlined above and described in Table 9.2 is the preferred method of intensity calibration.

Another less robust form of intensity calibration over a narrow wavelength range is flat-fielding. Flat-fielding techniques can be used if the major source of throughput fluctuations is the variation in the CTD's response from pixel to pixel [168]. Correct flat-fielding of a CTD requires uniform detector illumination, which can be achieved using a collimated source. Figure 9.5 is a result of such a flat-field attempt on a Thomson 7882 CCD. Note that the spots in the flat-field exposure are shadows of dust particles on both the outside and inside faces of the quartz window protecting the CCD chip. (The dust

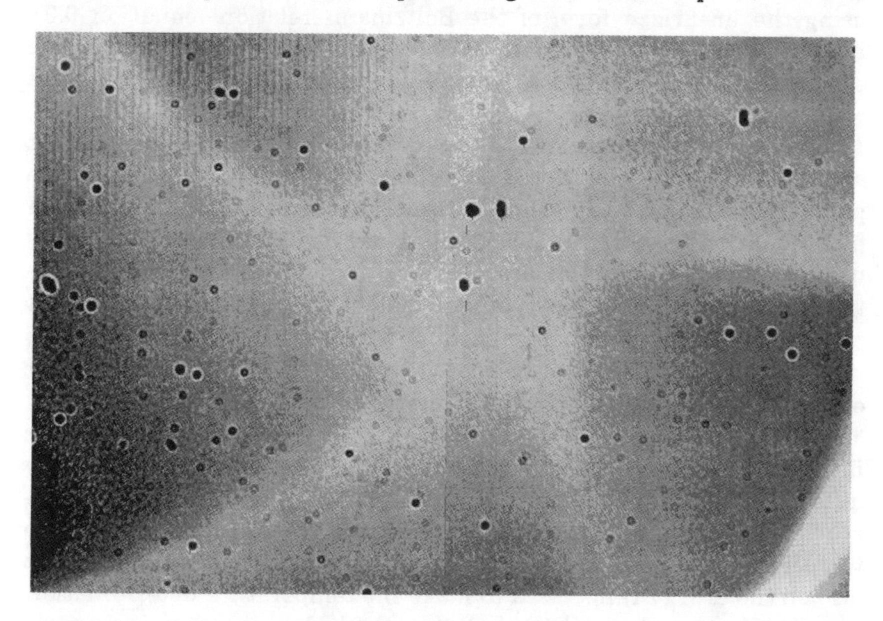

Figure 9.5 An example of flat-fielding using a collimated tungsten source. Here the lighter shades represent the regions of high intensity, and black represents the lowest intensity.

particles scatter only weakly, so that they are unnoticed when illuminated nonuniformly.) The position of these shadows will change depending on *f*/ of illumination, so flat-field corrections must be made using the same *f*/ of illumination as in the experimental optical train. These shadows give rise to some of the apparent noise in the calibration spectrum presented in Figure 9.6 (collected using the standard lamp-calibration method). The optical train in Figure 9.2B was used to collect these data. In Figure 9.6A the full CCD image is presented in 16-shade grayscale. Again, the lighter shades represent the regions of high intensity, with black representing the lowest intensity. A single column extraction of the image in Figure 9.6A is presented in 9.6B. Additional sources of noise can be found in the standard lamp's shot and flicker noise, scattering from the dust particles in the optical train. These would not be accounted for strictly when using flat-fielding. Consequently, the MgO-calibration procedure is preferred when an accurate accounting of system throughput is required, as in the case of Boltzmann excitation analysis.

9.4.2.3 Boltzmann Plot Construction Considerations

Having obtained an adequate wavelength range and an intensity calibration, the apparent excitation temperature can be measured using the linearized form of the Boltzmann relation (equation 9.3). Even at LTE, T_{ex} is not necessarily representative of the plasma's electron temperature unless the probed species is a major plasma constituent, namely Ar or Ar^+ for an argon-based plasma [2,4,22,46, 114,183,207]. If an absolute intensity calibration is performed that accounts for the solid angle of collection, then the ground state population can also be determined from the intercept of the Boltzmann plot (equation 9.3) [22,114,207–210]. However, the intercept is rarely used due to the difficulties associated with absolute intensity calibrations, where one must worry about matching solid angles and about polarization states between calibration and experimental sources. As a consequence, a Boltzmann plot is typically used as a means for obtaining excitation temperature information.

The application of the Boltzmann temperature technique is illustrated in Figure 9.4. The technique has been applied to two very different plasma systems (iron hollow-cathode lamp and high-voltage spark discharge sources), and one can see that the validity of the temperature determination depends not only on the type of source, but also on the energy range used to make the temperature determination [6]. The iron-neutral (Fe(I)) hollow-cathode lamp studies in Figures 9.4A (using a 2-m monochromator with a CCD) and 9.4C (using an echelle with a CCD) give differing temperature measurements depending on what energy range is used to construct the plot, with the results varying as much as 3700 K. In contrast, the Fe(I) spark

Table 9.2 Intensity Calibration Procedures

Intensity-Calibration Procedure for CTD Imaging Systems

1. Place MgO-frosted slide at the focus of the collection optics.

2. Illuminate the slide off-axis with the standard lamp.

3. Place light baffles along optical train to block stray light and indirect reflective paths.

4. Pre-fog CTD if charge trapping is a problem.

5. Expose CTD for an appropriate time to reach two-thirds of the saturation charge.

6. Subtract a stored pre-fog[a]-only frame from the calibration. Substitute a bias frame[a] for the pre-fog if charge trapping is not a problem.

7. Divide the fog- or bias-corrected, experimental-image frame pixel by pixel with the fog- or bias-corrected calibration frame in step 6. Multiply the result by the appropriate portion of the standard lamp's spectral irradiance profile.

Intensity Calibration Procedure for Multi-dispersive (Echelle) Systems

1. Follow steps 1–5 above.

2. Obtain a flat-field frame. Note: This should be pre-fog corrected if charge trapping is suspected.

3. Flat-field correct both experimental and calibration frames.

4. Divide the flat-field and fog-corrected experimental-image frame pixel by pixel with the identically corrected calibration frame. Multiply each order extraction by the appropriate portion of the standard lamp's spectral irradiance profile.

---

[a] Note that bias and pre-fog exposure times must match the experimental and calibration exposure times if the CTD dark current is significant. For liquid N_2-cooled CCDs, we have found that the dark current is negligible with exposure times as long as 10 minutes.

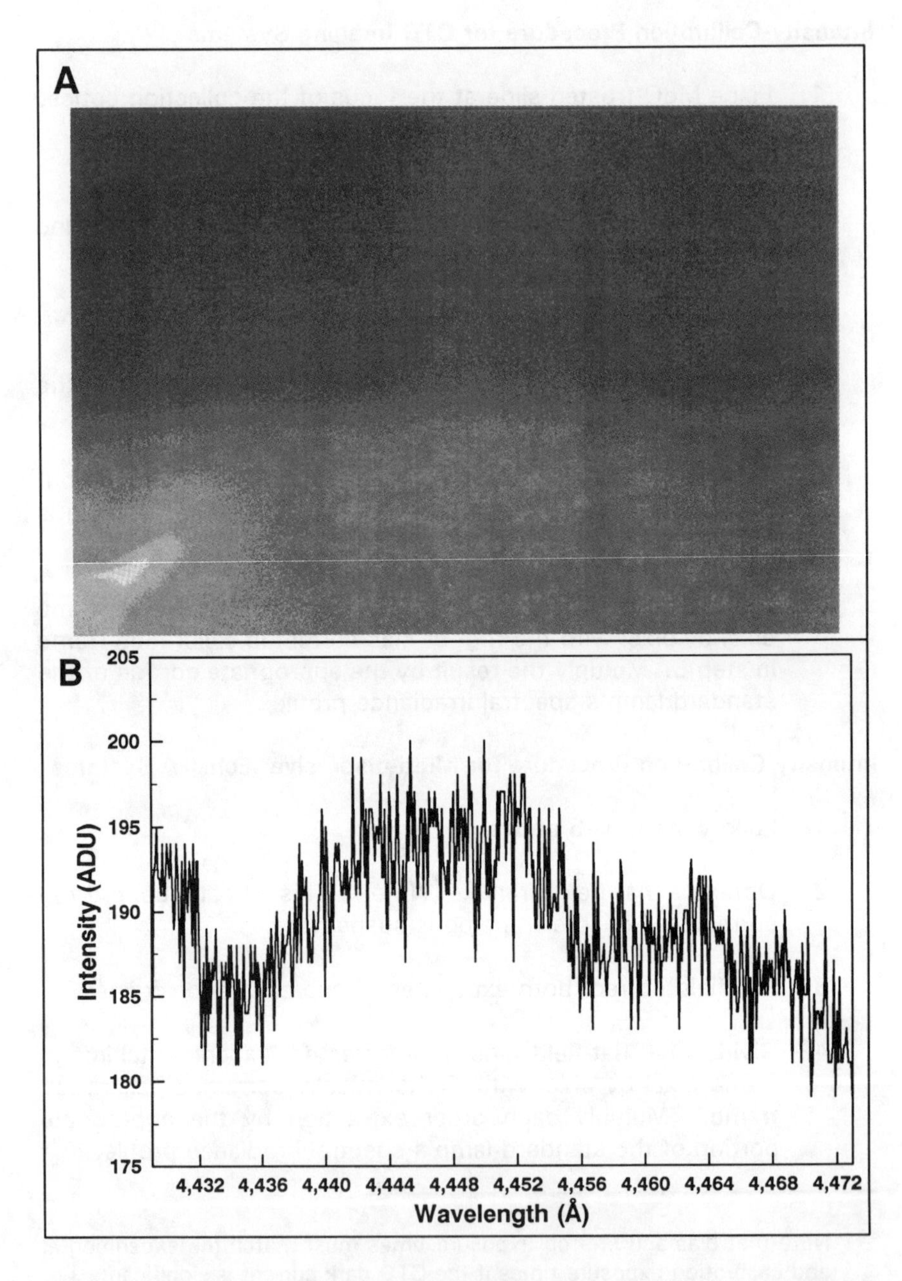

Figure 9.6 An example MgO/standard lamp calibration frame of a CCD. The optical train diagrammed in Figure 9.2B was used to collect these data.

measurements in Figure 9.4B (2m/CCD) and 9.4D (echelle/CCD) show excellent agreement when matching energy ranges are used for construction of the plot. A minimal variation of 1000 K, which is slightly outside the error bars, is observed when a wider energy range is used in the echelle measurement; therefore, one must take care when choosing the energy range over which to make the temperature measurement.

The reasons for the disagreement in the example systems (Figure 9.4) are twofold. First, a low electron density combined with a moderate excitation temperature is characteristic of a hollow-cathode source. Consequently, the frequency of collisions is not high enough for the hollow-cathode source to reach LTE. Instead the hollow-cathode source is somewhere between LTE and coronal equilibrium [2,175,180,211]. As a consequence of the non-LTE conditions in the hollow-cathode lamp, one can expect more disequilibrium and the non-LTE conditions to be more pronounced in the lower lying energy levels. The higher energy levels have more of a chance to be in LTE than the lower lying states [6,114,182,212]. Therefore, it is not surprising that the limited energy-level coverage of the 2m Boltzmann plot (Figure 9.4A, 4100 K) predicts a different temperature than the echelle measurements (Figure 9.4C, 8700 K) over a wider energy range. When matching energy ranges are used for comparison, the echelle (5000 K $\pm$ 400 K) and 2m (4100 K $\pm$ 400 K) excitation temperatures show better agreement. In contrast, the spark measurements (Figure 9.4B and 9.4D) show excellent agreement when temperatures calculated using the same energy range are compared: 6200 K $\pm$ 500 K from the echelle and 6400 K $\pm$ 800 K from the 2m measurements. Unlike the hollow-cathode measurements, the wider energy range measurement is within the error bars of the 2m measurement. Therefore, one can conclude that the high-voltage spark is fairly close to LTE, while the iron hollow-cathode source exhibits non-LTE type behavior.

The measurements in Figure 9.4 illustrate the shortcomings, checks for internal consistency, limitations, and power of Boltzmann excitation studies. When a source is at LTE, the Boltzmann plot is by far the easiest way to obtain temperature information with a reasonable degree of accuracy. When the source is not at LTE, the validity of the temperature measurement from the Boltzmann plot can quickly fall into question [110,112,114,187,189,195,197,207,210, 213–225].

9.4.3 Saha-Boltzmann Measurements

When Saha's ionization distribution is used with Boltzmann's excitation distribution, additional plasma information such as electron temperature, ionization temperature, electron density, and the degree of ionization can be obtained [2,114,180,183,187,189,226]. Saha's distribution is a statement of the chemical equilibrium between successive ionization states and provides a means for accessing the stage or degree of ionization that exists for systems in LTE [2,6,180, 183,187,207]. Saha's distribution (generalized form in equation 9.4) evaluates this equilibrium by minimizing the free energy for this reaction using both Maxwell's and Boltzmann's population distributions. The key assumption to this derivation is that complete equilibrium among the thermal (velocity) and excitation distributions exists. This presupposes that the individual species temperatures are equivalent [207]. The atomic (Q_a(T)) and ionic (Q_i(T)) partition functions have been explicitly stated, and the electron partition function (the cube of the inverse of the DeBroglie wavelength) is embedded in the constants. The concentration terms (N_e = electron density, N_{atom} = atom density, and N_{ion} = ion density) are expressed as number densities (m^{-3}). Due to the large energy difference (ionization potential, χ) between successive ionization states, it is seldom necessary to consider more than one ionization equilibrium reaction, which is generally the first ionization equilibrium for laboratory-produced plasmas [2,4,114,183,207]. The generalized form of Saha's distribution is given by

$$\frac{N_{ion}N_e}{N_{atom}} = \frac{2(2\pi m_e kT)^{3/2}}{h^3} \frac{Q_i(T)}{Q_a(T)} e^{-\chi/kT} \tag{9.4}$$

Total number densities such as those expressed in the Saha equation are extremely difficult to quantify in laboratory measurements without invoking Boltzmann's distribution to describe the partitioning of species into the various excited-states. A single excited-state number density can be probed optically using the intensity-density relationship defined in equation 9.1 (see Section 9.4.2). Therefore, one can modify Saha's equation (equation 9.4) by coupling it with the Boltzmann's distribution, so that both the atomic and ionic densities are expressed in terms of individual atomic (j) or ionic (k) excited state populations, which are experimentally obtainable. Further, these excited state populations can be replaced by emission intensities. This form of the coupled Saha-Boltzmann equation, equation 9.5, can be used to determine the plasma's LTE electron density and ionization (electron) temperature [2,183,187,189,195,226].

$$\frac{I_{ion,k}}{I_{atom,j}} = \frac{A_{ion,ki}\, g_{ion,k}\, \lambda_{atom}}{A_{atom,ji}\, g_{atom,j}\, \lambda_{ion}} \frac{2(2\pi m_e kT)^{3/2}}{N_e h^3} e^{-(\chi + E_{ion,k} - E_{atom,j})/kT} \tag{9.5}$$

To employ equation 9.5, either the plasma's electron density or temperature is required for most measurements [114,183,187,189,191, 195,222,226]. Typically, an independent electron-density measurement is made using a method such as H_β Stark broadening (Section 9.4.5.2) [183,189,191,195,222,224]. This electron density measurement coupled with a measurement of the intensity ratio of a particular ion- and atomic-emission line pair is used in equation 9.5 to calculate the ionization temperature (T). Likewise, a similar method is used when electron density information is desired using the electron (ionization) temperature as the independently measured quantity. However, one can simultaneously obtain both the ionization temperature and electron density from a coupled Saha-Boltzmann analysis by probing the behavior of several ion/atomic emission line ratios. This information is obtained from the linearized form of equation 9.5, explicitly stated in equation 9.6.

$$\ln\left[\frac{I_{ion,k}\, g_{atom,j}\, A_{atom,ji}\, \lambda_{ion}}{I_{atom,j}\, g_{ion,k}\, A_{ion,ki}\, \lambda_{atom}}\right] = -\left[\frac{\chi + E_{ion,k} - E_{atom,j}}{kT}\right] + \ln\left[\frac{2\,(2\pi m_e k T)^{3/2}}{N_e h^3}\right] \tag{9.6}$$

By plotting the ratio of several ionic- and atomic-emission line combinations as a function of their energy difference, the result, if in LTE, will be a line with a slope proportional to the electron (ionization) temperature and with electron density determined from the intercept. One can cover a substantial energy range with a measurement of just a few atomic and ionic emission lines [147]. Due to this large energy coverage, as illustrated in Figure 9.7, the temperature measurements made from the slope are less sensitive to model inaccuracies. Over the large energy range, disequilibrium in the form of nonlinearity can be readily observed. Figure 9.7A is a Boltzmann plot constructed from the observable Ar(I) emission lines present. Likewise, Figure 9.7B is a Boltzmann plot constructed from Ar(II) emission. Figure 9.7C is an illustration of the combined Saha-Boltzmann intensity ratio technique (equation 9.6) constructed from the Ar(I) and Ar(II) emission lines used in Figures 9.7A and 9.7B. However, uncertainties in the transition probabilities can become a significant factor, especially since ionic transition probabilities typically have uncertainties of $\pm 25\%$ or greater depending on the particular element being probed. This method is also highly susceptible to systematic errors and source disequilibrium. The intercept at this stage has not yielded reliable electron density values, but it has only been applied to the high-voltage spark discharge, where there are known problems associated with the time-integrated spark-data reduction. In this coupled Saha-Boltzmann method, LTE is a key assumption. Not only does the source have to be in excitational and collisional equilibrium, but it must also be in ionization equilibrium, unlike the uncoupled Boltzmann temperature analysis methods [114,183,207]. Consequently, not only must one be concerned with source disequilibrium, but also the experimental constraints on any such coupled Saha-Boltzmann measurement are more severe than those imposed in the Boltzmann temperature method.

In coupled Saha-Boltzmann techniques, the experimental constraints, such as large spectral bandpass and dynamic detection range, are doubly important. Since the measurement of at least two emission lines is necessary, single-channel detectors such as PMTs suffer from lack of adequate bandpass and cannot be used in non-steady-state,

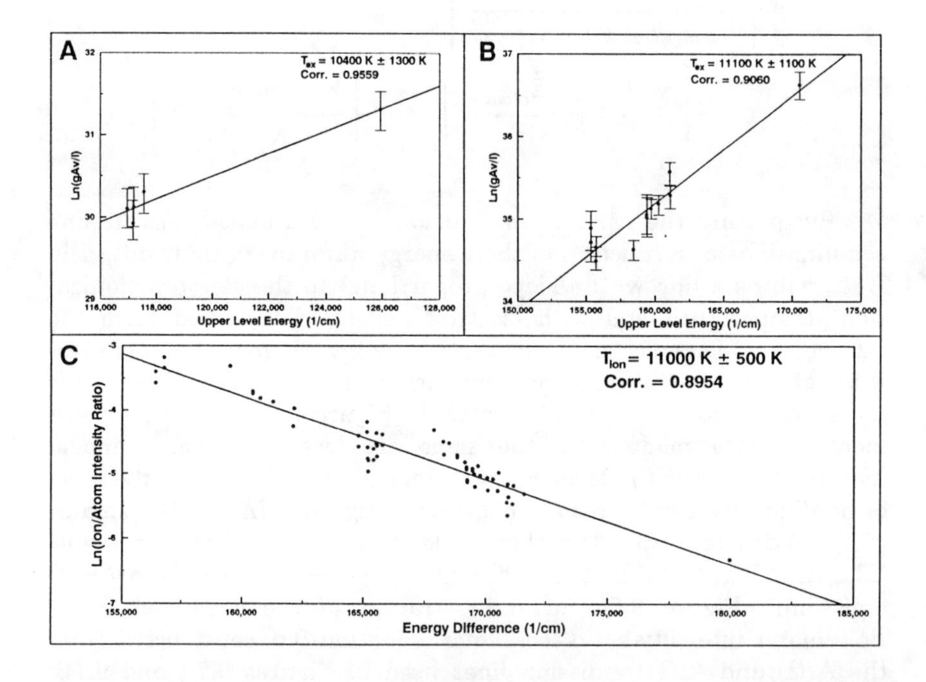

Figure 9.7 The Boltzmann plots (**A** and **B**) and the Saha-Boltzmann plot (**C**) from echelle spark emission measurements made on the discharge axis and in the cathode (pure Cu) space charge region.

irreproducible, or transient systems. For a sufficiently narrow wavelength window (roughly 5–25 nm), multichannel detectors (CTDs, vidicons, PDAs, MCPs, photographic emulsions, etc.) can be used in the Figure 9.1A configuration with success [46,114,147,191,222,226]. However, in some systems, ionic and atomic emission spectra differ too greatly, and as a consequence, both an ionic and an atomic emission transition may not be simultaneously observable within the limited accessible wavelength coverage. The dual dispersive configuration of Figure 9.1C offers several advantages over the single-dispersion systems. This experimental configuration is capable of resolving a wide wavelength range (~200 nm), necessary for the special dynamic range limitations that exist in such coupled Saha studies.

Not only does adequate spectral range coverage become a problem in such studies, but dynamic range of the detector also becomes important due to the potentially large concentration differences between successive ionization stages. Also, depending on the temperature of the source, upper energy levels in both stages might become increasingly difficult to probe since their populations will be relatively low in comparison with the lower-lying levels. The CTD-based systems have a distinct advantage over many of the other multichannel detectors due to their large dynamic detection range [19,51,62,77–80, 129,157,158,166]. Only the photoemissive-based detectors exceed the CTD in this respect, but they suffer from nonlinearity of response, geometrical instabilities, and relatively poor quantum efficiencies in comparison with CTDs [19,51,62,84,87]. Due to the limitations associated with both the detector and the plasma system, both single (Figure 9.1A) and dual (Figure 9.1C) dispersive CTD-based systems are well suited for intensity-ratio Saha-Boltzmann measurements, especially CID-based systems with increased dynamic detection range using random-access read techniques. Echelle/CID-based systems are ideal for measuring multiple intensity ratios and for dealing with spectrally difficult systems [78,129,157,158,166]. However, due to the increased experimental difficulty and equilibrium constraints (LTE), coupled Saha-Boltzmann techniques have been used only as a means of obtaining estimates, auxiliary information, or consistency checks [114,187,189,222].

9.4.4 Molecular Temperature Measurements

In addition to the coupled Saha-Boltzmann techniques, molecular temperature measurements round out the wide variety of Boltzmann excitation-temperature characterization techniques. In molecular Boltzmann studies, molecular band emission is used to construct the Boltzmann plot. Since the energy difference or separation of the individual *ro*-vibronic transitions is small relative to the difference in the atomic case, molecular temperature measurements are applicable to relatively cool plasmas, with temperatures of 5000 K or less [2,180,227]. Such plasmas include the increasingly important PECVD plasmas, which rely on these temperatures to perform some energetic chemical reactions; the temperatures cannot be so high as to completely fragment the molecular gases in the plasma. These plasmas have been well characterized in the literature and show significant variance from source to source [35,164,228–235]. This section will briefly explain the theory and assumptions implicit in temperature measurements exploiting the characteristic richness of the molecular emission in such systems and how CTDs can be used to enhance these measurements.

Just as in the atomic case (section 9.4.2), the population of the various rotational, vibrational, and electronic levels in a molecule can be described by Boltzmann's distribution of excited states, if the plasma source is in LTE or complete TE. Likewise, the number densities of these excited molecular states can be measured directly by monitoring the spectral intensity profile of a particular transition in the molecular band or a series of transitions if relative measurements are to be made [2,180,227,235]. Since measurement of actual population densities from intensity measurements requires an absolute intensity calibration and a set of reasonably accurate molecular transition probabilities, such measurements are difficult [2,22,114,180, 207–210,227,235]. (See section 9.4.2.2 for details.) In general, relative measurement techniques such as the Boltzmann plot are employed for temperature determination. Unlike the atomic case, three different processes (rotation, vibration, and electronic transition) must be accounted for when computing the molecular energy level and the intensity of a particular transition. Typically the Born-Oppenheimer approximation is made. This states that the three processes may be considered independently when computing the transition probability. In turn, the overall transition probability $A_{J'J''v'v''mn}$ can be factored into three independent terms $A_{J'J''}$, $A_{v'v''}$, and A_{mn}. While any particular transition can have only a specific transition probability, the three independent Einstein coefficients can be applied to families of transitions within the same band or manifold. The intensity of such a transition can represented with equation 9.7 [2,180,227].

$$I_{J'v'm} = \frac{C}{4\pi} A_{J'J''} A_{v'v''} A_{mn} h \nu^4 N_{J'v'm} l \quad \frac{W}{m^2\ srad} \tag{9.7}$$

This intensity relationship can be coupled with the Boltzmann distribution in several ways depending on the type of temperature desired and the spectral information available. If one is interested in measuring the rotational temperature of the emitting molecule, then only the spectral intensity distribution of a single band is required for the measurement. In addition, only the molecular rotational transition probabilities will factor into the linearized form of the molecular Boltzmann equation (equation 9.8).

$$\ln\left[\frac{\nu^4 A_{J'J''}(2J' + 1)}{I_{J'v'm}}\right] = \frac{B_v J'(J' + 1)}{kT_r} + constant \quad (9.8)$$

Here, as in Section 9.4.2, the rotational temperature (T_r) can be determined from the slope after adjusting for both the Boltzmann constant (k) and the rotational constant of the molecule (B_v) [2,180, 227,236,237]. Unlike the atomic case, the frequency dependence appears to be greater in the molecular formulation; however, the molecular and atomic cases have the same frequency dependence despite the explicit ν^4 term that appears in the molecular formulation. Due to the manner in which molecular constants are evaluated, it is a matter of convenience and convention to separate the frequency component instead of embedding it in the constants [2,180,227]. Again, unlike the atomic case, the energy is calculated based on stepping through the various components (spectral transitions) of the molecular emission band, since the bands are separated by roughly equivalent energy steps (though there is some variation in separation due to the dependence of the rotation constant B_v on vibrational quantum number v). Note that this form of the molecular Boltzmann equation deals only with rotational structure of the band, while the vibrational and electronic transition is held constant (i.e., same vibrational v′ → v″ level transition and m → n electronic transition). Therefore, the vibrational and electronic transition probabilities are embedded in the constant term of equation 9.8, which greatly simplifies the analysis.

This type of molecular Boltzmann analysis is well suited for high-resolution, small-bandpass experimental configurations, since one must be able to resolve the individual *ro*-vibronic lines within the band, typically requiring a resolution of 0.5 Å or better. If the vibrational temperature (T_v) is desired, then one must significantly increase the bandpass of the resolving system to probe other upper-level vibrational quantum numbers. Since there is coupling of the rotational and vibrational transitions, one must account for the rotational contribution to the vibronic spectrum. This vibrational form of the molecular Boltzmann plot is described by equation 9.9 [180,227].

$$\ln\left[\frac{\nu^4 g A_{J'J''} A_{v'v''}}{I_{J'v'm}}\right] + \frac{E_{J'}}{kT_r} = -\frac{E_{v'}}{kT_v} + constant \qquad (9.9)$$

An additional rotational analysis of one of the bands is necessary to determine T_r for substitution into equation 9.9 [180,227]. Again, as in similar Boltzmann methods, the slope yields the desired temperature (T_v) information. Both of these methods rely on the relative accuracy of the various transition probabilities being used. However, theoretical calculations can yield a consistent set of relative transition probabilities with better precision than can be obtained for most atomic systems [2]. In addition to Boltzmann-plot temperature determinations, equivalent temperature measurements can be made by fitting the experimental band-emission-intensity profile against the temperature-dependent synthesized-band spectrum with reasonable accuracy [227,235]. All temperature determinations based on these methods are ideally suited to the CTD-based detection schemes of Figure 9.1A and 9.1C, discussed in section 9.3.

Not only does one have to be concerned with adequate wavelength coverage; now both spectral resolution and dynamic detection range can also be significant factors when making molecular Boltzmann temperature measurements. CTDs are especially suited to these measurements due to their sensitivity (in some plasmas, molecular emission is extremely weak), dynamic range (the dynamic detection range is less of a constraint in molecular systems than in the atomic case), geometric stability (required for accurate wavelength calibration), and large information bandwidth (since molecular bands contain a multitude of lines due to the coupling of both the rotational and vibrational levels). Due to the high spectral resolution constraint and relatively small-wavelength window coverage, rotational temperature measurements based on a single molecular band are well suited to the experimental configurations illustrated in Figures 9.1A and 9.2. However, for vibrational temperature measurements, the narrow bandpass of such high-resolution imaging systems is not sufficient for the observation of more than two molecular emission bands for most systems. For these investigations, the dual dispersive CTD-based systems (Figures 9.1C and 9.3) are better [160,164]. In general, CTD-detection-based experimental schemes have proven to be quite useful in all of the Boltzmann-based temperature-diagnostic techniques due to their unique qualities. Likewise, these same qualities also make them quite advantageous for use in many of the line-shape plasma-diagnostic techniques discussed in the next section.

9.4.5 LineShape Diagnostic Techniques

In contrast with the Boltzmann techniques, lineshape-based diagnostic techniques are concerned with the structure of the emission profile and how plasma conditions modify that structure. Several plasma diagnostic techniques exploit the fact that the plasma alters the lineshape of the spectral emission lines. The predominant alteration is line broadening, although shifting also occurs. Spectral emission lines (atomic or molecular) can be broadened by several factors. Line broadening in plasmas can arise from thermal motion, collisions, and electric field interactions. Broadening due to thermal motion (Doppler) can be used to measure the translational temperature of the species being probed [2,21,180,181]. Likewise, the line broadening that arises due to elastic collisions can be modeled and (electron) density information can be extracted; however, due to the charged nature of a plasma, electrostatic repulsive and attractive forces must be accounted for in the model. Broadening due to both the collisional and electrostatic plasma environment is called Stark broadening [2,21,45,180,181]. Stark broadening measurement of electron density is the most common lineshape-based plasma diagnostic technique [20,115,187,238–273]. These two forms (Doppler and Stark) of spectral line broadening represent the primary mechanisms for spectral line broadening in plasma sources [2,21,180,181]. Both have found utility as nonperturbative probes of plasma behavior. Therefore, the following sections are devoted to how plasma information can be obtained by exploiting the line broadening that is naturally present in the energetic plasma environment; several other lineshape-based diagnostic techniques are discussed in Section 9.4.8.

9.4.5.1 Doppler-Broadening Techniques

The Doppler effect is simply the apparent shift in source wavelength as the source moves toward or away from a point of reference or "observer." In a plasma, emitters have a wide range of thermal energies, which results in the smearing of the shift. So what one observes is a symmetric broadening of the spectral line that depends only on the translational kinetic energy of the emitter [2,21,45,180, 181]. Doppler broadening gives rise to a Gaussian line shape. Temperature information is extracted by measuring the full-width at half maximum (FWHM, $2\delta\nu_D$) of the spectral line profile according to equation 9.10 [2,21,180,181].

$$\delta\nu_D = \frac{2\nu_o}{c}\sqrt{2kT\ln 2/m} \qquad (9.10)$$

The mass emitter (m) and central frequency (ν_0) are also required.

CTDs are ideally suited for lineshape studies such as this due to their multichannel nature. The entire spectral emission profile can be captured using a CTD oriented in the configurations in Figure 9.1A and 9.2. The only major complication to the Doppler-broadening temperature measurement is the requirement of precision measurement of narrow emission lines. For an argon plasma with a translational temperature of 10,000 K, the FWHM in the visible can range from roughly 0.04 to 0.09 Å, requiring high resolution for accurate measurement. In addition, linearity of detector response across the profile is also required. Again, CTDs outperform the other types of multichannel detectors.

For a single line, the wavelength range is sufficiently small that intensity calibration (see Section 9.4.2.2) is needed only to account for small pixel-to-pixel differences in quantum efficiency; however, such corrections will serve to degrade the SNR of the experimental data due to the causes outlined in Section 9.4.2.2. Again, CTDs have consistently better SNRs in comparison with the TV-type detectors such as vidicons [19,51,55,62]. If a sufficient number of data points of the line profile are obtained, then the CTD system's digital output can be conveniently fit to the Gaussian Doppler profile. Depending on accuracy of measurement, the familiar triangle approach may also be used for FWHM determination. One must be particularly careful in both of these approaches to account for the instrumental broadening due both to the discrete sampling effects of CTD detection and the dispersive optics [274–276]. The discrete sampling or diode registry problem can be partially accounted for using Fourier transform techniques [274–278]. In general, such Doppler-broadening investigations are feasible for low atomic or molecular weight species in high-temperature, low-electron-density plasmas. As the electron density increases, Stark broadening becomes significant and likewise the spectral profile becomes more difficult to analyze due to the additional broadening phenomena [2,21,180].

9.4.5.2 Stark-Broadening Techniques

In contrast to Doppler broadening, Stark broadening is more prevalent at high electron densities and has only a weak temperature dependence [2,20,21,180,181]. It is the smearing of the Stark effect due to the distribution of the local electric fields, which can be quite large in a dense plasma [2,20,177]. The degree of the line broadening is indicative of the local concentration (density) of charged perturbers. Typically, Stark broadening is due to the electrons, since their thermal velocities and collision frequencies are greater than that of the ions [2,20]. However, the actual broadening observed will be a convolution of the electrostatic, collisional impacts of both the electrons and ions

[2,20,21,240,241,249,250,262–266,278–281]. Consequently collisional and electrostatic dynamics of both species must be included in the theoretical model or be accounted for independently [20,21,239,249, 251,262,263,272,273,281–285]. The result of such models is that the FWHM of the spectral profile (in absence of other types of broadening, such as physical and observational) is roughly proportional* to the plasma's electron density [2,20,21,180,181]. In addition, the various atoms (energy levels) respond differently to the electrostatic interactions such that there are two distinct types of Stark broadening, namely linear and quadratic. Of these two, the linear effect gives rise to a greater broadening of the spectral line than the quadratic. Consequently Stark-broadening electron-density studies are conducted using spectral emission lines that are susceptible to linear Stark broadening; of these the Balmer H_β (4861.332 Å $4d^2D_{3/2} \rightarrow 2p^2P^o_{1/2}$) emission line is the most common [115,187,238–264].

With the great interest in H_β, numerous theoretical H_β Stark-broadened line profiles have been tabulated for a wide range of electron densities and temperatures [20,239,242,249]. By measuring the plasma's H_β emission profile and using a method of comparative fitting with the theoretical profile database, one can determine the plasma electron density with a relatively (with respect to other available electron density measurement techniques) high degree of accuracy ($\sim$5%) [2,20,21,240,249]. In addition, a rough estimate of the plasma's electron temperature can also be made, due to the weak temperature dependence of the Stark phenomenon. In contrast to the Doppler profile, the typical H_β profile can have a FWHM on the order of a few nanometers for atmospheric pressure plasmas, which eases the constraints on the fitting and measurement processes [147,240,250, 255,257,263,264,286].

As was the case for Doppler measurements, CTDs are ideally suited for Stark measurements, due to similar considerations. However, unlike the Doppler measurements, the breadth of the Stark H_β profile makes intensity calibration necessary due to the relatively large wavelength coverage and the variations in quantum efficiency over that range. With adequate spectral resolution, which is controlled both by pixel size and the reciprocal linear dispersion of the spectrometer, a large number of profile points can be measured so the discrete sampling effects of the CTD become less of an issue [147,274–276]. With adequate dispersion, both Doppler and instrumental broadening

---

*For the linear Stark broadening of hydrogen lines, the proportionality is FWHM = $C(Ne, T)\, N_e^{2/3}$. For quadratic Stark broadening, the direct proportionality is observed in the absence of significant ion broadening.

can be neglected when observing high-density plasmas such as the high voltage spark and most of the laser-produced plasmas [147,240, 249,260,263,264]. For such investigations, a detector with multichannel capability, high sensitivity, response linearity, and large dynamic detection range is desirable.

In these plasma investigations, CTDs have outperformed their single-channel and multichannel vidicon counterparts [147,240, 247–250,255,256,260–264,266,269,286]. For example, consider the results presented in Table 9.3 and Figure 9.8. Figure 9.8 shows the results of three H_β Stark-broadening investigations for the characterization of a unidirectional high-voltage-spark discharge: a CCD (A), a photographic emulsion (B), and a PMT (C), with the data in A intensity normalized. The three different detectors were used in conjunction with the same dispersion and collection optical train for the observation of the Stark-broadened H_β emission in a 4% hydrogen-doped, analytical spark discharge [147,255].

As can be seen in Table 9.3, the multichannel capability of both the CCD and photographic emulsion drastically reduces the actual experimental data-acquisition time. Comparable spatial coverage obtained using the PMT would take roughly 64 times longer than the photographic emulsion measurements and 140 times longer than the CCD measurements. In addition, the CCD's multidimensionality and geometrical stability allow one to characterize simultaneously the entire spatial extent of the discharge with a high degree of integrity, whereas the single-channel PMT integrates over the exit slit height. If the slit height is varied, the spatial resolution of the PMT can be increased with the sacrifice of significantly longer data-acquisition time. The CCD's SNR also is better than the PMT's, though less than the read-noise limit due to experimental constraints. From the results tabulated in Table 9.3 and graphically presented in Figure 9.8, one can see that the CCD clearly outperforms its counterparts in all categories, especially in useful spectral resolution.

In addition, when compared with a vidicon used for characterizing a laser-produced Ar plasma, the CCD offers better SNR and requires less data processing. Similar vidicon investigations on a laser-produced Ar plasma required extensive image processing due to irregularities in the vidicon response [248,286]. For the CCD investigation, variation in quantum efficiency across the region can be accounted for using a linear approximation and proper scaling factors with sufficient accuracy for spectral fitting purposes, removing the need for intensity calibration [147]. With the excellent SNR, sensitivity, spatial resolution, response linearity, and dynamic range, spectral lineshape analysis, such as H_β Stark broadening, becomes a point-and-shoot operation. The major difficulty in such investigations becomes one of

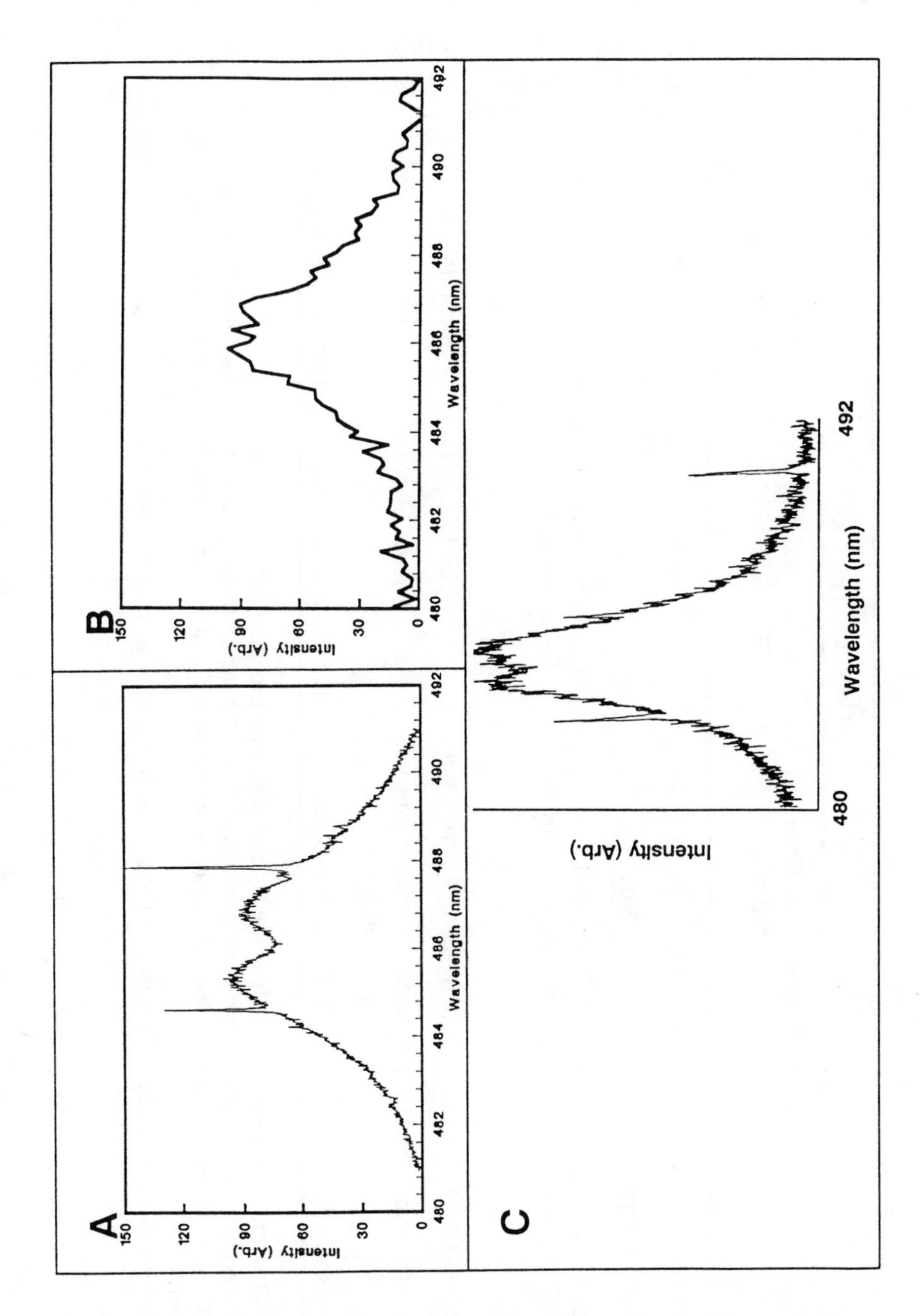

Figure 9.8 Direct detector comparison for observation of time-resolved Stark broadened H_β emission in the high voltage spark discharge.

Table 9.3 Comparison of Detector Performance in H_β Stark-broadening Spark Studies

Characteristics	Detector Type		
	Photographic Emulsion (103-aD)	PMT (EMI 9781R)	CCD (Thomson 7882) 384x576
Data acquisition time	~1 to 10 min	16 min	10 min[a]
Response linearity	No	Yes	Yes
Detector sensitivity (480-490 nm)	(~35% variation) Depends on fluence	QE ≈ 12% ±0.5% Sensitive to RFI	QE ≈ 25% ±1%
S/N	6	12	18[b]
Data-processing time	45 min. dev., 20 min. cal.	Seconds	Seconds
Wavelength resolution	1.5 Å densitometry interval; 0.2 Å instrumental	0.6 Å	0.05 Å
Spatial resolution	150 μm × 50 μm	50 μm × 2 mm	50 μm × 23 μm[c]

[a] Two CCD frames were used to capture the H_β profile (5 min exposure time for each).
[b] Due to the weakness of the H_β emission and experimental timing stability, only 10% of the full-well capacity of the CCD was utilized. Therefore the S/N could be as high as 53 (shot-noise limit) if the entire full-well capacity is utilized.
[c] The spectrometer slit width of 50 μm, not the pixel size, sets the horizontal resolution.

what model to use for data analysis and whether to consider ion broadening, Doppler, and/or instrumental effects in that analysis.

In addition to the linear H_β Stark measurements previously mentioned, quadratic Stark broadening can be used to measure plasma electron density when doping of the plasma with hydrogen is not viable or is too perturbative [1,20,21,187, 191,265]. In contrast to linear Stark broadening, the quadratic variant asymmetrically splits the energy levels, which gives rise to a symmetrically broadened (Lorentzian), shifted (usually red-shifted) peak [2,20,21,180,181,259]. As in the linear case, several such quadratic Stark theoretical profiles have been evaluated for a number of elements and excitation levels. However, the quadratic theoretical database is much less reliable than that of H_β, with a typical accuracy of 20% or worse [20,21,238,261]. In addition, the asymmetry in the profile introduced by ion broadening is more severe in the quadratic case and complicates the electron-density determination. Also, the extent of broadening is on the order of a few angstroms in contrast to the few nanometers of the linear H_β effect [147,187,238,240,259,261,267,279,280]. Despite these limitations, quadratic Stark broadening of Ar(I) lines in the near infrared (8115 Å and 8103 Å) have been successfully used for electron-density characterization studies, showing excellent agreement with H_β measurements [147,238]. The increased quantum efficiency (as high as ~90%) of CTDs in the red region of the spectrum has also been exploited in these quadratic Stark-broadening Ar(I) CCD studies [147]. In these investigations, the same arguments as in the H_β measurements can be used to expound on the merits of CTD detection for such lineshape studies.

9.4.6 Thomson Scattering Methods

Unlike the Boltzmann and line-broadening techniques, the Thomson-scattering technique exploits the interaction of the plasma's electron cloud with a high-intensity radiation field provided by a laser [23,24, 287]. The laser photons are scattered from the free electrons in the plasma, and the electromagnetic oscillations of the laser light serve to accelerate the electrons in the plasma [23, 24,288]. The scattered spectrum is broadened due to the thermal distribution that exists in the plasma. The degree of this effect is dependent *only* on the plasma electron temperature [23,24,287,289]. In addition, the electron velocity distribution can also be determined by analyzing the scattered profile without any equilibrium assumptions, unlike the Boltzmann measurements [24,287]. Likewise, the number of scattered photons observed is a function only of the plasma electron density. As a consequence of this orthogonality, Thomson scattering is an excellent means for simultaneously measuring both the plasma's electron density and temperature [23,24,287,290,291]. As in most scattering measurements,

Thomson scattering is a fairly improbable event, so high photon fluxes are required to produce even weak signals.

Two types of Thomson scattering, coherent and incoherent, can be used for plasma N_e and T_e measurements [23,24,287]. One can effectively probe both types of scattering, coherent and incoherent, by selectively changing the scattering observation angle [24,287,288, 292–296]. In incoherent scattering, the Thomson event appears as a weak broad feature centered about the Rayleigh peak [23,24,287–289, 292,297,298]. For coherent scattering, one observes two weak satellite peaks off the Rayleigh scattered line [23,24,287,288,292,294,299]. Due to these differences in the scattering peak structure, the two types of Thomson scattering pose differing experimental problems and require different data analysis.

When observing coherent scattering, one will be looking for two weak satellite peaks in the presence of a huge Rayleigh peak signal. Here, the CID is likely to be better suited for the measurement due to its random-access read capabilities, thereby enabling the detection of weak signals in the presence of the strong Rayleigh background [24,287,300]. This Rayleigh factor in turn can be somewhat avoided by choosing an observational bandpass sufficiently off the Rayleigh line [287]. Now only the issue of the scattered light (due to scattering of the Rayleigh line by the grating) in the spectrometer comes into play. By using a triple monochromator with a subtractive dispersion section in the Figure 9.1A configuration, the stray-light background from the grating scatter can be minimized. The echelle configuration in Figure 9.1C will not be suitable for these scattering measurements due to grating scatter and poor throughput [159,160]. Other techniques, such as those outlined in Chapter 7 dealing with the application of CTD for characterizing Raman systems, can be used. Wavelength resolution is essential for these types of measurements, so again the configuration in Figure 9.1A is well suited. In addition to the intense Rayleigh line, the background of the plasma continuum emission is another complication [24,287,289,291,300–302]. This can be the major limiting factor; the plasma continuum may swamp the scattered signal, so that detector dynamic reserve and SNR are reduced [24,287,289,291,300, 301]. Once again, these are areas in which the CTD offers advantages.

In contrast to the coherent scattering measurements that are akin to Raman measurements, incoherent Thomson scattering shows no external peak structure, posing a much different set of problems. Since the Thomson-scattered feature is superimposed about the Rayleigh peak, subtraction of the Rayleigh lineshape is necessary for proper integration of the Thomson-scattered peak [287,289,291,298]. Another complication arises due to the necessity for an absolute intensity calibration [24,290,291,303]. Here the calibration is not straightforward since the scattering measurements are dependent on polarized light. Accurate polarization calibration of the plasma continuum is extremely difficult to obtain [24,290]. Problems associated with absolute intensity calibration can be circumvented by calibrating from the Rayleigh peak [287,291,298,304–308]. The weakness of the broad Thomson scattering, due to the small numbers of electrons and their small scattering cross section, leads to the question of whether the scattered signal is detectable above the plasma continuum [24,287,301]. Because this profile is not discrete, relative intensity calibration is necessary. Also, the plasma's continuum emission may serve to degrade the SNR by swamping the weak scattered signal [309]. Therefore, as in the coherent measurements, CTDs offer detector sensitivity and excellent SNR for making these measurements. In fact, CTDs outperform their photoemissive-based counterparts in this respect and are likely to find utility in making the difficult incoherent Thomson-scattering measurements [287,291,295, 305,310].

9.4.7 Continuum-Based Measurements

In contrast to Thomson measurements in which the plasma continuum emission is a hindrance, continuum-based plasma characterization techniques use this phenomenon for plasma temperature and density determinations. There are a variety of continuum-based plasma-diagnostic techniques [2,180,191,311,312]. Of these, the blackbody-comparison, absolute-continuum-intensity, and continuum-to-line-emission-ratio methods are the most commonly employed [191, 311–315]. The blackbody-continuum investigations are generally conducted in the far red to infrared portions of the spectrum, where line emission is less frequent and equilibrium conditions are better met [311]. In this technique, the plasma's blackbody radiation is measured over a large bandpass and is fit with Planck's blackbody radiation curve for temperature determination. The accuracy of such determinations is suspect and depends on the assumptions regarding the origin of the continuum emission used in the temperature calculation [311,315]. This comparative fitting method requires only that a relative intensity calibration be performed, if a sufficient experimental bandpass is employed. Greater accuracy requires an absolute intensity calibration [311,313,315]. Since these investigations are typically performed in the red portion of the spectrum, the CTD's excellent red sensitivity and multidimensional format makes it ideally suited for such measurements. In the comparative fitting technique, a large wavelength bandpass is desirable, so an echelle/CTD (Figure 9.1C) is ideal; however, to the authors' knowledge, no such observations using CTDs have been made. In addition to interest in emission in the red portion of the spectrum, continuum emission in the extreme ultraviolet (~100-nm range) is particularly useful in characterizing argon plasmas due to the optically thick nature of the Ar(I) and Ar(II) resonances [70]. Again, the CTD's response in this region of the spectrum makes it desirable for such investigations. In general, these blackbody and ultraviolet-based continuum measurements are less preferred than the continuum-ratio measurements [188,311,312,314].

9.4.8 Other Plasma Diagnostic Techniques

A number of other plasma diagnostic techniques could benefit from CTD detection. As examples, both line reversal and interferometric techniques deserve mention. In the line reversal method, one measures the line profile of an optically thin plasma emission line under normal conditions and under the illumination of either an intense blackbody or laser source [2,316–318]. Under blackbody illumination, the ratio of these two integrated line-profile measurements is exponentially dependent on the electron temperature [2,316–318]. In the case of laser illumination, the relationship between the ratio and electron temperature becomes increasingly complex [317]. Since the line profile is to be measured, a relative intensity calibration is generally required. As in the lineshape section of this chapter (9.4.5), the experimental configuration in Figure 9.1A is optimal for this type of measurement. In addition, arguments and methods outlined in that section apply here. Unlike the Stark- and Doppler-broadening methods, CTD detection has not, to the authors' knowledge, been explored in line reversal investigations. Most of these measurements have employed PMTs and are infrequently used due to doubts about accuracy and potential problems in high-temperature sources such as the ICP and high-voltage spark [2,316–318]. The multichannel format, sensitivity (especially in the red), geometrical stability, excellent SNR, and dynamic detection range of the CTD could be exploited for these types of measurements.

In contrast to all of the methods previously discussed, interferometer-based techniques are concerned with fringe patterns and not line structure. Multichannel detectors have been successfully employed to measure electron density and Doppler shifts using interferometric observation techniques [315,319–321]. The electron density measurement is made based on the dependence of the plasma refractive index on the electron plasma frequency, ω_p [2,288,322,323]. In these measurements, a Mach-Zehnder interferometer typically is used and the fringe shift is measured. The electron density (cm^{-3}) can then be determined from equation 9.11 [288,307,321,322].

$$\int_0^l N_e \, dl = 2.25 \times 10^{13} \frac{\delta}{\lambda}; \quad \delta = \frac{\omega_p^2 l \lambda}{2c^2} \tag{9.11}$$

Here, δ is the fringe shift that is measured directly from the experimental image. The electron density is calculated from the plasma frequency. The imaging capabilities of the system in Figure 9.1B are utilized for fringe observation. In one investigation, a pyroelectric vidicon was successfully employed for the imaging [321]; however, CTD-based detection has several advantages over the vidicon systems, since CTDs have characteristically better SNRs, larger linear dynamic ranges, and better imaging fidelity. Consequently CTDs could find applicability in this type of measurement.

In contrast to the interferometric plasma electron density measurements, CTDs have been successively employed in interferometric Doppler shift measurements [320]. In fact, a Fabry-Perot CCD combination has been used to measure the Doppler shifts in the upper atmosphere that are indicative of the wind velocities present. Here, the CTD is used in the Figure 9.1B, direct-imaging configuration and the Airy image produced by the interferometer is recorded by the array. The two-dimensional character of the CTD results in a reduction of data-acquisition and analysis time. In addition, the ability to simultaneously observe several diffraction orders enhances the SNR. In these measurements, the line profile is reconstructed from the interferogram and is used to precisely measure the shift of the central maximum of the line due to the movement of the emitter with respect to the data-acquisition apparatus. The amount of this shift is proportional to the velocity of the source. The Fabry-Perot interferometer has a greater spectral resolution than do grating spectrometers; hence, this interference-detection combination has several advantages, the main one being better spectral resolution over the conventional grating spectrometer systems [319,320].

9.5 Conclusions

With CTD detection, many plasma diagnostic techniques can effectively be employed under a wide variety of plasma conditions and experimental configurations. Compared with previous technologies, CTDs allow the simultaneously recorded wavelength range to increase significantly. As the bandwidth of spectrometer/detector combinations become larger, the amount of information available also increases. For example, the bandpass of the echelle/CCD combination used in gathering the data of Figure 9.3 allows the experimenter to use simultaneously several of the diagnostic techniques discussed in this chapter [160,164]. The quality of data obtainable on a routine basis from these combinations of detectors and dispersion systems has helped to overcome several difficult experimental problems. In addition, as larger format CTD arrays become available, higher bandpass, spatially resolved studies become feasible and may eventually replace the current high-bandpass echelle systems. With these high-bandpass systems, plasma analysis may become a point-and-shoot operation, allowing the experimenter to design the experiment after the fact through the choice of data-analysis schemes. In this situation, use of CTDs for plasma analysis becomes more of a computer and chemometrics problem than a basic science, engineering, or experimental problem.

Acknowledgments

We would like to thank Steven W. Rynders and Duane L. Miller for contributions of CCD data and to figure generation. In addition we would like to acknowledge Dean L. Olson and Duane L. Miller for their help in editing this document. Some of the results presented in this chapter were partially supported by grants from the National Science Foundation, the University of Illinois Research Board, and the Department of Energy (Office of Basic Energy Sciences).

List of Symbols

Symbol	Description	Units
A_{ij}	Transition probability	(Hz)
$A_{J'J''}$	Rotational component of molecular transition probability	
$A_{v'v''}$	Vibrational component of molecular transition probability	
A_{mn}	Electronic component of molecular transition probability	
$A_{atom,ji}$	Atomic transition probability	(Hz)
$A_{ion,ki}$	Ionic transition probability	(Hz)
B_v	Molecular rotational constant	(cm^{-1})
c	Speed of light	(2.9979×10^8 m/s)
C	Constant factor in molecular intensity relation	
δ	Interference fringe shift	(unitless)
$\delta\nu_D$	FWHM of Doppler profile	(Hz)
e	Charge of electron	(1.60219×10^{-19} C)
E_i	Energy of excited level i	(J)
E_j	Energy of excited level j	(J)
E_o	Ground state energy	
$E_{ion,k}$	Energy of ionic excited level k	(J)
$E_{atom,j}$	Energy of atomic excited level j	(J)
g	Statistical weight of molecular system	(unitless)
g_o	Statistical weight of ground state	(unitless)
g_i	Statistical weight of state i	(unitless)
g_j	Statistical weight of state j	(unitless)
$g_{atom,j}$	Statistical weight of atomic excited state j	(unitless)
$g_{ion,k}$	Statistical weight of ionic excited state k	(unitless)

Symbol	Description	Units
h	Planck's constant	(6.626176×10^{-34} Js)
I_{ij}	Intensity of transition from state i to j	($Wm^{-2}srad^{-1}$)
$I_{ion,k}$	Intensity of ionic transition from state k	($Wm^{-2}srad^{-1}$)
$I_{atom,j}$	Intensity of atomic transition from state j	($Wm^{-2}srad^{-1}$)
$I_{J'v'm}$	Intensity of molecular transition	($Wm^{-2}srad^{-1}$)
J	Total angular momentum quantum number	(unitless)
J'	Rotational quantum number	(unitless)
k	Boltzmann's constant	(1.3805×10^{23} J/K)
λ	Generic wavelength symbol	(m)
λ_{ij}	Wavelength of transition from excited i to j	(m)
λ_{atom}	Wavelength of atomic transition	(m)
λ_{ion}	Wavelength of ionic transition	(m)
l	Plasma thickness	(m)
m	Emitter mass	(kg)
m_e	Mass of electron	(9.109534×10^{-31} kg)
ν_o	Central Doppler frequency	(Hz)
N_e	Electron density	(m^{-3})
N_i	Population density of excited level i	(m^{-3})
N_j	Population density of excited level j	(m^{-3})
N_o	Ground state population density	(m^{-3})
N_{atom}	Total atom number density	(m^{-3})
N_{ion}	Total ion number density	(m^{-3})
$N_{J'v'm}$	Population density of the molecular excited level J'v'm	(m^{-3})
$Qa(T)$	Atomic partition function	(unitless)
$Qi(T)$	Ionic partition function	(unitless)
$Tapp$	Apparent excitation temperature	(Kelvin)
Te	Plasma electron temperature	(Kelvin)
Tex	Boltzmann excitation temperature	(Kelvin)
Tr	Molecular rotational temperature	(Kelvin)
Tv	Molecular vibrational temperature	(Kelvin)
v'	Molecular vibration quantum number	(unitless)
χ	First ionization potential	(J)

References

1. Meek, J.M.; Craggs J.D., Eds. *Electrical Breakdown of Gases*. Wiley, New York, 1978.
2. Thorne, A.P. *Spectrophysics*. Chapman, New York, 1988.
3. Lifshitz, E.M.; Pitaevskii, L.P. *Physical Kinetics*. Pergamon Press, New York, 1981.
4. Oxenius, J. *Kinetic Theory of Particles and Photons: Theoretical Foundations of Non-LTE Plasma Spectroscopy*. Springer-Verlag, Berlin, 1986.
5. Nowak, S.; Van der Mullen, J.A.M; Van Lammeren, A.C.A.P.; Schram, D.C. *Spectrochim. Acta* **1989**, *44B,* 411.
6. Van Der Mullen, J.A.M. *Spectrochim. Acta* **1989**, *44B*, 1067.
7. Van Der Mullen, J.A.M. *Spectrochim. Acta* **1990**, *45B*, 1.
8. Van Der Mullen, J.A.M. *Spectrochim. Acta* **1990**, *45B*, 233.
9. C.S. Rann, A.N. Hambly, *Anal. Chem.* **1965**, *37*, 879.

10. Walters, J.P.; Goldstein, S.A. In *Sampling, Standards and Homogeneity*, ASTM ST P 540, American Society for Testing and Materials, 1973, p. 45.
11. Walters, J.P.; Goldstein, S.A. *Spectrochim. Acta* **1984**, *39B*, 693.
12. Olesik, J.W.; Walters, J.P. In *Multichannel Image Detectors Vol. 2*, Talmi, Y., Ed.. American Chemical Society: Washington, D.C., 1983, pp 31–56.
13. Scheeline, A. *Prog. Analyt. Atom. Spectrosc.* **1984**, *7*, 21.
14. Cousins, J.C.; Scheeline, A.; Coleman, D.M. *Appl. Spectrosc.* **1987**, *41*, 954.
15. Scheeline, A.; Coleman, D.M. *Anal. Chem.* **1987**, *59*, 1185A.
16. Kolczynski, J.D.; Pomeroy, R.S; Jalkian, R.D.; Denton, M.B. *Appl. Spectrosc.* **1989**, *43*, 887.
17. Mork, B.J.; Scheeline, A. *Spectrochim. Acta* **1989**, *44B*, 1297.
18. Boumans, P.W.M. *Inductively Coupled Plasma Emission Spectroscopy, Part 1--Methododology, Instrumentation, and Performance.* Wiley, New York, 1987.
19. Busch, K.W.; Busch, M.A. *Multielement Detection Systems for Spectrochemical Analysis*. Wiley, New York, 1990.
20. Griem, H.R. *Spectral Line Broadening by Plasmas*. Academic Press, New York, 1974.
21. Wiese, W.L. In *Plasma Diagnostic Techniques*. Huddlestone, R.H.; Leonard, S.L., Eds. Academic Press, New York, 1965, pp. 265–317.
22. Reif, I.; Fassel, V.A.; Kniseley, R.N. *Spectrochim. Acta* **1973**, *28B*, 105.
23. Sheffield, J. *Plasma Scattering of Electromagnetic Radiation*. Academic Press, New York, 1975.
24. Scheeline, A.; Zoellner, M.J. *Appl. Spectrosc.* **1984**, *38*, 245.
25. Huang, M.; Yang, P.Y.; Hanselman, D.S.; Monnig, C.A.; Hieftje, G.M. *Spectrochim. Acta* **1990**, *45B*, 511.
26. Huang, M.; Hanselman, D.S.; Jin, Q.; Hieftje, G.M. *Spectrochim. Acta* **1990**, *45B*, 1339.
27. Scheeline, A.; Walters, J.P. *Anal. Chim. Acta* **1977**, *95*, 59.
28. Scheeline, A.; Walters, J.P. In *Contemporary Topics in Analytical and Clinical Chemistry*. Hercules, D.M.; Hieftje, G.M.; Synder, L.R.; Evenson, M.A., Eds.; Plenum, New York, 1982, Vol. 4, pp. 295–372.
29. Prost, M.M. *Spectrochim. Acta* **1982**, *37B*, 541.
30. Scheeline, A.; Walters, J.P. *Anal. Chem.* **1975**, *48*, 1519.
31. Mork, B.J.; Scheeline, A. *Spectrochim. Acta* **1987**, *42B*, 1063.
32. Monnig, C.A.; Gebhart, B.D.; Marshall, K.A.; Hieftje, G.M. *Spectrochim. Acta* **1990**, *45B*, 261.
33. Or, T.W.; Cong, P.C.; Pfender, E. *Plasma Chem. Plasma Proc.* **1992**, *12*, 189.
34. Wagatsuma, K.; Hirokawa, K. *SIA Surf. Interf. Anal.* **1986**, *8*, 37.
35. Kushner, M.J. *J. Appl. Phys.* **1987**, *62*, 2803.
36. Newton, I. *Phil. Trans.* **1672**, *80 (Feb.)*, 3075.
37. Harrison, G.R.; Lord, R.C.; Loofbourow, J.R. *Practical Spectroscopy*. Prentice-Hall, Englewood Cliffs, N.J., 1948, pp, 141–165.
38. Harvey, C.E. *Spectrochemical Procedures*. Applied Research Laboratories, Sunland, Calif., 1950, pp. 47–81.
39. Nachtrieb, N.H. *Principles and Practice of Spectrochemical Analysis*. McGraw-Hill, New York, 1950, pp. 102–124.

40. Sawyer, R.A. *Experimental Spectroscopy*, 3rd ed. Dover, New York, 1963, pp. 192–214.
41. Slavin, M. *Emission Spectrochemical Analysis*. Wiley-Interscience, New York, 1971, pp. 124–152.
42. Mika, J.; Torok, T. *Analytical Emission Spectroscopy: Fundamentals*. Floyd, P.A., Trans. Crane, Russak & Company, New York, 1974, pp. 473–496.
43. Poehler, T.O. In *Physical Optics and Light Measurements, Vol. 26: Methods of Experimental Physics*. Malacara, D., Ed. Academic Press, San Diego, 1988, pp. 291–334.
44. Malmstadt, H.V.; Franklin, M.L.; Horlick, G. *Anal. Chem.* **1972**, *44(8)*, 63A.
45. Demtroder, W. *Laser Spectroscopy: Basic Concepts and Instrumentation*, corrected 3rd printing. Springer-Verlag, Berlin, 1981.
46. Blades, M.W.; Horlick, G. *Spectrochim. Acta* **1981**, *36B*, 861.
47. Rybarczyk, J.P. Ph.D. Dissertation, University of Missouri-Columbia, 1981.
48. Hieftje, G.M. *Anal. Chem.* **1972**, *44(6)*, 81A.
49. Hieftje, G.M. *Anal. Chem.* **1972**, *44(7)*, 69A.
50. Aiello, P.J.; Enke, C.G. In *Multichannel Image Detectors, Vol. 2*, Talmi, Y., Ed. American Chemical Society, Washington, D.C., 1983, pp. 57–73.
51. Busch, K.W.; Malloy, B. In *Multichannel Image Detectors, Vol. 1*, Talmi, Y., Ed. American Chemical Society, Washington, D.C., 1983, pp. 27–58.
52. Felkel Jr., H.L.; Pardue, H.L. In *Multichannel Image Detectors, Vol. 1*, Talmi, Y., Ed. American Chemical Society, Washington, D.C., 1983, pp. 59–96.
53. Catchpole, C.E. In *Photoelectronic Imaging Devices, Vol. 2*, Biberman, M.L.; Nudelman, S., Eds. Plenum Press, New York, 1971, pp. 167–190.
54. Hall, J.A. In *Photoelectronic Imaging Devices, Vol. 2*, Biberman, M.L.; Nudelman, S., Eds. Plenum Press, New York, 1971, pp. 483–514.
55. Talmi, Y. *Anal. Chem.* **1975**, *47(7)*, 658A.
56. Talmi, Y. *Anal. Chem.* **1975**, *47(7)*, 699A.
57. Weber, P.G. *Rev. Sci. Instrum.* **1983**, *54*, 1331.
58. Baker, O.J.; Steed, A.J. *Appl. Opt.* **1968**, *7*, 2190.
59. Golightly, D.W.; Kniseley, R.N.; Fassel, V.A. *Spectrochim. Acta* **1970**, *25B*, 451.
60. Daneilsson, A.; Lindblom, P. *Appl. Spectrosc.* **1976**, *30*, 151.
61. Siegmund, O.H.W.; Malina, R.F. In *Multichannel Image Detectors, Vol. 2*. Talmi, Y., Ed. American Chemical Society, Washington, D.C., 1983, pp. 256–275.
62. Talmi, Y. In *Multichannel Image Detectors, Vol. 1*. Talmi, Y., Ed. American Chemical Society, Washington, D.C., 1983, pp. 1–25.
63. Read, P.D.; Powell, J.R.; van Breda, I.G.; Lyons, A.; Ridley, N.R. *Proc. SPIE* **1986**, *627*, 645.
64. Benjamin, R.D.; Terry, J.L.; Moos, H.W. *Rev. Sci. Instrum.* **1987**, *58*, 520.
65. Timothy, J.G. *Opt. Eng.* **1985**, *24*, 1066.
66. Timothy, J.G.; Joseph, C.L.; Wolff, S.C. *Proc. SPIE* **1982**, *331*, 301.
67. Timothy, J.G. *Proc. SPIE* **1984**, *501*, 89.

68. Slater, D.C.; Timothy, J.G.; Morgan, J.S.; Kasle, D.B. *Proc. SPIE* **1990**, *1243*, 35.
69. Bashkin, S.; Stoner Jr., J.O. *Atomic Energy-Level and Grotrian Diagrams, 2: Sulfur(I) - Titanium(XXII).* North-Holland, New York, 1978.
70. Minnhagen, L. *Arkiv Fysik* **1963**, *25*, 203.
71. Striganov, A.R.; Sventitskii, N.S. *Tables of Spectral Lines of Neutral and Ionized Atoms*. Plenum, New York, 1968.
72. Stern, R.A.; Catura, R.C.; Blouke, M.M.; Winzenread, M. *Proc. SPIE* **1982**, *331*, 583.
73. Janesick, J.R.; Elliott, T.; Collins, S.; Blouke M.M.; Freeman, *J. Opt. Eng.* **1987**, *26*, 692.
74. Sims, G.R.; Griffin, F.; Lesser, M.P. *Proc. SPIE* **1989**, *1071*, 31.
75. Janesick, J.; Elliott, T.; Bredthauer, R.; Cover, J.; Schaefer, R.; Varian, R. *Proc. SPIE* **1989**, *1071*, 115.
76. Vallerga, J.; Lampton, M. *Proc. SPIE* **1987**, *868*, 25.
77. Epperson, P.M.; Sweedler, J.V.; Bilhorn, R.B.; Sims, G.R.; Denton, M.B. *Anal. Chem.* **1988**, *60*, 327A.
78. Sweedler, J.V.; Jalkian, R.D.; Pomeroy, R.S; Denton, M.B. *Spectrochim. Acta* **1989**, *44B*, 683.
79. Sweedler, J.V.; Jalkian, R.D.; Denton, M.B. *Appl. Spectrosc.* **1989**, *43*, 953.
80. Denton, M.B. *Proc. SPIE* **1990**, *1318*, 107.
81. Howell, N.G.; Morrison, G.H. *Anal. Chem.* **1977**, *49*, 106.
82. Furuta, N.; McLeod, C.W.; Haraguchi, H.; Fuwa, K. *Appl. Spectrosc.* **1980**, *4*, 211.
83. Fultz, M.L.; Durst, R.A. *Talanta* **1983**, *30*, 933.
84. Talmi, Y., Busch, K.W. In *Multichannel Image Detectors, Vol. 2*, Talmi, Y., Ed. American Chemical Society, Washington, D.C., 1983, pp. 1–29.
85. Goldberg, J.M.; Sacks, R.D. *Appl. Spectrosc.* **1983**, *37*, 531.
86. Uchida, T.; Itoh, S.; Coleman, D.M.; Minami, S. *Appl. Spectrosc.* **1990**, *44*, 391.
87. Hall, J.A. *Proc. SPIE* **1976**, *78*, 14.
88. Horlick, G.; Codding, E.G. *Appl. Spectrosc.* **1975**, *29*, 167.
89. Horlick, G. *Appl. Spectrosc.* **1976**, *30*, 113.
90. Yates, D.A.; Kuwana, T. *Anal. Chem.* **1976**, *48*, 510.
91. Chuang, F.S.; Natusch, D.F.S.; O'Keefe, K.R. *Anal. Chem.* **1978**, *50*, 525.
92. Vogt, S.S. *Proc. SPIE* **1981**, *290*, 70.
93. Bortoletto, F. *Proc. SPIE* **1983**, *290*, 90.
94. Küveler, G.; Wöhl, H. *Astron. Astrophys.* **1983**, *122*, 69.
95. Blades, M.W.; Horlick, G. *Spectrochim. Acta* **1981**, *36B*, 861.
96. Winge, R.K.; Fassel, V.A.; Edelson, M.C. *Spectrochim. Acta* **1988**, *43B*, 85.
97. Zhu, J.; Mann, C.K.; Vickers, T.J. *Appl. Spectrosc.* **1988**, *42*, 1567.
98. Brett, L.; Stahl, R.G.; Timmins, K.J. *J. Anal. Atom. Spectrom.* **1989**, *4*, 333.
99. Moulton, G.P.; O'Haver, T.C.; Harnly, J.M. *J. Anal. Atom. Spectrom.* **1989**, *4*, 673.
100. Lepla, K.C.; Horlick, G. *Appl. Spectrosc.* **1989**, *43*, 1187.
101. Brushwyler, K.R.; Furuta, N.; Hieftje, G.M. *Talanta* **1990**, *37*, 23.

102. Kendall-Tobias, M. *Am. Lab.* **1989**, *21*, 102.
103. Edmonds, T.E.; Horlick, G. *Appl. Spectrosc.* **1977**, *31*, 536.
104. Blades, M.W.; Horlick, G. *Spectrochim. Acta* **1981**, *36B*, 881.
105. Choot, E.H.; Horlick, G. *Spectrochim. Acta* **1986**, *41B*, 889.
106. Blades, M.W.; Horlick, G. *Appl. Spectrosc.* **1980**, *34*, 696.
107. Blades, M.W. *Appl. Spectrosc.* **1983**, *37*, 371.
108. Furuta, N.; Horlick, G. *Spectrochim. Acta* **1982**, *37B*, 53.
109. Choot, E.H.; Horlick, G. *Spectrochim. Acta* **1986**, *41B*, 935.
110. Caughlin, B.L.; Blades, M.W. *Spectrochim. Acta* **1984**, *39B*, 1583.
111. Caughlin, B.L.; Blades, M.W. *Spectrochim. Acta* **1985**, *40B*, 579.
112. Caughlin, B.L.; Blades, M.W. *Spectrochim. Acta* **1985**, *40B*, 1539.
113. Burton, L.L.; Blades, M.W. *Appl. Spectrosc.* **1986**, *40*, 256.
114. Walker, Z.; Blades, M.W. *Spectrochim. Acta* **1986**, *41B*, 761.
115. Chan, Shi-Kit; Montaser, A. *Spectrochim. Acta* **1989**, *44B*, 175.
116. Clary, M.C.; Klaasen, K.P.; Snyder, L.M.; Wang, P.K. *Proc. SPIE* **1979**, *203*, 98.
117. Gursky, H. *Proc. SPIE* **1981**, *290*, 2.
118. Tower, J.R.; Cope, A.D.; Pellon, L.E.; McCarthy, B.M.; Strong, R.T.; Moldovan, A.G.; Levine, P.A.; Kinnard, K.F.; Elabd, H.; Hoffman, D.M.; Kramer, W.M.; Longsderff, R.W.; Kennerly, R.E.; Calvin, W.M. *Proc. SPIE* **1985**, *570*, 172.
119. Waddell, P.; Christian, C. *Opt. Eng.* **1987**, *26*, 734.
120. Blouke, M.M.; Corrie, B.; Heidtmann, D.L.; Yang, F.H.; Winzenread, M.; Lust, M.L.; Marsh IV, H.H.; Janesick, J.R. *Opt. Eng.* **1987**, *26*, 837.
121. Tseng, H.; Nguyen, B.T.; Fattahi, M. *Proc. SPIE* **1989**, *1071*, 1705.
122. Boksenberg, A.; *Phil. Trans. R. Soc. Lond. A* **1982**, *307*, 531.
123. Schlemmer, H.H.; Mächler, M. *J. Phys. E: Sci. Instrum.* **1985**, *18*, 914.
124. Boulade, O.; Lemaitre, G.; Vigroux, L. *Proc. SPIE* **1986**, *627*, 268.
125. Boulade, O.; Biagio, B.D.; Jouan, R.; Lecomte, A.; Montiel, P.; Revest, D.; Rio, Y.; Testard, O.; Vigroux, L. *Astron. Astrophys.* **1986**, *163*, 301.
126. Mork, B.J.; Scheeline, A. *Appl. Spectrosc.* **1988**, *42*, 1332.
127. Niemczyk, T.M.; Gobeli, G.W. *Proc. SPIE* **1990**, *1318*, 33.
128. Mermet, J.M.; Carré, M.; Lemarchand, A. *Proc. SPIE* **1990**, *1318*, 88.
129. Denton, M.B. *Proc. SPIE* **1990**, *1318*, 107.
130. Savoye, E.D.; Engstrom, R.W.; Zimmerman, H.S. *Proc. SPIE* **1979**, *203*, 59.
131. Yamazaki, K.; Haba, K.; Hirokura, S.; Kawahata, K.; Noda, N.; Matsuura, K; Sato, K.; Tanahashi, S.; Taniguchi, Y.; Toi, K.; Fujita, J. *J. Nuclear Materials* **1984**, *128*, 186.
132. Balmer, J.E.; Lampert, W.; Roschger, E.; Hares, J.D.; Kilkenny, J.D. *Rev. Sci. Instrum.* **1985**, *56*, 860.
133. Baumgart, J.S. *Proc. SPIE* **1989**, *1155*, 546.
134. Lucht, H. *Rev. Sci. Instrum.* **1990**, *61*, 1849.
135. Balinenko, I.N.; Zvegintsev, Y.Y.; Ivanov, K.N.; Kuz'min, G.A.; Lozovoi, V.I.; Malyarov, A.V.; Naumov, S.K.; Pischelin, E.V.; Postovalov, V.E.; Prokhorov, A.M.; Sornakov, D.B.; Schelev, M.Y. *Proc. SPIE* **1990**, *1243*, 87.
136. Sweedler, J.V.; Denton, M.B. *Appl. Spectrosc.* **1989**, *43*, 1378.
137. Wingard, M.A.; Dessy, R.E. *Appl. Spectrosc.* **1990**, *44*, 1444.

138. Roesler, F.L.; Harlander, J. *Proc. SPIE* **1990**, *1318*, 234.
139. Conrady, A.E. *Applied Optics and Optical Design*. Dover, New York, 1957.
140. Zajac, A.; Hecht, E. *Optics*. Addison-Wesley, Reading, 1976.
141. Pedrotti, F.L.; Pedrotti, L.S. *Introduction to Optics*. Prentice-Hall, Englewood Cliffs, N.J., 1987.
142. Reader, J. *J. Opt. Soc. Am.* **1969**, *59*, 1189.
143. Walters, J.P.; Goldstein, S.A. *Spectrochim. Acta* **1976**, *31B*, 201.
144. Walters, J.P.; Goldstein, S.A. *Spectrochim. Acta* **1976**, *31B*, 295.
145. Hsu, W.-H.; Sainz, M.A.; Coleman, D.M. *Spectrochim. Acta* **1989**, *44B*, 109.
146. Pomeroy, R.S; Denton, M.B. *Appl. Spectrosc.* **1989**, *43*, 887.
147. Bye, C.A. Ph.D. Thesis, University of Illinois, Champaign-Urbana, 1993.
148. Wood, D.L.; Dargis, A.B.; Nash, D.L. *Appl. Spectrosc.* **1975**, *29*, 310.
149. Felkel Jr., H.L.; Pardue, H.L. *Anal. Chem.* **1977**, *49*, 1112.
150. Felkel Jr., H.L.; Pardue, H.L. *Anal. Chem.* **1978**, *50*, 602.
151. Felkel Jr., H.L.; Pardue, H.L. *Clin. Chem.* **1978**, *24*, 602.
152. York, D.G.; Jenkins, E.B.; Zucchino, P.; Lowrance, J.L.; Long, D.; Songaila, A. *Proc. SPIE* **1981**, *290*, 202.
153. White, R.E. *Pub. Astro. Soc. Pac.* **1984**, *96*, 488.
154. Ramsey, L.W.; Huenemoerder, D.P. *Proc. SPIE* **1986**, *627*, 282.
155. Goodrich, R.W.; Veilleux, S. *Pub. Astro. Soc. Pac.* **1988**, *100*, 1572.
156. Hsiech, C.; Petrovic, S.C.; Pardue, H.L. *Anal. Chem.* **1990**, *62*, 1983.
157. Bilhorn, R.B.; Denton, M.B. *Appl. Spectrosc.* **1989**, *43*, 1.
158. Sims, G.R.; Denton, M.B. *Talanta* **1990**, *37*, 1.
159. Scheeline, A.; Bye, C.A.; Miller, D.L.; Rynders, S.W. *Proc. SPIE* **1990**, *1318*, 44.
160. Scheeline, A.; Bye, C.A.; Miller, D.L.; Rynders, S.W.; Owen Jr., R.C. *Appl. Spectrosc.* **1991**, *45*, 334.
161. Keliher, P.N. *Research/Development* **1976**, *27(6)*, 26.
162. Lowen, E.G. In *The Echelle Story*. Bausch & Lomb, Spectrametrics Inc., Andover, Md., 1970.
163. ARL SpectraSpan V Product Literature on Direct Current Plasma Emission Spectrometers, ARL, Sunland, Calif., 1986.
164. Rynders, S.W. Ph.D. Thesis, University of Illinois, Champaign-Urbana, 1991.
165. Krupa, R.J.; Owen, R.C. United States Patent No: 4,995,721, 1991.
166. Pomeroy, R.S; Sweedler, J.V.; Denton, M.B. *Talanta* **1991**, *37*, 15.
167. Dantzler, A.A. *Appl. Opt.* **1985**, *24*, 4504.
168. Ponz, D.; Brinks, E.; D'Odorico, S. *Proc. SPIE* **1986**, *627*, 707.
169. Tody, D. *Proc. SPIE* **1986**, *627*, 733.
170. Gehren, T.; Ponz, D. *Astron. Astrophys.* **1986**, *168*, 386.
171. Horne, K. *Pub. Astro. Soc. Pac.* **1986**, *98*, 609.
172. Miller, D.L. Ph. D. Thesis, University of Illinois, Champaign-Urbana, 1994.
173. Boyd, T.J.M.; Sanderson, J.J. *Plasma Dynamics*. Barnes & Noble, New York, 1969.
174. Shohet, J.L. *The Plasma State*. Academic Press, New York, 1971.

175. Davidson, R.C. *Methods in Nonlinear Plasma Theory*. Academic Press, New York, 1972.
176. Nicholson, D.R. *Introduction to Plasma Theory*. Wiley, New York, 1983.
177. Chen, F.F. *Introduction to Plasma Physics and Controlled Fusion, Vol. 1: Plasma Physics*, 2nd Ed. Plenum Press, New York, 1985.
178. Krall, N.A.; Trivelpiece, A.W. *Principles of Plasma Physics*. San Francisco Press, San Francisco, 1986.
179. McWhirter, R.W.P. In *Plasma Diagnostic Techniques*, Huddlestone, R.H.; Leonard, S.L., Eds. Academic Press, New York, 1965, pp. 201–264.
180. Marr, G.V. *Plasma Spectroscopy*. Elsevier, New York, 1968.
181. Donskoi, A.V.; Goldfarb, V.M.; Klubnikin, V.S. *Physics and Technology of Low-Temperature Plasmas*, Dresvin, S.V. (Russian); Eckert, H.V. (English), Eds.; Cheron, T., Trans. Iowa State University Press, Ames, 1977.
182. Griem, H.R. *Phys. Rev.* **1963**, *131*, 1170.
183. Blades, M.W.; Caughlin, B.L.; Walker, Z.H.; Burton, L.L. *Prog. Analyt. Spectrosc.* **1987**, *10*, 57.
184. Richter, J. *Z. Astrophysik* **1965**, *61*, 57.
185. Boumans, P.M.J.M. *Theory of Spectrochemical Excitation*. Adam Hilger, London, 1966.
186. Weisbach, M.F. *J. Quant. Spectrosc. Radiat. Transfer* **1971**, *11*, 1225.
187. Kalnicky, D.J.; Fassel, V.A.; Kniseley, R.N. *Appl. Spectrosc.* **1977**, *31*, 137.
188. Neufeld, C.R. *J. Appl. Phys.* **1979**, *50*, 3218.
189. Alder, J.F.; Bombelka, R.M.; Kirkbright, G.F. *Spectrochim. Acta* **1980**, *35B*, 163.
190. Chien, Y.K.; Benenson, D.M. *IEEE Trans. Plasma Sci.* **1980**, *PS-8*, 411.
191. Batal, A.; Jarosz, J.; Mermet, J.M. *Spectrochim. Acta* **1982**, *37B*, 511.
192. Gunter, W.H.; Visser, K.; Zeeman, P.B. *Spectrochim. Acta* **1982**, *37B*, 571.
193. Meubus, P. *Can. J. Phys.* **1982**, *60*, 886.
194. Gunter, W.H.; Visser, K.; Zeeman, P.B. *Spectrochim. Acta* **1983**, *38B*, 949.
195. Raaijmakers, I.J.M.M.; Boumans, P.W.J.M.; Van Der Sijde, B.; Schram, D.C. *Spectrochim. Acta* **1983**, *38B*, 697.
196. Bond, D.J. *J. Quant. Spectrosc. Radiat. Transfer*. **1983**, *29*, 429.
197. Haddad, G.N.; Farmer, A.J.D. *J. Phys. D: Appl. Phys.* **1984**, *17*, 1189.
198. Blades, M.W.; Lee, N. *Spectrochim. Acta* **1984**, *39B*, 879.
199. Nick, K.-P.; Richter, J.; Helbig, V. *J. Quant. Spectrosc. Radiat. Transfer*. **1984**, *32*, 1.
200. Eddy, T.L. *J. Quant. Spectrosc. Radiat. Transfer* **1985**, *33*, 197.
201. Yang, P.; Barnes, R.M.; Mostaghimi, J.; Boulos, M.I.; *Spectrochim. Acta* **1989**, *44B*, 657.
202. Smith, T.R.; Denton, M.B. *Appl. Spectrosc.* **1989**, *43*, 1385.
203. Bezman, R.; Faulkner, L.R. *Anal. Chem.* **1971**, *43*, 1749.
204. Michael, P.R.; Faulkner, L.R. *Anal. Chem.* **1976**, *48*, 1188.
205. Knowles Middleton, W.E.; Sanders, C.L. *J. Opt. Soc. Am.* **1951**, *41*, 419.
206. Sims, G.R.; Denton, M.B. *J. Opt. Eng.* **1987**, *26*, 1008.
207. Blades, M.W. *Spectrochim. Acta* **1982**, *37B*, 869.
208. Drawin, H.W. *High Pressure---High Temperatures* **1970**, *2*, 359.

209. Olsen, H.N. *J. Quant. Spectrosc. Radiat. Transfer* **1963**, *3*, 305.
210. Hirabayashi, A.; Nambu, Y.; Hasuo, M.; Fujimoto, T. *Phys. Rev. A* **1988**, *7*, 77.
211. Sobelman, I.I.; Vainshtein, L.A.; Yukov, E.A. *Excitation of Atoms and Broadening of Spectral Lines*. Springer, Berlin, 1981.
212. Drawin, H.W. In *Progress in Plasmas and Gas Electronics*. Rompe, R.; Steenbeck, M., Eds. Akademic Verlag, Berlin, 1974, Vol. 1.
213. Burton, L.L.; Blades, M.W. *Spectrochim. Acta* **1990**, *45B*, 139.
214. Mermet, J.M. *Spectrochim. Acta* **1975**, *30B*, 383.
215. Wilson, R. *J. Quant. Spectrosc. Radiat. Transfer* **1962**, *2*, 477.
216. Gusinow, M.A.; Gerado, J.B.; Hill, R.A. *J. Quant. Spectrosc. Radiat. Transfer* **1969**, *9*, 383.
217. Leonard, S.L. *J. Quant. Spectrosc. Radiat. Transfer* **1972**, *12*, 619.
218. Swallom, D.W.; Scholz, P.D. *J. Quant. Spectrosc. Radiat. Transfer* **1972**, *2*, 107.
219. Blades, M.W.; Hieftje, G.M. *Spectrochim. Acta* **1982**, *37B*, 191.
220. Montaser, A.; Fassel, V.A.; Larsen, G. *Appl. Spectrosc.* **1981**, *35*, 385.
221. Kornblum, G.R.; De Galan, L. *Spectrochim. Acta* **1977**, *32B*, 71.
222. Walker, Z.; Blades, M.W. *Spectrochim. Acta* **1987**, *42B*, 1077.
223. Oxenius, J. *J. Quant. Spectrosc. Radiat. Transfer* **1990**, *44*, 157.
224. Lovett, R.J. *Spectrochim. Acta* **1982**, *37B*, 969.
225. Schram, D.C.; Raajmakers, I.J.M.; Van der Sidje, B.; Schenkebars, H.J.W.; Boumans, P.W.J.M. *Spectrochim. Acta* **1983**, *38B*, 1545.
226. Suh, S.Y.; Sacks, R.D. *Spectrochim. Acta* **1981**, *36B*, 1081.
227. Mavrodineanu, R.; Boiteux, H. *Flame Spectroscopy*. Wiley, New York, 1965.
228. Ho, P.; Buss, R.J.; Leohman, R.E. *J. Mater. Res.* **1989**, *4*, 873.
229. Matsuda, A.; Yokoyama, S.; Tanaka, K. *Appl. Phys. Lett.* **1988**, *53*, 1489.
230. Downey, S.W.; Mitchell, A.; Gottscho, R.A. *J. Appl. Phys.* **1988**, *63*, 5280.
231. Broadbent, E.K.; Morgan, A.E.; Flanner, J.M.; Coulman, B.; Sadana, D.K.; Burrow, B.J.; Ellwanger, R.C. *J. Appl. Phys.* **1988**, *64*, 6721.
232. Economou, D.J.; Park, S.-K.; Williams, G.D. *J. Electrochem. Soc.* **1989**, *36*, 188.
233. Heintze, M.; Veprek, S. *Appl. Phys. Lett.* **1989**, *54*, 1320.
234. Toyoshima, Y.; Arai, K.; Matsuda, A.; Tanaka, K. *Appl. Phys. Lett.* **1990**, *6*, 1540.
235. Ishihara, S.-I.; Otsuka, A.; Nagata, S. *Appl. Phys. Lett.* **1989**, *55*, 2396.
236. Broida, H.P. *Nat. Bur. Stand. (USA)* **1954**, *523*, 23.
237. Marr, G.V. *Can. J. Phys.* **1957**, *35*, 1265.
238. Griem, H.R. *Phys. Rev.* **1962**, *128*, 515.
239. Kepple, P.; Griem, H.R. *Phys. Rev.* **1962**, *173*, 317.
240. Wiese, W.L.; Kelleher, D.E.; Paquette, D.R. *Phys. Rev. A* **1972**, *6*, 1132.
241. Kelleher, D.E.; Wiese, W.L. *Phys. Rev. Lett.* **1973**, *31*, 1431.
242. Vidal, C.R.; Cooper, J.; Smith, E.W. *Astrophys. J. Suppl. Series No. 214* **1973**, *25*, 37.
243. Ramette, J.; Drawin, H.W. *Z. Naturforsch.* **1976**, *31a*, 401.
244. Wiese, L.L.; Augis, J.A. *J. Appl. Phys.* **1977**, *48*, 4528.
245. Barreto, E.; Jurenka, H.; Reynolds, S.I. *J. Appl. Phys.* **1977**, *48*, 4510.
246. Jarosz, J.; Mermet, J.M.; Robin, J.P. *Spectrochim. Acta* **1978**, *33B*, 55.

247. Drawin, H.W.; Ramette, J. *Z. Naturforsch.* **1979**, *34a*, 1041.
248. Dhali, S.K.; Williams, P.F.; Crumley, R.J.; Gundersen, M.A. *IEEE Trans. Plasma Sci.* **1980**, *PS-8*, 164.
249. Ehrich, H.; Kelleher, D.E. *Phys. Rev. A* **1980**, *21*, 319.
250. Coulaud, G.; Fleurier, C.; Ranson, P.; Chapelle, J. *Phys. Rev. A* **1980**, *21*, 851.
251. Smith, E.W.; Talin, B.; Cooper, J. *J. Quant. Spectrosc. Radiat. Transfer* **1981**, *26*, 229.
252. Oberauskas, J.; Serapinas, P.; Salkauskas, J.; Svedas, V. *Spectrochim. Acta* **1981**, *36B*, 799.
253. Capelle, B.; Mermet, J.M.; Robin, J.P. *Appl. Spectrosc.* **1982**, *36*, 102.
254. Williams, R.R.; Coleman, G.N. *Spectrochim. Acta* **1983**, *38B*, 1171.
255. Scheeline, A.; Kamla, G.J.; Zoellner, M.J. *Spectrochim. Acta* **1984**, *39B*, 677.
256. Blades, M.W.; Lee, N. *Spectrochim. Acta* **1984**, *39B*, 879.
257. Goode, S.R.; Deavor, J.P. *Spectrochim. Acta* **1984**, *39B*, 813.
258. Caughlin, B.L.; Blades, M.W. *Spectrochim. Acta* **1985**, *40B*, 987.
259. Abbas, A.; Basha, T.S.; Abdel-Aal, Z.A. *Jap. J. Appl. Phys.* **1988**, *27*, 804.
260. Nowak, S.; Van der Mullen, J.A.M.; Van der Sijde, B.; Schram, D.C. *J. Quant. Spectrosc. Radiat. Transfer* **1989**, *41*, 177.
261. Bakshi, V.; Kearney, R.J. *J. Quant. Spectrosc. Radiat. Transfer* **1989**, *42*, 405.
262. Bakshi, P.; Kalman, G. *J. Quant. Spectrosc. Radiat. Transfer* **1990**, *44*, 93.
263. Stehlé, C. *J. Quant. Spectrosc. Radiat. Transfer* **1990**, *44*, 135.
264. Ashkenazy, J.; Kipper, R.; Caner, M. *Phys. Rev. A* **1991**, *43*, 5568.
265. Preston, R.C. *J. Phys. B* **1977**, *10*, 523.
266. Kelleher, D.E. *J. Quant. Spectrosc. Radiat. Transfer* **1981**, *25*, 191.
267. Wiese, W.L.; Knojevic, N. *J. Quant. Spectrosc. Radiat. Transfer* **1982**, *28*, 185.
268. Czernichowski, P.A.; Chapelle, J. *Acta Physica Polonica* **1983**, *A63*, 67.
269. Arata, Y.; Miyake, S.; Matsuoka, H. *J. Quant. Spectrosc. Radiat. Transfer* **1984**, *32*, 343.
270. Vaessen, P.H.M.; Van Engelen, J.M.L.; Bleize, J.J. *J. Quant. Spectrosc. Radiat. Transfer* **1985**, *33*, 51.
271. Stehlé, C. *Phys. Rev. A* **1986**, *34*, 4153.
272. Marasinghe, P.A.B.; Lovett, R.J. *Spectrochim. Acta* **1986**, *41B*, 349.
273. Stamm, R.; Calisti, A.; Kaftandjian, P.; Talin, B. *J. Quant. Spectrosc. Radiat. Transfer* **1990**, *44*, 19.
274. McGeorge, S.W.; Salin, E.D. *Spectrochim. Acta* **1986**, *41B*, 327.
275. Winge, R.K.; Fassel, V.A.; Eckels, D.E. *Appl. Spectrosc.* **1986**, *40*, 461.
276. Lepla, K.C.; Horlick, G. *Appl. Spectrosc.* **1990**, *44*, 1259.
277. Ng, R.C.L.; Horlick, G. *Spectrochim. Acta* **1981**, *36B*, 529.
278. Ng, R.C.L.; Horlick, G. *Spectrochim. Acta* **1981**, *36B*, 543.
279. Jones, D.W.; Wiese, W.L. *Phys. Rev. A* **1984**, *30*, 2602.
280. Jones, D.W.; Wiese, W.L.; Woltz, A.L. *Phys. Rev. A* **1986**, *34*, 450.
281. Hey, J.D.; Griem, H.R. *Phys. Rev. A* **1975**, *12*, 169.
282. Greene, R.L.; *J. Quant. Spectrosc. Radiat. Transfer* **1983**, *30*, 409.
283. Iglesias, C.A. *J. Quant. Spectrosc. Radiat. Transfer* **1983**, *30*, 55.

284. Dufty, J.W.; Boercker, D.B.; Iglesias, C.A. *J. Quant. Spectrosc. Radiat. Transfer* **1990**, *44*, 115.
285. Kelleher, D.E.; Oza, D.H.; Cooper, J.; Greene, R.L. *J. Quant. Spectrosc. Radiat. Transfer* **1990**, *44*, 101.
286. Zerkle, D.K. Ph.D. Thesis, University of Illinois, Champaign-Urbana, 1992.
287. Evans, D.E.; Katzenstein, J. *Rep. Prog. Phys.* **1969**, *32*, 207.
288. Holzhauer, E. *Infrared Phys.* **1976**, *16*, 135.
289. Bessenrodt-Weberpals, M.; Kempkens, H.; Uhlenbusch, J. *Plasma Phys. Cont. Fus.* **1986**, *28*, 279.
290. Desoppere, E.; Van Oost, G. *Revue Phys. Appl.* **1983**, *18*, 803.
291. Bretz, N.; Dimock, D.; Foote, V.; Johnson, D.; Long, D.; Tolnas, E. *Appl. Opt.* **1978**, *17*, 192.
292. Lovberg, R.H. In *Modern Optical Methods in Gas Dyanmic Research.* Dosanjh, D.S., Ed. Plenum Press, New York, 1971, pp. 155–176.
293. Berney, A. *Plasma Phys.* **1973**, *15*, 699.
294. Ramsden, S.A.; Davies, W.E.R. *Phys. Rev. Lett.* **1966**, *16*, 303.
295. Baldis, H.A.; Walsh, C.J.; Benesch, R. *Appl. Opt.* **1982**, *21*, 297.
296. Huang, M.; Hieftje, G.M. *Spectrochim. Acta* **1985**, *40B*, 1387.
297. Kronast, B.; Pietrzyk, Z.A. *Phys. Rev. Lett.* **1971**, *26*, 67.
298. Döbele, H.F.; Kirsch, K. *Phys. Lett.* **1974**, *46A*, 352.
299. Kito, Y.; Sakuta, T.; Kamiya, A. *J. Phys. D* **1984**, *17*, 2283.
300. Downing, J.N.; Eisner, M. *Phys. Fluids* **1975**, *18*, 991.
301. Nagayama, Y.; Sakuma, K.; Toyama, H. *Jap. J. Appl. Phys.* **1982**, *21*, 1056.
302. Vriens, L.; Adriaansz, M. *J. Appl. Phys.* **1974**, *45*, 4422.
303. Phillips, P.E.; Nielsen, P. *Plasma Phys.* **1978**, *20*, 1265.
304. Huang, M.; Marshall, K.A.; Hieftje, G.M. *Spectrochim. Acta* **1985**, *40B*, 1211.
305. Fresse, K.B.; Massey, R.S.; Gribble, R.; Smith, J.D. *Proc. SPIE* **1981**, *288*, 280.
306. Sethian, J.D.; Ekdahl, C.A. *Rev. Sci. Instrum.* **1978**, *49*, 729.
307. Chiang, W.T.; Murphy, D.P.; Griem, H.R. *Phys. Lett.* **1979**, *72A*, 341.
308. Cobble, J.A. *Rev. Sci. Instrum.* **1985**, *56*, 73.
309. Kimberlin, S.R.; Chan, P.W.; Hazelton, R.C.; Yadlowsky, E.J. *J. Appl. Phys.* **1978**, *49*, 2700.
310. Huang, M.; Hanselman, D.S.; Yang, P.; Hieftje, G.M. *Spectrochim. Acta* **1992**, *47B*, 765.
311. Kimmitt, M.R.; Prior, A.C.; Roberts, V. In *Plasma Diagnostic Techniques*, Huddlestone, R.H.; Leonard, S.L., Eds. Academic Press, New York, 1965, pp. 399–430.
312. Batal, A.; Jarosz, J.; Mermet, J.M. *Spectrochim. Acta* **1981**, *36B*, 983.
313. Tracy, D.H.; Myers, S.A. *Spectrochim. Acta* **1982**, *37B*, 1055.
314. Lamoureux, M.; Möller, C.; Jaeglé, P. *J. Quant. Spectrosc. Radiat. Transfer* **1985**, *33*, 127.
315. Zangers, J.; Meiners, D. *J. Quant. Spectrosc. Radiat. Transfer* **1989**, *42*, 25.
316. Weisbach, M.F. *J. Quant. Spectrosc. Radiat. Transfer* **1971**, *11*, 1225.

317. Stormberg, H.-P.; Schäfer, R. *J. Quant. Spectrosc. Radiat. Transfer* **1985**, *3*, 27.
318. Kavetsky, A.; O'Mara, B.J. *J. Quant. Spectrosc. Radiat. Transfer* **1985**, *33*, 93.
319. Küveler, G.; Wöhl, H. *Astron. Astrophys.* **1983**, *122*, 69.
320. Abreu, V.J.; Skinner, W.R. *Appl. Opt.* **1989**, *28*, 3382.
321. Brannon, P.J.; Gerber, R.A.; Gerardo, J.B. *Rev. Sci. Instrum.* **1982**, *53*, 1403.
322. Baessler, P.; Kock, M. *J. Phys. B* **1980**, *13*, 1351.
323. Konjević, N.; Ćirković, L.; Labat, J. *Fizika* **1970**, *2*, 121.

10
CTD Detectors in Atomic Emission Spectroscopy

Robert S. Pomeroy

Department of Chemistry, United States Naval Academy
Annapolis, Maryland

10.1 Atomic Emission Spectroscopy for Chemical Analysis

Atomic emission spectroscopy (AES) is based on the sampling and excitation of a sample and the subsequent collection, separation, and detection of light emitted from the excited species. AES is perhaps the most commonly used technique for sensitive, selective, and rapid determination of a large number of metals and nonmetals in a wide variety of materials. Fields of application span agronomy, air quality analysis, ceramics, environmental science, forensic science, geochemistry, materials science, medicine, metallurgy, and water analysis. In addition, AES enjoys ease of automation, high sample throughput, low cost, and multielemental capability while maintaining good precision and accuracy over a large range of sample concentrations.

Currently atomic emission spectroscopic instruments range in price from $5000 for simple flame photometric instruments to $250,000 for complete systems using an inductively coupled plasma (ICP) as the source. Currently 12 companies produce ICP-based spectroscopic systems with total sales of emission-based instruments on the order of $50 million per year [1].

In many laboratories, the emission instrument is the central elemental analysis tool. Demands for higher sample throughput and lower cost per analysis are pushing AES instrumentation to new levels of sophistication. The trend is toward instruments that (1) require less user input, that is, have a high degree of automation, (2) provide multielemental analysis both at trace and percent levels, and (3) generate an organized report of the analysis. The trend toward reduced operator intervention requires some means of quality control. This typically is accomplished by the periodic analysis of standards and presupposes, to a certain extent, the composition of the sample. An alternative approach incorporates a database of spectroscopic information and proceeds with the analysis by applying known spectroscopic principles. This places two demands on the development of future instruments: (1) better ways to collect the database information and (2) better ways to manage the information. These considerations are where charge-transfer device (CTD) technology has had and will continue to have the biggest impact on AES instrumentation.

This chapter describes basic background information on atomic emission spectroscopy, providing the framework for an understanding of how and why CTD detectors are incorporated into spectroscopic systems. Brief descriptions follow of the spectroscopic systems developed to date. The discussion then addresses the use of CTD detectors in atomic spectroscopy and the application of these spectroscopic systems to various modes of elemental analysis. The principal advantage of these systems is the complete acquisition of all the spectroscopic data over a large spectral region while the performance characteristics at any one wavelength equal or surpass those of commercially available single-channel systems. Automation is a central theme in the development of new analytical instrumentation, so active control and "intelligent" instruments are the focus for the future. The rapid developments of computer technology combined with the multichannel data acquisition of CTD-based spectroscopic systems are revolutionizing atomic emission spectroscopy.

10.2 Background

The application of CTD technology to AES requires special consideration of the design of the spectrometer and of the mode of detector operation in order to obtain the unique advantages CTD detectors offer over other means of optical detection. A brief discussion of AES instrumentation provides the basis for this discussion. The interested reader can obtain more information from the excellent texts by Boumans [2,3].

In AES, the liquid or gaseous sample is typically introduced to the excitation source by direct aspiration; solids may be sampled by laser ablation or by an electrical discharge such as an arc or a spark. The source must then vaporize, atomize and excite the constituents of the sample. The spectrum emitted from such an experiment carries with it information about the sample, the source, and the medium in which the excitation is supported. The emitted radiation is collected and introduced to the entrance of a wavelength-dispersing system, the spectrometer. The spectrometer then disperses the light into its various components and focuses the light at the focal plane. The intensity of the emitted radiation at a given wavelength is proportional to the concentration of a particular species in the source and is monitored by the detector system. Commonly used detectors include

(1) photographic emulsion, (2) a series of fixed detectors mounted on the focal plane (a direct reader), and (3) a movable grating system that moves the spectrum past a fixed single-channel detector (a scanning monochromator).

The emission spectra produced by modern sources can be quite complex and extend from the vacuum ultraviolet to the near-infrared region of the electromagnetic spectrum. Consequently, stringent demands are placed on the spectrometer. Since a considerable amount of broadband radiation is generated by the source, stray light is a concern. Another important consideration in the design of a spectrometer system is the compromise between wavelength coverage and spectral resolution. Typically, a wavelength region from the near ultraviolet through the visible portion of the spectrum is chosen, as this is where most of the analytes have strong emission transitions. Ideally, the spectral resolution needs to be high enough to isolate the radiation emitted by the species of interest from other radiation sources. In practice, the spectral bandpass of spectroscopic instruments associated with high-temperature atomic emission sources needs to be less than 0.02 nm; analyses performed with spectrographs with lower resolution suffer from increased noise, analytical bias, poorer sensitivity, and nonlinear response curves. While important, the throughput of the spectrograph is not the primary concern, because source fluctuation is the dominant source of noise.

In principle, the intensity of the emitted radiation from an excited element is directly proportional to its concentration. In addition, the identification of an emitting species can be deduced since the radiation wavelength is specific to the analyte. Under ideal conditions, one needs only to monitor at a single position on the focal plane to determine if a particular species is present and at what level. Quantification entails comparing the emission intensity from a series of standards with that of the unknown. In the practical design of an emission system, however, there is a compromise between spectral resolution and spectral coverage. As a result, for samples rich in emission lines there can be a problem with spectral interferences; the emission from the analyte of interest may overlap the emission from other species present. The spectrochemical analysis then requires that the analyst eliminate or compensate for the interference by chemical pretreatment, selective sampling, or source excitation, or by the use of an alternate emission line that suffers no such interference. The sample composition itself also affects the analysis, because the physical and chemical characteristics of the sample matrix influence both the sample introduction and the source characteristics (such as temperature). Here again, the analyst needs to either eliminate the matrix effect by standard addition, or employ selective sampling and/or excitation, or must duplicate the sample matrix in the preparation of the standards (matrix matching).

10.2.1 Acquisition of the Emission Spectrum

The spectral information is usually obtained by either a sequential approach, in which the information is encoded in time such as with a scanning monochromator, or by a simultaneous method, in which the information is encoded spatially as with a polychromator (direct reader). Scanning monochromators typically employ a Czerny-Turner optical configuration in which a moving grating brings individual spectral lines into focus at a fixed position. This approach allows any wavelength to be observed but observes only one wavelength at a time. For transient sources such as an arc or spark discharge, where an internal standard line must be monitored for quantitative work, scanning monochromators are ineffective. For temporally stable sources like the inductively coupled plasma (ICP) or the direct current plasma (DCP), scanning spectrometers have worked well. However, in situations where there is a limited amount of a complex sample, it may be impossible to monitor all the spectral lines of interest before the sample is exhausted.

Polychromators generally use a Pashen-Runge optical configuration that employs a concave grating to focus the emission spectrum on the circumference of a circle, the Rowland circle. By placing the detection system on the focal plane, it is possible to obtain information simultaneously from multiple wavelengths. Photographic emulsions were initially popular with these systems because they allowed a wide spectral range to be simultaneously monitored with an integration time that helped to smooth out source fluctuations. Analytical emission spectroscopy often must deal with a large dynamic range of photon fluxes—intense emission from strong emitters at percent levels in the sample to extremely low fluxes for trace quantities of weak emitters—requiring multiple exposures (and time to develop the emulsions) in order to evaluate the analysis.

The photographic emulsion has been replaced by the photomultiplier tube (PMT) due to its high sensitivity, better precision, wide dynamic range, and ease of electronic data acquisition. This required a compromise—either the spectrum is scanned with a PMT, in which case analysis time becomes an issue, or a series of PMTs, in which case data are acquired only at discrete wavelengths, greatly reducing the flexibility while at the same time increasing the complexity and cost. Hybrid systems have been attempted, but these do not offer true simultaneous coverage of the entire emission spectrum. They do exhibit gains in analysis speed and flexibility, but at the cost of increased instrumental complexity.

10.2.2 Advent of Charge-Transfer Device Detectors

As the chemical community demanded more and more of AES instrumentation, it became clear that what is desired is the electronic equivalent of a photographic emulsion. Initial attempts tried to adapt multichannel array detectors designed for the television industry, such as the vidicon [4], the silicon intensified-target tube [5], and the image-dissector tube [4]. Unfortunately, the difficulties of trying to cope with the demands of ultraviolet response, high sensitivity, pixel selection, an extremely large linear dynamic range, and the large numbers of detector elements prevented the widespread use of any of these detectors in AES.

Solid-state array detectors based on silicon integrated-circuit technology are the latest multichannel detector technology. One class of silicon array detectors is the CTD. These detectors are solid-state, multichannel, integrating detectors with negligible dark currents (when cooled), high quantum efficiencies over a wide spectral region, low read noise, and a large dynamic range and are available in a variety of formats [6–9]. This class of detectors can be further divided into the subclasses of charge-injection devices (CIDs) and charge-coupled devices (CCDs), described in detail in Chapters 2 and 3 of this volume. The CID differs from the CCD primarily by the mode in which the photogenerated charge is measured. Some of the more salient characteristics of CIDs and CCDs will be presented against a backdrop of their application to AES.

10.2.3 CID versus CCD Charge Readout

The architectures and readout modes of CIDs and CCDs have been described in detail in Chapters 2 and 3. As mentioned in Chapter 3, the CID allows the charge contained in an individual detector element to be determined nondestructively (the nondestructive readout or NDRO mode) and allows a detector element to be independently addressed. In addition, CIDs do not suffer from charge blooming so that if the charge capacity of the detector element is exceeded, the excess charge does not spill into adjacent elements. These characteristics allow the extension of the dynamic range of the detector and enables the measurement of low fluxes by extending the integration time. By varying the integration time from detector element to detector element, the user can obtain the optimum signal-to-noise ratio (SNR) for each spectral line during the course of the measurement with no previous knowledge of photon fluxes.

In contrast, CCDs exhibit lower read noise than CIDs by about one order of magnitude, which makes them superior low-light-level detectors. However, the architecture of the CCD requires that charge from the entire device must be shifted out to obtain the desired information, even if the information represents only a small portion of the acquired data. Just as for a CID, the simple dynamic range of the CCD is not sufficient for a single exposure to be used; however, since the readout for the CCD is both destructive and lacks random access, the technique of variable integration time cannot be used to extend the range. Another disadvantage present in many CCDs is charge blooming. This is disastrous in AES where one detector element can be illuminated by a background feature with many magnitudes greater intensity than the nearby spectral feature of interest. In response to this problem, new CCD architectures that resist blooming have been developed, and antiblooming CCDs are now commercially available. Several advantages of CCDs include the greater variety of formats, larger array sizes, large number of manufacturers offering devices, and higher quantum efficiencies over extended wavelength areas. The choice of CID or CCD depends on a number of factors and could change as device characteristics improve.

10.2.4 Camera Systems

The camera systems built for operating CIDs and CCDs in spectroscopic and scientific imaging applications are designed as still cameras, as opposed to video cameras that read out the array at fixed (relatively short) time intervals. A mechanical shutter is used to control the exposure time and several-minute exposures are routinely used to achieve very good low-light-level sensitivity. Charge, which is thermally generated in the silicon, becomes significant at long exposure times, so some form of cooling is necessary. If the arrays are not cooled, a significant portion of the dynamic range is consumed by the dark current, and more important, SNR is degraded by the dark-current shot noise. The pixel intensity is digitized directly to 12 to 16 bits of precision. To achieve very low noise levels, the rate of pixel readout is much slower than in a video camera, hence the term *slow scan* is often applied. Pixel rates ranging from 50 to 500 kHz are typical, so the time required to read an entire array can vary from 1 to 20 s depending on the size of the array and the readout speed.

10.3 Echelle Spectroscopic Systems

Unfortunately, CTDs are not readily available in geometries suitable for use in conventional optical spectrometers. In fact, while CTDs are available in sizes of hundreds to millions of detector elements, most of the larger devices are rectangular arrays and each detector element is relatively small. Conventional linear spectroscopic dispersion systems are not suitable; however, the echelle spectrograph [10] is capable of achieving the spectral resolution necessary for AES in a compact instrument; its two-dimensional output of the spectral information makes it an ideal match to the two-dimensional formats of CTDs. The output from an echelle spectrograph is generated by a coarsely ruled grating designed to disperse the light in very high grating orders. The orders are separated by means of a cross dispersing element producing a series of grating orders stacked vertically, one above the other. More detailed information on the design and application of echelle spectrographs is detailed in Chapter 3 of this book, in Ken and Mariana Busch's book [11], and in publications from the Bonner Denton group [12,13].

The two-dimensional output of echelle spectrographs is an obvious match to the format of CIDs and CCDS. However, CTDs provide a large number of pixels in a relatively small area of silicon, requiring considerable attention to the design of a compatible polychromator with a flat focal plane format. Prior to the application of CTDs to AES, the commercially available echelle spectrographs (designed for film) did not generate a sufficiently small spectral display for use with the sizes of CTDs that are typically available. Consequently, custom optical systems had to be designed to utilize the CTD detectors effectively. Descriptions of five different approaches to designing an echelle system are presented below, emphasizing the unique aspects of each.

10.3.1 The University of Arizona Echelle/CID Spectrometer

The first CID-detector-based systems were built at the University of Arizona by Denton and colleagues and have been described in detail elsewhere [12,13]. The basic design consists of an echelle spectrometer using a quartz prism for cross dispersion. An image size that matches the size of the CID is achieved from a relatively high-angular-dispersion-grating/prism combination by using a camera-mirror focal length that is 7.5 times shorter than the collimating mirror focal length (see Figure 10.1). The detector is a General Electric CID17BAS (presently manufactured by CIDTEC, Liverpool, NY) operated at 50 kHz and cooled to liquid nitrogen temperatures. The device has a full-well capacity of 600,000 electrons, exhibits negligible dark current, and has a read noise of about 100 electrons based on 100 nondestructive readouts (NDROs, see Chapter 3). The camera is operated by a Photometrics Ltd. camera controller (Model CC183, Tucson, AZ) with custom firmware and resides in an Intel Multibus-based computer system. The dispersion obtained with these systems is approximately 0.005 nm/pixel at 300 nm with a spectral range from 200 to 550 nm.

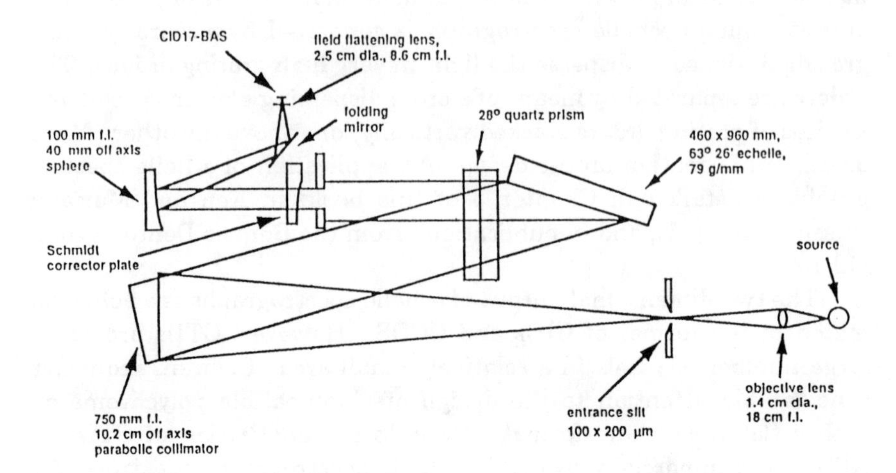

Figure 10.1 Optical diagram of the University of Arizona Echelle/CID spectrometer. (Reprinted with permission from reference 12.)

10.3.2 The Thermo Jarrell-Ash Echelle/CID Spectrometer

A commercial system also using a CID17BAS detector system similar to the one described above has been introduced by Thermo Jarrell-Ash [14]. The optical system is designed to cover the wavelength region from 170 to 800 nm with an improved light-gathering efficiency compared with the University of Arizona instrument. The detector quantum efficiency was enhanced in the UV by the application of Metachrome II, a wavelength-conversion phosphor available from Photometrics Ltd., Tucson, Arizona. The optical system uses only a single pass through a prism cross disperser to a custom echelle grating, and the light is brought into focus on the detector with a torroidal camera mirror (Figure 10.2). This optical arrangement reduces the number of surfaces, thereby cutting down on the amount of stray light and increasing the throughput of the spectrograph. The result is a spectroscopic system with sensitivity and wavelength coverage rivaling those of conventional instruments while also exhibiting simultaneous, continuous wavelength coverage.

The spectrometer and detector performances are described in detail by Pilon [14,15]. The trade-off with this system is the reduction in dispersion (0.02 nm/pixel) to gain wavelength coverage. The primary constraint is the size of the detector and the number of pixels

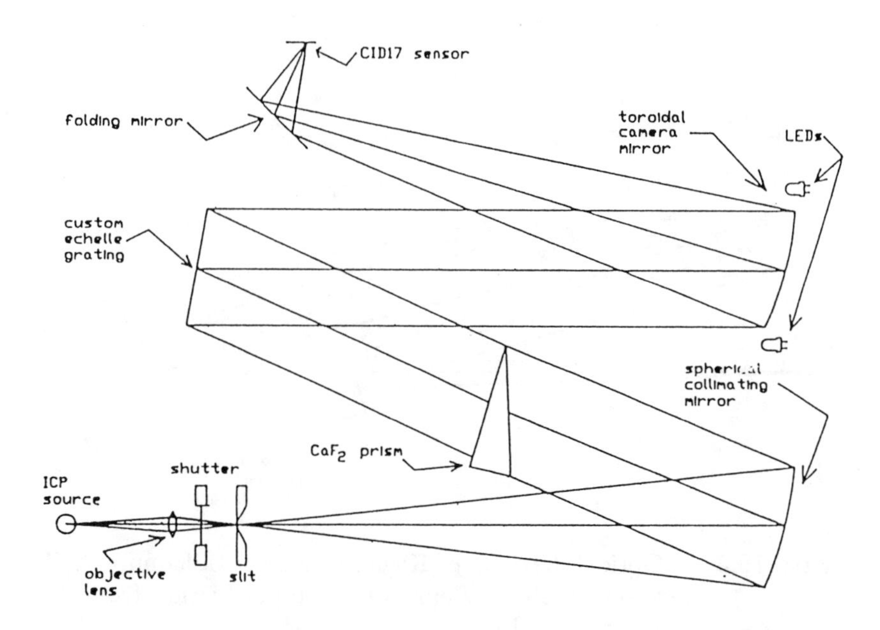

Figure 10.2 Optical diagram of Thermo Jarrell-Ash echelle/CID spectrometer. (Reprinted with permission from reference 14.)

available. The advent of CIDs with larger formats will allow for the design of systems with the same benefits of throughput and spectral coverage and will exhibit increased resolution.

10.3.3 Kodak Antiblooming Echelle/CCD Spectrometer

Bilhorn [16] at Eastman Kodak introduced a CCD-based echelle system that also uses an echelle grating with a prism for cross dispersion. An image size compatible with the size of the CCD is produced by using a more coarsely ruled echelle grating. The coarser ruling produces more echelle orders and a lower angular dispersion within an order, so that a longer focal length camera mirror can be used to produce the same size image. The optical system includes an off-axis parabolic collimator, a calcium fluoride prism, and a spherical camera mirror. The echelle grating is illuminated in the plane that is perpendicular to the plane of diffraction in order to minimize the amount of light lost due to shadowing by the deep echelle grooves (Figure 10.3). The CCD camera is base on an Eastman Kodak KAF 1300L CCD. The CCD was introduced in January 1990 as a 1.3-million-pixel, antiblooming, full-frame imager. The camera system consists of a liquid-nitrogen-cooled camera head (model CH210), camera electronics module (model

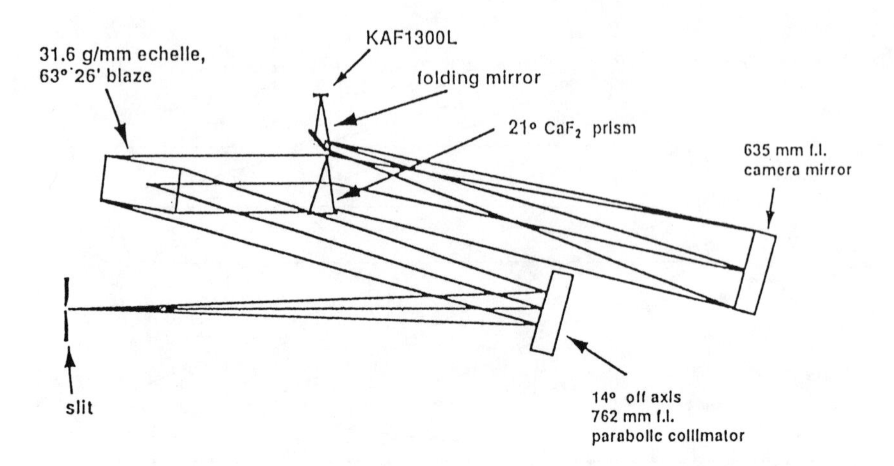

Figure 10.3 Optical diagram of Kodak echelle/antiblooming-CCD spectrometer. (Reprinted with permission from reference 16.)

CE200), and a VME-bus camera controller (all from Photometrics, Ltd.). The camera controller board is used with a Sun 3e computer system (Sun Microsystems, Mountain View, California). The Multibus-based computer system was also configured by Photometrics (Model DIPS3000). The spectrometer covers a wavelength range extending from approximately 180 to 800 nm and displays a dispersion of 0.0025 nm/pixel at 300 nm.

10.3.4 Perkin-Elmer Echelle/CCD Spectrometer

At the 1992 Pittsburgh Conference [17,18], Perkin-Elmer described a novel solid-state detector that utilizes a series of CCD subarrays located in specific locations in order to acquire the emission from the most prominent spectral lines for most elements. Each subarray can be accessed separately, allowing a variable integration time for each subarray of detectors. This allows sensitive low noise detection with a CCD without the requirement of reading out an entire device at a series of integration times. The spectrometer divides the emission from an ICP as it leaves the echelle grating using a Schmidt reflector; that creates two discrete wavelength regions: a UV region, 167 to 380 nm, and a visible region, 380 nm to 867 nm (Figure 10.4). The Schmidt reflector eliminates spherical aberration, and the grating generates little stray light. This system overcomes the limitations of the CCD full frame readout and the limited size of current CID

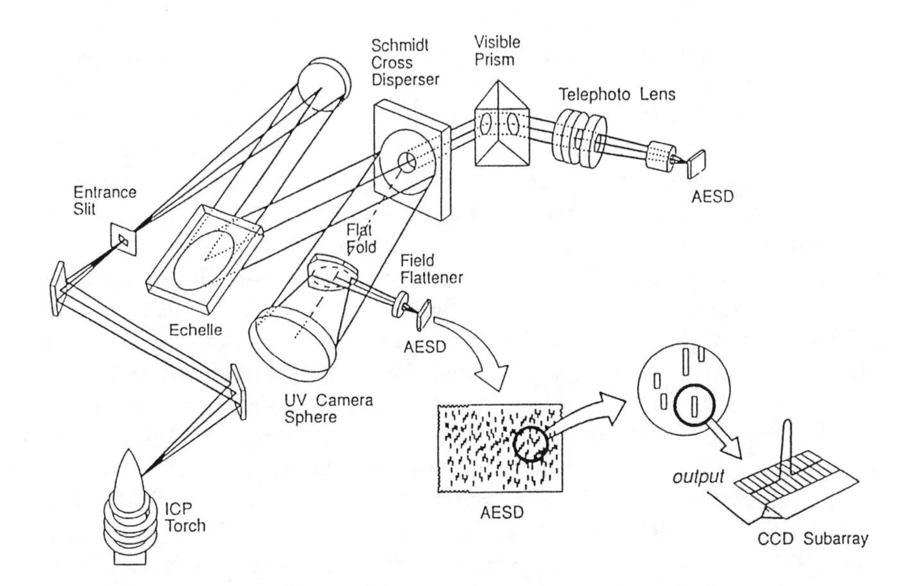

Figure 10.4 Optical diagram of Perkin-Elmer echelle/custom-CCD spectrometer. (Figure courtesy of Perkin-Elmer.)

detectors. The resultant system mimics the random access integration of the CID and acquires the spectral radiation at a higher resolution, 0.0025 nm/pixel. There is some additional complexity in the design of the instrument and not all of the radiation is acquired; however, this system effectively overcomes the present shortcomings in available CTD technology and promises nearly all the advantages of the systems discussed earlier. This instrument is in a development stage and was introduced in the spring of 1993.

10.4 Chemical Analysis with CTD-Based Systems

10.4.1 Qualitative Elemental Analysis with a Multichannel Detector

Many times the object of an elemental analysis is simply to identify some or all of the elements present in the sample. Traditional qualitative elemental analysis by emission spectroscopy usually is performed by observing the emission at one or more spectral lines to confirm the presence or absence of an element. When the question is posed about only a few elements, the analysis is easily handled with a slew-scanning monochromator or a fixed-slit polychromator (provided that channels are installed for the specific elements). When a large number of elements must be included in the survey, the problem becomes more difficult. At times the analyst has little information about the sample and is presented with a "good" and a "bad" material and is asked to find the differences. In these cases it is often desirable to perform a qualitative analysis for as many elements as possible.

The availability of the entire emission spectrum from a single exposure in an echelle polychromator equipped with a two-dimensional array detector makes qualitative surveys for a large number of elements relatively straightforward. Effective qualitative analysis then hinges on the system's ability to identify the wavelengths of spectral lines in the acquired spectrum. An echelle polychromator with a solid-state two-dimensional detector array contains no moving parts and exhibits stable wavelength registration with a calibration that is easy to establish and maintain. Verification of the calibration simply involves checking to see that the spectral lines from a source such as a mercury lamp fall on the correct pixels. Spectral lines for other elements are referenced to a mercury emission line, typically the 253.65-nm line. This approach allows the location of a particular spectral line for an element by simply calculating an X and a Y offset. Additionally, if the coordinates of the mercury line do change, simply updating these coordinates recalibrates the system, since the relative coordinates of the spectral lines for all of the other elements remain unaffected [13].

Qualitative analysis of a sample can be carried out with varying degrees of sensitivity by controlling the exposure time of a single exposure. A single image of the entire spectrometer focal plane is checked for the presence of emission at the locations corresponding to a particular element. With either the CID- or antiblooming CCD-echelle spectrometer system, a relatively long exposure (1 min) results in saturation of the intense lines from any element present at high concentration. Trace elements will show a number of spectral lines at intermediate intensities, and ultratrace elements will show only a few of the most intense spectral lines at relatively low signal levels. Even if a spectral line is saturated, the qualitative analysis is unaffected since neither detector blooms. Shorter exposure times (down to 0.2 s) are used to find only the major components of a sample.

10.4.2 Quantitative Elemental Analysis

Quantitative elemental analysis requires some form of standardization, either internal or external. Spectral line intensities are measured in terms of emission intensity per unit time, so that values derived from different exposure times can be directly compared. The differences in operation between the CID and CCD, specifically NDRO and the pseudorandom access capabilities of the CID, lead to different means of implementation [13].

Operationally, the differences between echelle spectroscopic systems based on CIDs and those based on CCDs center around their differences in available modes of signal readout. The CID can employ random access integration (RAI), a combination of the NDRO of the CID with the ability to pseudorandomly access individual detector elements. Nondestructive readouts (NDROs) can be used to evaluate the amount of photogenerated charge and select an optimum integration time, approximately 80% of the detector's full-well capacity. When this optimum exposure level is reached the detector can be read out and the integration recorded. Read windows, subarrays within the detector array where spectral information of interest falls, are then sampled by the computer in a looping structure. As particular spectral features have integrated to an optimum level they are removed from the loop, and the looping continues until data for all the spectral lines have been recorded at the optimum level or until a preset upper limit on the integration time is reached. This upper limit is governed by the background emission and the stability of the source [19].

Since the CCD is incapable of the RAI mode, its operation relies on the acquisition of several separate exposures: short exposures to gather information on intense emitters, long exposures for weak emitters, similar to the approach used with photographic emulsions. In light of this need for several exposures, it becomes apparent that antiblooming CCDs are required. Without the antiblooming CCD, faint features are overcome by the blooming from intense lines despite the array's having been exposed to the source for the length of time required to obtain a satisfactory SNR [20]. As the CID detector with RAI readout acquires data only at wavelengths where information is present, it acquires much less extraneous data than the CCD and places much less of a demand on the computer system and memory storage. Computers and memory are not expensive, however, and so in principle there are no real advantages to one device for detection over the other.

10.5 Expert Systems

10.5.1 Automated Qualitative Analysis

The qualitative analysis procedure can be automated by replacing the decision making of the analyst with a computer algorithm. A flowchart of one possible algorithm is shown in Figure 10.5 [21]. The process begins by acquiring an emission spectrum. For plasma-emission spectroscopy, a blank-subtracted emission spectrum is used so that emission features due to the argon plasma and solvent are removed. The corrected spectrum is produced by the digital subtrac-

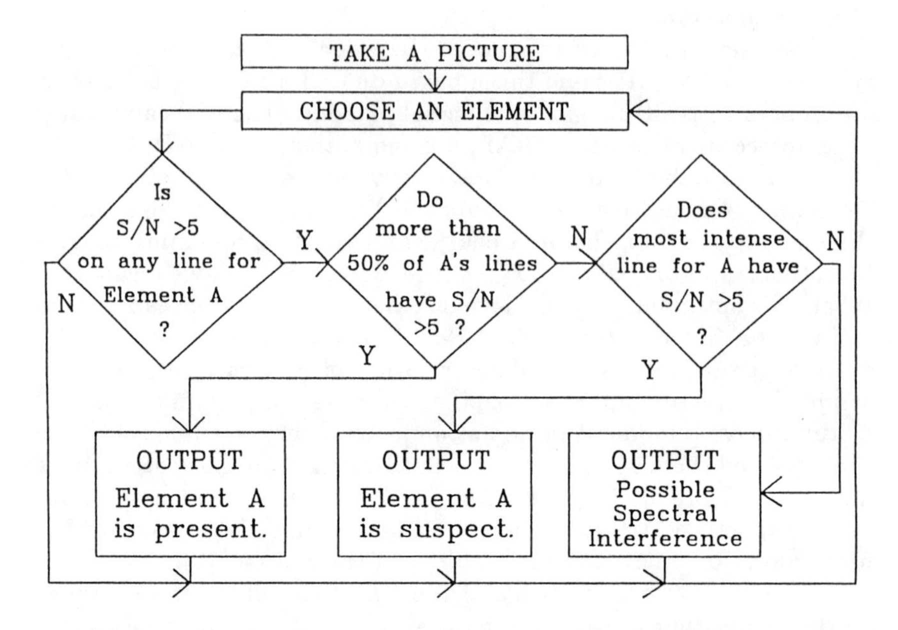

Figure 10.5 Flow diagram for the automated qualitative analysis procedure. (Reprinted with permission from reference 22.)

tion of two images taken with the same exposure time: one of the sample solution and one of the blank, typically a 5% nitric acid solution.

The algorithm employs a database containing information on up to 10 spectral lines per element for the elements that can be determined by emission spectroscopy in the ultraviolet. The lines included in the database are the strongest emitters for each element (for a particular source type), and the position of each spectral line is stored in coordinates relative to the mercury calibration wavelength. Data files exist for approximately 60 elements, and these are searched sequentially to determine if the various criteria shown in the diagram are met. The "best emitter" is the spectral line that produces the highest signal level at the detector and incorporates detector sensitivity, spectrometer throughput, and source emissivity factors.

The algorithm checks for the presence of each element one at a time by first calculating the SNR for each of the spectral lines. A subarray of pixels centered on the spectral line location is used for both spectral line and background intensity measurements. The width of the subarray corresponds to the width of an echelle order and the length corresponds to approximately eight times the half-width of a spectral line. The resulting subarray is rectangular with a 4:1 aspect ratio. The two groups of pixels at either end of the subarray are used as the spectral background and to estimate the noise in the baseline. The spectral line intensity is calculated from the sum of the pixel intensities at the center of the subarray after the background contribution has been subtracted. The noise value used is the standard deviation of the mean of all of the background pixels.

Once the SNR has been calculated for all of the spectral lines for an element, a check is made to determine if any of the SNRs exceed a predetermined value, typically 5. If none do, the algorithm moves on to the next element. The algorithm next checks to see if more than half of the lines have an SNR greater than 5. If more than half of the lines do, it is concluded that the element is definitely present. If fewer than half of the spectral lines have an SNR of 5 or more, the algorithm checks to see if the SNR on the most intense emitter is over 5. If this is the case, the conclusion is that the element is probably present but at such a low concentration that only the most intense lines have an SNR that is greater than 5. For these cases the algorithm reports that the presence of the element is suspected. If the most intense emitter does not have an SNR of 5 or more, then the line with the high SNR must be a less intense line. The only way this can possibly occur is if there is a spectral interference. The algorithm this reports that a spectral interference is suspected and outputs the SNR and the wavelength of the line.

One further refinement is required to ensure against false positives. A false positive theoretically can occur in one of two ways. If enough spectral interferences occur so that half of the spectral lines in an element database yield an SNR > 5, the algorithm will report that the element is present. The likelihood of this occurrence is kept low by making sure that each element database contains multiple spectral lines and by using a system with high resolution. A false positive can also be reported if the noise calculated for a given spectral line is anomalously low. Since the standard deviation is calculated from a relatively small number of pixel intensities, this outcome occurs frequently enough to be troublesome. The difficulty is avoided by tabulating noise values from all of the spectral lines found with an SNR > 5. Any spectral lines with an SNR > 5 and a noise value more than 2 sigma from the mean value are ignored.

An example of a typical automated qualitative analysis output for an element found to be present in the sample is shown in Table 10.1. The most intense emitter is placed first in the element's database and the rest of the lines are placed in random order. The SNR value chosen as a cutoff value is shown in the output. Five has been found

Table 10.1 Automated Qualitative Analysis Report for the Element, Chromium, Found to Be Present in a Sample[a]

Line (N=6)	Wavelength	SNR	Signal	Bkgd. Devn.[b]	Abs X	Abs Y
0	359.35	670.31	670	1.09	175	61
1	192.47	192.47	192	1.72	130	33
2	305.65	305.65	306	1.55	199	32
3	287.35	287.35	287	1.96	242	61
4	138.44	138.44	138	1.55	39	33
5	460.74	460.74	460	0.95	94	61

S/N Threshold = 5

Number of Lines = 6 Number Found = 6

[a] Synthetic solution containing 10 ppm each Cr, Ni, Al, Mg, and Sr analyzed by direct current plasma emission.

[b] Standard deviation of the background.

to be a conservative value that still provides high sensitivity. In Table 10.1 all of the spectral lines of chromium were detected at an SNR > 5, so the element was reported as being present in the sample. The other information in the table (and stored in the database) includes the line wavelengths and the coordinates of the spectral line relative to the mercury reference wavelength.

Table 10.2 shows an example of the output for an element not present in the sample. Even though an SNR > 5 is reported for the mercury line at 435.8 nm (line number 6 in the table), none of the other lines have an SNR > 5. This is not a case of the element being present at a low concentration, since the most intense emitter, the 253.65-nm line (line 0 in the table), does not have an SNR > 5. The example demonstrates how the automated, qualitative analysis algorithm can be used to identify spectral interferences. By simply aspirating a single-element solution, all other elements that have a spectral line exhibiting an SNR > 5 must be spectral interferences. The routine also provides a measure of the severity of the interference, since the actual SNRs are reported.

Table 10.2 Automated Qualitative Analysis Report for the Element, Mercury, Not Found to Be Present in a Sample[a]

Line	Wavelength	SNR	Signal	Bkgd. Devn.[b]	Abs X	Abs Y
0	253.65	−0.98	−4.90	5.08	214	168
1	404.66	−1.00	−4.20	4.22	151	36
2	365.00	0.05	0.24	5.30	160	54
3	313.20	287.3	287	1.96	242	61
4	312.57	0.83	3.18	3.81	97	89
5	296.73	1.21	5.82	4.80	151	104
6	435.83	8.50	38.10	4.49	126	26
7	366.33	−1.56	−4.56	2.93	232	52
8	365.48	0.38	1.44	3.80	186	52
9	546.00	−0.02	−0.86	4.24	272	2

S/N Threshold = 5

Number of Lines = 10; Number Found = 1

[a] Same solution as in Table 10.1. Note that the element is reported as not found even though one of the spectral lines gave an SNR greater than 5. Since the most intense emitter did not give an SNR greater than 5, the algorithm assumes a spectral interference has occurred.

[b] Standard deviation of the background.

The sensitivity of the automated qualitative analysis algorithm is demonstrated by analyzing a dilution series of a six-element standard solutions. The series spans the concentration range from 5000 to 5 ng/g and contains the elements Ca, Cr, Mn, Fe, Ni, and Pb. Table 10.3 summarizes the results of the analysis. At the highest concentration all of the components are reported as being present and also of equal importance, and no false positives are reported. The greatest risk of a false positive occurs when the spectrum contains many intense spectral lines. As the solution becomes more dilute, the reported results change from "present" to "suspected to be present" to finally "not found." The trend observed matches that predicted by the

Table 10.3 Automated Qualitative Analysis Results for a Dilution Series of a Six-Element Standard

	Present	Suspect	Not Found
5000 ppb	Ca,Cr,Mn, Fe,Ni,Pb	---	---
500 ppb	Ca,Cr,Mn, Fe,Ni	Pb	---
50 ppb	Ca,Cr,Mn	Fe,Ni,Pb	---
5 ppb	Ca	Cr,Mn	Pb,Fe,Ni

emission intensities of the spectral lines and the limits of detection for these elements with this instrument. For example, Pb, Fe, and Ni are reported as not found at a concentration of 5 ng/g, which is below the instrument detection limit for these elements. Also note the sensitivity that is still exhibited for the strong emitter Ca at the 5 ng/g level.

Even though the system has a sensitivity close to that of a traditional instrument, qualitative analysis detection limits are achieved by the automated qualitative analysis algorithm with virtually no false positives. The high level of discrimination is in large part due not only to precise background correction, which is possible because of the simultaneous nature of the instrument, but also to the fact that spectral information from more than one emission line is being used.

A sample application for the automated qualitative analysis routine is illustrated with several examples. Figure 10.6 shows the results for 10 elements from the qualitative analysis of three stream water samples taken near a mining operation in southern Arizona. One sample is from upstream of the mining operation; one sample is from the runoff from the operation, and one sample is from downstream of the operation. By being able to simultaneously screen for the presence of many elements, it is possible to quickly determine which elemental species serve as good chemical markers of the mining output. For example, the concentration of iron, cobalt, and copper in the mining effluent is very high, but no significant difference exists between the upstream and downstream water samples. This is due to the instability of iron, cobalt, and copper in the water column at the natural pH of the stream water. Manganese and zinc can serve as markers because they are undetected in the water sampled upstream,

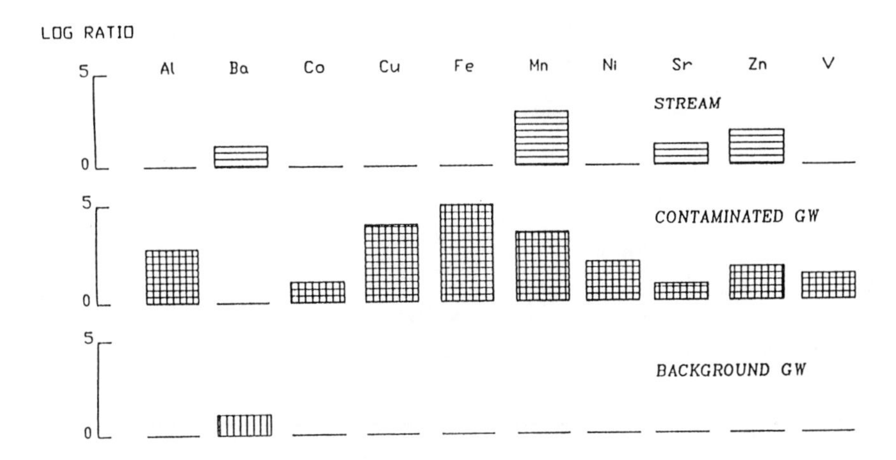

Figure 10.6 Automated qualitative analysis results for a stream water analysis in the vicinity of a mining operation. (Reprinted with permission from reference 22.)

but their presence is still detectable downstream. The decrease in concentration is due to the dilution of the mining effluent by the stream.

Another strength of the automated qualitative analysis algorithm is that little or no prior knowledge about the sample is necessary. This is beneficial in applications such as the screening of unidentified laboratory-waste solutions. The composition of the waste solutions must be determined so that the proper disposal method can be chosen. The results of an automated qualitative analysis of an unidentified aqueous waste solution from a university teaching laboratory is shown in Table 10.4. Once these results were reported to the laboratory that generated the waste, it was able to identify the processes that generated the waste. The likely source of the manganese was a still used to make triply distilled water; the tin and cadmium were used in several electrochemical experiments. Boron, calcium, magnesium, aluminum, and silicon were reported in the qualitative analysis and were verified as present by independent measurements. The likely source of these elements is the glass container in which the waste was collected. Two important points are illustrated by this example: the automated qualitative analysis was successful in analyzing an uncharacterized sample, and sensitivity comparable to conventional quantitative analysis was achieved as demonstrated by the detection of ions leached from the glass container.

Table 10.4 Results of Automated Qualitative Analysis of Unlabeled Laboratory Waste

Elements Found	Elements Suspected
Manganese	Magnesium
Boron	Cadmium
Calcium	Barium
Aluminum	Silicon
Tin	
Carbon	

Automated qualitative analysis is not limited to solution samples analyzed by plasma emission. With use of a spark source for excitation, it is possible to use automated qualitative analysis for rapid screening of metal alloys. The results for samples of low-alloy steels and aluminum samples are shown in Tables 10.5 and 10.6 [22]. We note that there were no false positives reported even in the spectrally rich spark spectra. All elements were either positively identified, suspected to be present (trace alloying components whose concentration was so low as not to provide an SNR > 5 on 50% or more of the emission lines), or not detected. Those elements listed as not detected are present at a concentration near or below the limit of detection for the instrument; in the cases of sulfur, beryllium, and gallium, no databases containing spectral line information for those elements currently exist.

An additional refinement that can be added to this automated qualitative analysis routine to aid in the identification of elements would be to examine the relative emission intensity of the spectral lines for an element. Spectral lines would not only have to be found in a sample, but their relative intensities would have to be consistent given the sample type. Fortunately, this has not turned out to be necessary since the multiline approach appears to be sufficiently robust. If relative line intensities were entered, the algorithm would be more complex in order to consistently avoid reporting false negatives; spectral interferences would have to be corrected by an iterative process before the ratios for an element could be correctly calculated. Additionally, variations in sample-matrix composition significantly alter the source excitation temperature which affects

Table 10.5 Automated Qualitative Analysis for Steel Standards[a]

Standard	Elements Present	Elements Detected	Elements Suspected	Not Detected
SS 406/1	Fe,C, Si,Mn, P,S,Cr,Mo, Ni,Co,Cu,V	Fe,C,Si, Mn,Cr,Mo, Ni,Cu,V	Co: 0.0006%	P: 0.0009% S: No database
SS 407/1	Fe,C,Si, Mn,P,S, Cr,Mo, Ni,Cu,V	Fe,C,Si, Mn,Cr,Mo, Ni,Cu,V	P: 0.030%	S: No database
SS 408/1	Fe,C,Si, Mn,P,S, Cr,Mo, Ni,Cu,V	Fe,C,Si, Mn,Cr,Mo, Ni,Cu,V	P: 0.037%	S: No database
SS 409/1	Fe,C,Si, Mn,P,S, Cr,Mo,Ni, Co,Cu,V	Fe,C,Si, Mn,Cr,Mo, Ni,Cu,V	Co:0.014% P: 0.025%	S: No database
SS 410/1	Fe,C,Si, Mn,P,S, Cr,Mo, Ni,Cu,V	Fe,C,Si, Mn,Cr,Mo, Ni,Cu,V	P: 0.072%	S: No database

[a] Bureau of Analyzed Samples, Ltd., Newham Hall, Middleborough, England.

spectral-line-intensity ratios (particularly atom-line to ion-line ratios). Therefore, a robust qualitative analysis scheme that requires emission-line intensity ratios to fit a predetermined pattern would have to take into consideration source excitation conditions as well. This is in contrast to ICP-mass spectrometry, where the iterative approach is quite successful because of relatively small variability in isotope ratios [23].

10.5.2 Semiquantitative Analysis

Although qualitative screening is useful as an initial means of analysis, many applications require a better estimate of the concentration of each constituent. Simply using the SNR obtained from the automated qualitative analysis routine provides a crude estimate, but this method

Table 10.6 Automated Qualitative Analysis for Aluminum Standards[a]

Standard	Elements Detected	Elements Suspected	Not Detected
SS 1100-AN	Al,Mg,Si, Fe,Cu,Zn,Ti	Cr,Mn	Ni: 0.003% Ga: No database
SS 2011-X	Al,Mg,Si, Fe,Cu,Cr,Zn	Mn,Cr,Ni	Ti: 0.005%
SS 2618-F	Al,Mg,Si, Fe,Cu,Ni	Mn,Zn,Ti	Cr: 0.001%
SS 2324-B	Al,Mg,Si, Fe,Cu,Mn	Cr,Zn,Ti	Ni: 0.009% Ga,Be: No database

[a] Aluminum Company of America, Alcoa Center, Pennsylvania.

is not highly reliable since the sample matrix can have such a pronounced effect on emission line intensities. In emission sources, the presence of easily ionized elements in the sample can lead to as much as a 20% enhancement of atom line emission intensities and a corresponding depression of the emission intensity from ion lines [24]. Changes in the nebulization conditions can also be caused by changes in the sample matrix, and these can lead to erroneous results. For example, a change in surface tension will alter the aerosol particle-size distribution, which in turn will alter the mass per unit time of sample transported into the plasma. These problems associated with the composition of the sample matrix can be overcome, at least partially, by using an internal standard. An internal standard in this context is simply an element whose concentration in the sample is known or which can be spiked into the sample solution at a known concentration. Measuring analyte line intensities relative to the internal standard compensates for changes in sample aerosol introduction rate and sample excitation conditions. The internal standard can also serve as an ionization buffer making the various samples and standards under analysis more uniform in composition.

The semiquantitative analysis algorithm was developed to give not only the qualitative composition of a sample but also a better estimate of the concentration of each component. A semiquantitative type of analysis is advantageous when speed of analysis is more important than high accuracy. The semiquantitative algorithm is also used as the first step in a matrix-dependent spectral-line-selection routine for true quantitative analysis as described in section 10.5.3. An internal standard element is used and is ideally chosen to match the spectroscopic, physical, and chemical behavior of the analyte element or elements. Using an echelle polychromator with a two-dimensional multichannel array, internal standard and analyte lines can be monitored simultaneously so that there is no increase in analysis time and the effects of source drift are reduced. The ratio of the analyte response to that of the internal standard compensates for fluctuations in sample transport parameters and ideally corrects for nonspectral interferences that affect the analyte and internal standard similarly. Factors used in choosing an appropriate internal standard are discussed by Ahrens and Taylor [25] and by Slavin [26].

In solid-sample analysis, the matrix material serves as the internal standard. In liquid samples, an element is typically added to both the sample and the standard solutions, which is used initially to determine the correspondence between emission-intensity ratio and concentration. In liquid solutions, the internal standard can also serve as a spectroscopic buffer, a material used to partially swamp out the effects of variations in the concentration of easily ionized elements between samples. Once the equations for the emission-ratio working curves are obtained, semiquantitative information can be obtained over an extended period without the need to recalibrate. Figure 10.7 shows the results of the semiquantitative evaluation of the magnesium concentration in a solution employing a lutetium internal standard and DCP excitation; concentrations were measured daily over a 15-day period using three replicate analyses per day. The solid line represents the average magnesium concentration determined by 10 replicate analyses obtained prior to the 15-day experiment. Care was exercised to keep the instrument and analysis conditions as uniform as possible over the test period; however, no calibration was performed for any of the analyses. As the figure shows, reproducible semiquantitative results were obtained.

Table 10.7 shows a comparison of the reported (Thorn Smith, Royal Oak, Mich.) concentration values for alloying components in steels versus those obtained with the semiquantitative algorithm using the dirct current plasma (DCP) for excitation. Iron emission at 259.940 nm was used as the internal standard. Although the accuracy achieved may not be acceptable in some applications, the advantage of not having to prepare calibration standards more than offsets the reduction in accuracy for other applications. Examples of applications for which the accuracy of semiquantitative analysis might be sufficient include quality assurance where the technique could be used to ensure that the concentration of certain elements does not exceed some specified value. Another potential application is as a screening

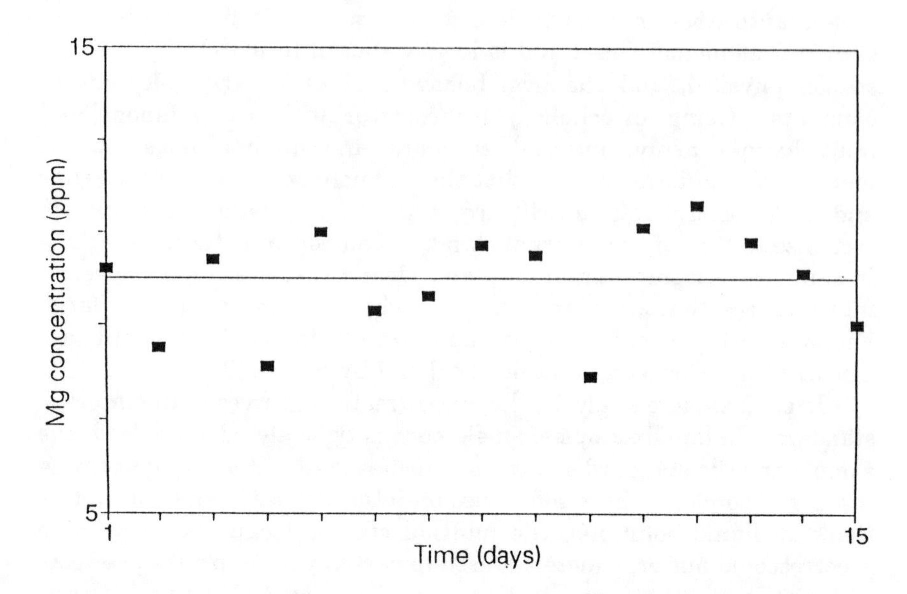

Figure 10.7 Long-term study for semiquantitative determinations of magnesium concentrations using lutetium as an internal standard in the direct current plasma. (Reprinted with permission from reference 22.)

technique for rapid and reliable identification or sorting of materials. Other frequently encountered applications include problem solving, where differences in composition are being sought, and the screening of raw materials for new manufacturing processes or from new vendors for potential contaminations.

10.5.3 Matrix-Dependent Spectral Line Selection

A question that is raised when using the semiquantitative analysis algorithm, particularly when analyzing materials such as steels that produce an emission spectrum with a very large number of intense spectral lines, is: Do inaccuracies arise simply as random variations, or do uncorrected spectral interferences cause true biases? The solution to this dilemma, and the next step in the evolution of the intelligent atomic emission spectrometer that performs completely automated analysis, is the development of an algorithm for choosing the best spectral lines for a particular analysis and for accurately correcting for any remaining spectral interferences. Thus far, an algorithm has been described for qualitatively identifying the composition of a sample and then roughly determining the concentration of

Table 10.7 Semiquantitative Analysis of Steel Standards[a]

Standard #		% Mo	% Mn	% Si	% Ni	% Cr
36	Reported	0.18	0.96	0.24	0.03	0.95
	Found	0.21	0.89	0.27	0.03	1.05
38	Reported	0.0	1.22	0.01	0.0	0.0
	Found	0.0	1.18	0.006	0.0	0.0
40	Reported	0.23	0.68	0.25	1.81	0.70
	Found	0.19	0.73	0.22	1.77	0.67
42	Reported	0.19	0.83	0.27	0.14	1.07
	Found	0.23	0.88	0.25	0.12	1.21
47	Reported	0.0	0.05	0.74	10.41	18.35
	Found	0.0	0.02	0.69	9.67	17.63
53	Reported	0.01	0.34	0.02	0.01	0.01
	Found	0.003	0.28	0.05	0.02	0.006

[a] Thorn Smith, Royal Oak, Michigan.

each component; all of this information is available for the next step.

Our proposed approach for matrix-dependent spectral line selection is that put forth by Boumans and co-workers [27–29]. A "true detection limit" is defined as the sum of the conventionally determined detection limit and a selectivity term that accounts for additional signal (noise) introduced by the presence of interfering spectral lines. Following the notation used by Boumans, the true detection limit is given by

$$c_{l,true} = \frac{2}{5}\frac{S_{Ij}}{S_A}C_I + C_{l,conv} \tag{10.1}$$

where

$C_{l,true}$ = true limit of detection
$C_{l,conv}$ = conventional limit of detection
C_I = concentration of an interfering species
S_{Ij} = sensitivity of the interferant j on analyte A
S_A = sensitivity of the analyte A

The best spectral line for a determination is the one that gives the lowest detection limit. The determination of which of the prominent lines will give the best detection limit in a particular sample requires the sensitivity, S_A, of the analyte lines and the sensitivity of all the potential interferants, S_{Ij}, as well as the concentration of the interferants. As a matter of convenience, the sensitivity ratios rather than the individual sensitivities are calculated and stored. Boumans calls these quotients Q values:

$$Q_{Ij} = \frac{S_{Ij}}{S_A} \tag{10.2}$$

The Q ratios are more reliable because they are independent of spectrometer throughput and detection efficiency; however, source excitation conditions and spectrometer resolution do have an impact. The Q values are easily determined using a multichannel detector-equipped polychromator. Simply by running a series of pure standards (100 μg/g, for example) and monitoring the intensities at all of the spectral lines for the other elements, the Q values can be determined quickly.

Sensitivity and Q values have been tabulated in our laboratories for the major alloying components in steel (Fe, Mn, Cr, Ni, Mo, and V) and also for the lanthanide elements. An example of the tabulated values is shown in Table 10.8, which shows the sensitivity and Q values for nickel in a steel matrix. The sensitivity is given in units of analog-to-digital units (ADUs) per ppm and the Q values are the sensitivity ratio $\times$ 10^6. The values marked with footnote b were obtained using a straight-line fit to the background data from only one side of the analyte spectral line (the normal procedure is to use both sides). This was necessary because interferences occurred in the spectral line wing, and higher-order background fits have yet to be implemented on the system.

An example of the application of the matrix-dependent spectral-line selection is found in the analysis of a monazite sand. Monazite sands are rich in rare-earth elements, as shown in Table 10.9 which lists the typical composition. Lanthanides are spectrally rich, and consideration of spectral interferences is essential. For example, the large amount of cerium present creates a problem for the determination of other rare earth elements. Typically in the analysis of neodymium and lanthanum, the Nd line at 430.357 nm and the La line at 433.373 nm are the most sensitive lines and are commonly used for quantitative determinations. However, with the high cerium content

Table 10.8 Sensitivity and Q Values for Ni in a Steel Matrix: Analysis by Direct Current Plasma on the University of Arizona Echelle/CID[a]

Wavelength (nm)	Ni Sensitivity ADU/ppm	*Q* Fe	*Q* Mo	*Q* Mn	*Q* Cr	*Q* V
349.30	5.1	38	1262[b]	18	55	243[b]
341.47	9.3	4918	6728	62	101[b]	14500
352.45	5.3	1492[b]	5011[b]	121	1869[b]	37
339.30	4.3	1952[b]	1113[b]	211	10013	32
344.63	4.2	318	7909[b]	0	8211[b]	342
347.25	4.1	3332	285	589	56367	615[b]

[a] Background correction is determined using a straight-line fit to the background on only one side of the spectral line.
b. Value subject to spectral interference on the wing of the analytical line.

of the monazite (44% CeO), these lines suffer interference. Figures 10.8a and 10.8b show working curves for lanthanum (33.749 nm) and neodymium (415.608 nm) in a cerium matrix. The solid points represent data obtained in a matrix low in cerium (25 μg/g) and the open points represent data obtained in a matrix of 1000 μg/g Ce. All of the points fall on the same line, demonstrating that these lines are essentially interference free. However, automatically using the interference-free spectral line for an analysis, that is, selecting analyte lines on the basis of values alone and ignoring the actual concentration of the interferant, ignores the inherently higher sensitivity of the line being rejected. Consequently the true detection limit calculation includes the interferant concentration term (C_{Ij}). The determination of Nd in a cerium matrix illustrates the impact of interferant concentration on optimum analyte line selection. At high Ce concentrations, the line at 415.6 nm is the best line; however, as the Ce concentration decreases, the line at 430.4 nm, which has a conventional limit of detection 10 times lower than the line at 415.6 nm, becomes the best line for analysis.

The situation is shown graphically in Figure 10.9. The crossover point occurs at a Ce concentration of approximately 700 μg/g and is dictated primarily by the resolution of the instrument. The 430.4-nm

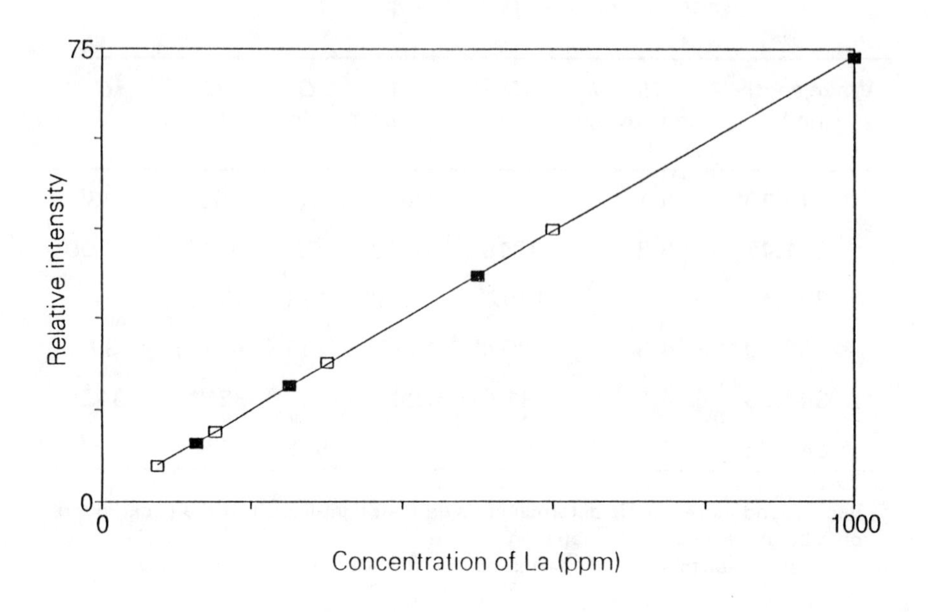

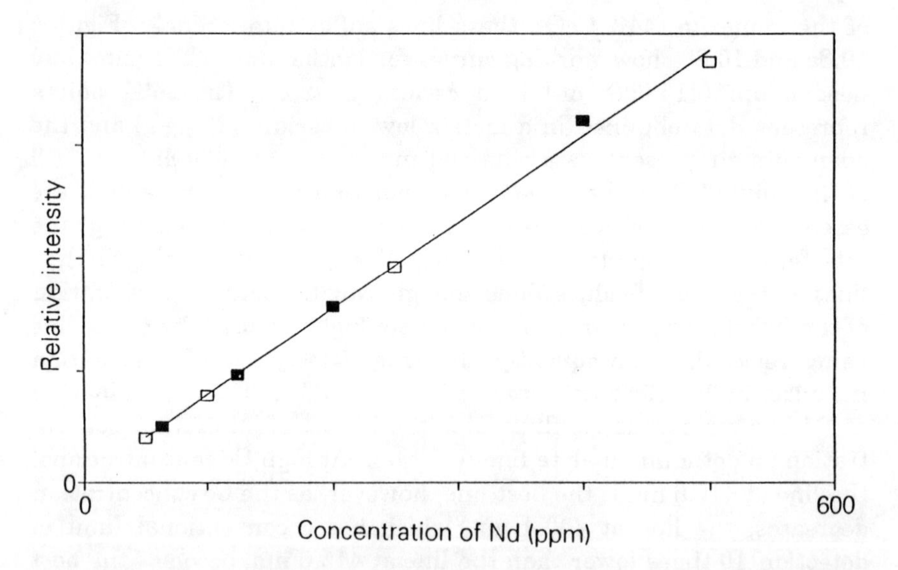

Figure 10.8 Lanthanum 333.749 nm (a) and neodymium 415.608 nm (b) working curves determined in the presence of cerium. Solid markers: 25 ppm concomitant Ce; open markers: 1000 ppm concomitant Ce.

Table 10.9 Monazite Sand Typical Composition, Percent as Oxides[a]

Ce	44.0	Yb	0.25
La	18.1	Tb	0.25
Nd	13.1	Er	0.19
Pr	4.8	Ho	<0.1
Y	3.3	Lu	<0.1
Sm	3.0	Tm	<0.1
Gd	2.2	Th	6.2
Du	0.9	U	0.6
Eu	<0.5		

[a] Prof. Q. Fernando, University of Arizona, Department of Chemistry, personal communication.

spectral line is the more sensitive when the Ce concentration in the matrix is low; however, the 415.6-nm line is the best line at high Ce concentrations because it suffers from less spectral interference.

10.5.4 Element-Specific Chromatographic Detection

Element-specific detection of chromatographic eluates places additional constraints on atomic emission analysis since the detection must take place within a relatively short time compared with conventional emission experiments. The emission must be sampled several times during the elution of a peak in order to maintain an adequate representation of the peak shape. Also, in many chromatographic applications the organic solvents give rise to highly structured backgrounds. In these high-background situations the excellent spectral registration of the CID-based system allows for extremely precise spectral subtractions, reducing the influence of the eluent. The multiwavelength capability and the RAI of the CID make it possible to acquire simultaneously the sample and background emission under computer control so as to maximize the SNR and maintain a sufficient data acquisition rate to ensure proper delineation of the chromatographic peak [30].

Utilizing the ability of the CID to perform nondestructive reads, the computer monitors the databases, and when the signal reaches a preset threshold value the time and signal are recorded; next the

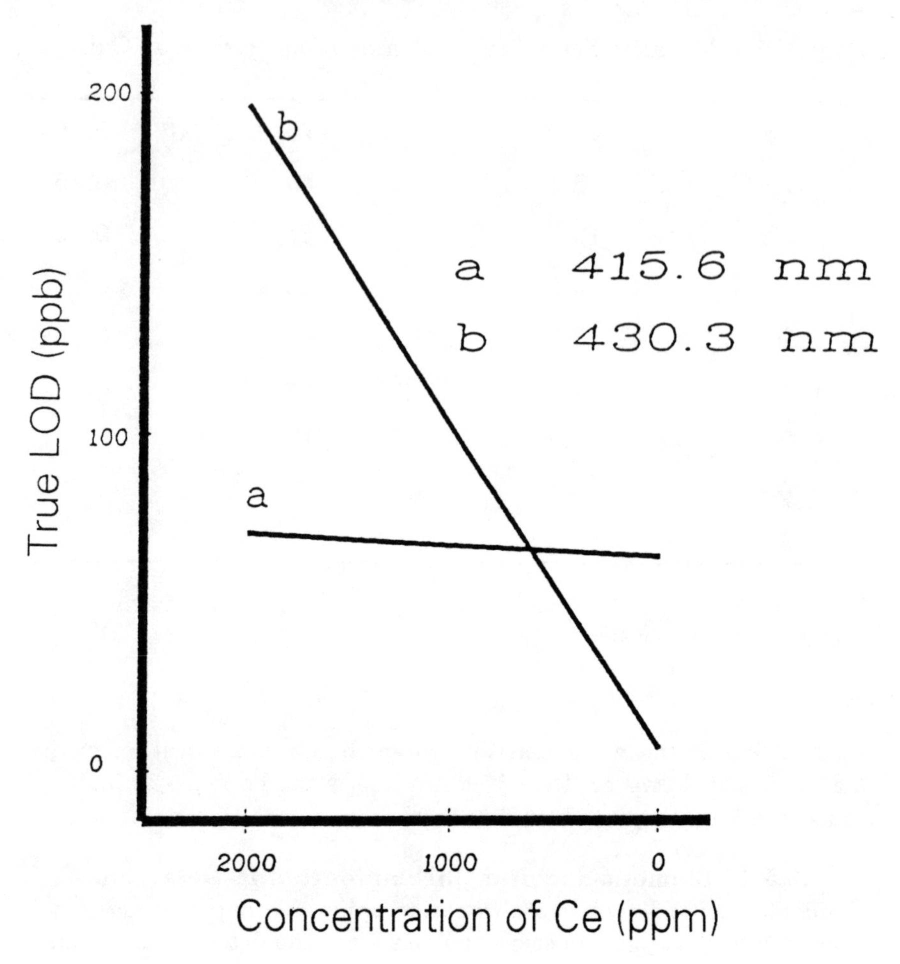

Figure 10.9 The limit of detection (LOD) for neodymium at two spectral lines as a function of cerium concentration in the sample matrix. (Reprinted with permission from reference 22.)

detector is cleared, and the integration is restarted. This approach allows for the data to be acquired when the SNR is optimal, not at some predetermined time interval (which would be necessary with the CCD). Comparison of the two approaches to data acquisition shows a 20-fold increase in the sensitivity by using the random-access-integration capability of the CID over the fixed-integration time method of the CCD [31]. Future developments of atomic emission detection of chromatographic eluates employing CID technology will allow optimal temporal resolution as well as the multiwavelength advantage of AES with CTD detection. (Note that the Perkin-Elmer system would have the same advantages.)

10.6 Expert Systems in the Near Future

Speed, flexibility, automation, and reliability can be achieved through the proper use of the spectral information available in an atomic emission spectrum. In the very near future, the combination of automated qualitative analysis, semiquantitative analysis, and automatic spectral line selection will be used routinely for analysis. It is easy to envision an instrument that will use a single continuous process to first qualitatively identify all of the components present in a sample, then make estimates of the concentrations of each component and, on the basis of the estimates, choose the very best spectral lines for the determination of the concentrations of the analytes of interest. An analytical report for the analysis of a series of similar samples might contain information on differences in matrix composition that result in different spectral lines being used in addition to the concentrations of the analytes. Since the "true" detection limit for an element depends on the matrix composition of the sample, concentrations might be reported relative to the detection limit in the sample, rather than relative to detection limits in pure water as is sometimes done in ICP-AES.

Information relating to variation in matrix composition that is discovered by the "intelligent" atomic emission spectrometer has an impact on the accuracy of qualitative determinations by plasma emission as well as being of potential importance in problem-solving or process-control applications. Unanticipated variations in matrix composition may signal a problem in the sample preparation steps. Assumptions regarding the degree to which the matrix composition of standards matches that of the samples may be invalidated. In these examples, the sample preparation procedure or the matrix composition of the standards may need to be altered to give the best possible analytical results. The time-consuming and expensive procedure of retesting samples that produce unexpected results could be avoided if there were a high degree of confidence in the original analysis.

A whole range of possibilities exists for improving the accuracy of analysis (removing sample-specific biases) through making use of spectral features that have diagnostic value. The emission lines for the elements normally determined by atomic emission can be used to determine sample-matrix composition, but other spectral features can be used to monitor sample-introduction conditions and analyte-excitation conditions. For example, in argon plasmas, argon atom to argon ion intensity ratios indicate excitation temperature; hydrogen or OH emission correlates with aqueous sample introduction rates;

carbon and hydrogen emission correlates with the concentration of organic species in the sample, and so on. [32]. The use that is made of this additional information, if provided, is currently the subject of active research. At the very least, abnormal conditions would be signaled so that corrective actions could be taken by an operator. More sophisticated approaches would apply correction factors when the correspondence is known. Another approach would seek to alter analysis conditions (powers, flow rates, etc.) to maintain the analytical source at constant conditions (excitation, sample introduction).

Effective use of the large database available makes realizable the intelligent atomic emission instrument. The advantage of this type of spectroscopic hardware is that even in situations where there is a large variation among samples, analysis by automated expert systems can still be effective. The system can either compensate or advise the analyst about the situation and in time even make recommendations for continued analysis. The next step to be taken is to make rational use of all the additional data. The last year has seen the introduction of CTD-based instruments from a number of instrument houses; as the number of users of such systems increases, the database and software to use the wealth of information effectively will soon follow.

References

1. Personal communication, Perkin-Elmer.
2. Boumans, P.M.J.M., Ed. *Inductively Coupled Plasma Emission Spectroscopy, Part 1 Methodology, Instrumentation, and Performance.* Wiley, New York, 1987.
3. Boumans, P.W.J.M. *Inductively Coupled Plasma Emission Spectroscopy - Part 2 Applications and Fundamentals.* Wiley, New York, 1987.
4. Felkel, H.L.; Pardue, H.L. *Anal. Chem.* **1978**, *50*, 602.
5. Aldous, K.M.; Mitchell, D.G.; Jackson, K.W. *Anal. Chem.* **1975**, *45*, 1034.
6. Bilhorn, R.B.; Epperson, P.M.; Sweedler, J.V.; Denton, M.B. *Appl. Spectrosc.* **1987**, *41*, 1125.
7. Epperson, P.M.; Sweedler, J.V.; Bilhorn, R.B.; Sims, G.R.; Denton, M.B. *Anal. Chem.* **1988**, *60*, 327A.
8. Bilhorn, R.B.; Sweedler, J.V.; Epperson, P.M.; Denton, M.B. *Appl. Spectrosc.* **1987**, *41*, 1114.
9. Sweedler, J.V.; Bilhorn, R.B.; Epperson, P.M.; Sims, G.R.; Denton, M.B. *Anal. Chem.* **1988**, *60*, 282A.
10. Keliher, P.N.; Wolhers, C.C. *Anal. Chem.* **1976**, *48*, 333A.
11. Busch, K.W.; Busch, M.A. *Multielement Detection Systems for Spectrochemical Analysis.* Wiley, New York, 1990.
12. Bilhorn, R.B.; Denton, M.B. *Appl. Spectrosc.* **1989**, *43*, 1.
13. Bilhorn, R.B. Ph.D. Dissertation, University of Arizona, 1987.
14. Pilon, M.J.; Denton, M.B.; Schleicher, R.G.; Moran, P.M.; Smith Jr., S.B. *Appl. Spectrosc.* **1990**, *44*, 1613.
15. Pilon, M.J. Ph.D Dissertation, University of Arizona, 1991.

16. Bilhorn, R.B.; Pomeroy, R.S; Denton, M.B. In *Computer Enhanced Analytical Spectroscopy*, P.C. Jurs, Ed. Plenum Press, New York, 1992.
17. Barnard, T.W.; Crockett, M.I.; Ivaldi, J.C.; Lundberg, P.L.; Young, E.F. Presented at the 1992 Pittsburgh Conference, New Orleans, paper 990.
18. Barnard, T.W.; Crockett, M.I.; Ivaldi, J.C.; Lundberg, P.L.; Ziegler, E.M. Presented at the 1992 Pittsburgh Conference, New Orleans, paper 991.
19. Bilhorn, R.B.; Denton, M.B. *Appl. Spectrosc.* **1990**, *44*, 1538.
20. Sweedler, J.V.; Jalkian, R.D.; Pomeroy, R.S; Denton, M.B. *Spectrochim. Acta* **1989**, *44B*, 683.
21. Pomeroy, R.S; Kolczynski, J.D.; Denton, M.B. *Appl. Spectrosc.* **1991**, *45*, 1120.
22. Pomery, R.S; Denton, M.B. *Appl. Spectrosc.* **1991**, *45*, 1111.
23. Ekimoff, D.; Van Nordstrand, A.M.; Mowers, D.A. *Appl. Spectrosc.* **1989**, *43*, 1252.
24. Olesik, J.W.; Williamson, E.J. *Appl. Spectrosc.* **1989**, *43*, 1223.
25. Ahrens, L.H.; Taylor, S.R. *Spectrochemical Analysis*, 2nd edn. Addison Welley, Reading, Pa., 1950.
26. Slavin, W. *Emission Spectrochemical Analysis*. Wiley-Interscience, New York, 1971.
27. Boumans, P.W.J.M.; Vrakking, J.J.A.M. *Spectrochim. Acta* **1987**, *42B*, 819.
28. Boumans, P.W.J.M.; Vrakking, J.J.A.M. *Spectrochim. Acta* **1985**, *40B*, 1085.
29. Boumans, P.W.J.M.; Maessen, F.J.M.J. *Spectrochim. Acta* **1988**, *43B*, 173.
30. Pomeroy, R.S; Baker, M.E.; Kolczynski, J.D.; Denton, M.B. *Appl. Spectrosc.* **1991**, *49*, 198.
31. Bilhorn, R.B.; Pomeroy, R.S; Denton, M.B. In *Trace Metal Analysis and Speciation*, I.S. Krull, Ed. Elsevier Science, Amsterdam, 1991.
32. Pomeroy, R.S; Sweedler, J.V.; Denton, M.B. *Talanta* **1990**, *37*, 15.

Index